T0337891

Robot Manipulator Redundancy Resolution

WILEY-ASME PRESS SERIES LIST

Robust Adaptive Control for Fractional-Order Systems with Disturbance and Saturation	Shi	November 2017
Robot Manipulator Redundancy Resolution	Zhang	October 2017
Combined Cooling, Heating, and Power Systems: Modeling, Optimization, and Operation	Shi	August 2017
Applications of Mathematical Heat Transfer and Fluid Flow Models in Engineering and Medicine	Dorfman	February 2017
Bioprocessing Piping and Equipment Design: A Companion Guide for the ASME BPE Standard	Huitt	December 2016
Nonlinear Regression Modeling for Engineering Applications	Rhinehart	September 2016
Fundamentals of Mechanical Vibrations	Cai	May 2016
Introduction to Dynamics and Control of Mechanical Engineering Systems	To	March 2016

Robot Manipulator Redundancy Resolution

Yunong Zhang and Long Jin
Sun Yat-sen University
Guangzhou, China

This Work is a co-publication between ASME Press and John Wiley & Sons Ltd

This Work is a co-publication between ASME Press and John Wiley & Sons Ltd

Registered Office(s)
John Wiley & Sons, Inc., 111 River Street, Hoboken, NJ 07030, USA
John Wiley & Sons Ltd, The Atrium, Southern Gate, Chichester, West Sussex, PO19 8SQ, UK

Editorial Office
The Atrium, Southern Gate, Chichester, West Sussex, PO19 8SQ, UK

For details of our global editorial offices, customer services, and more information about Wiley products visit us at www.wiley.com.

Wiley also publishes its books in a variety of electronic formats and by print-on-demand. Some content that appears in standard print versions of this book may not be available in other formats.

Library of Congress Cataloging-in-Publication Data
Names: Zhang, Yunong, 1973- author. | Jin, Long, 1988- author.
Title: Robot manipulator redundancy resolution / Yunong Zhang, Long Jin, Sun Yatsen University.
Description: Hoboken, New Jersey : Wiley, [2017] | Includes bibliographical references and index. |
Identifiers: LCCN 2017018502 (print) | LCCN 2017019566 (ebook) | ISBN 9781119381426 (pdf) | ISBN 9781119381433 (epub) | ISBN 9781119381235 (cloth)
Subjects: LCSH: Robots–Control systems. | Manipulators (Mechanism) | Redundancy (Engineering)
Classification: LCC TJ211.35 (ebook) | LCC TJ211.35 .Z44 2017 (print) | DDC 629.8/933–dc23
LC record available at https://lccn.loc.gov/2017018502

Cover image: © kynny/Gettyimages

Set in 10/12pt WarnockPro by SPi Global, Chennai, India
Printed and bound in Malaysia by Vivar Printing Sdn Bhd

10 9 8 7 6 5 4 3 2 1

To our parents and ancestors, as always

Contents

List of Figures

List of Tables

Preface

In recent decades, robotics has attracted considerable attention in scientific research and engineering applications. Much effort has been contributed to robotics and different types of robots have thus been developed and investigated. Among these robots, redundant robot manipulators have played a more and more important role in numerous fields of engineering applications, and they have been widely applied in industrial automation for performing repetitive dull work, such as welding, painting, and assembly. For a redundant robot manipulator, it possesses more degrees-of-freedom (DOF) than the minimum number required to perform a given end-effector primary task. One important issue in operating such robotic systems, redundancy resolution, has been widely studied, which is related to the motion planning and control of redundant manipulators. That is, given the desired Cartesian paths of the end effector, the corresponding joint trajectories need to be obtained online or in real-time t. Various redundancy-resolution schemes have thus been developed and investigated for the motion planning and control of redundant robot manipulators. By resolving the redundancy properly (or say, by using a specific redundancy-resolution scheme), redundant robot manipulators can avoid joint physical limits, while conducting the given end-effector primary path-tracking task.

In general, the redundancy-resolution problem can be solved at the joint-velocity level or at the joint-acceleration level, resulting in the corresponding velocity-level and acceleration-level redundancy-resolution schemes. In this book, focusing on redundancy resolution, we present and investigate different methods and schemes for the motion planning and control of redundant robot manipulators. Specifically, in view of the fact that the Jacobian matrix of a robot manipulator is actually varying with time during the motion-task execution, the problem of time-varying matrix pseudoinversion, as a new issue, is involved in the pseudoinverse-based scheme formulation. By computing the time-varying pseudoinverse of the Jacobian matrix (of the robot manipulator), discrete-time zeroing dynamics (ZD) models, as a new approach to the time-varying Jacobian matrix pseudoinversion, are applied to the redundant-manipulator kinematic control. Then, considering that calculating the inverse of Jacobian matrix is less efficient, three types of inverse-free simple solution based on the gradient dynamics (GD) method and ZD method, are thus presented and investigated to avoid the Jacobian inversion. Recent progress in our 16-year study shows the advantages of unifying the treatment of various schemes of manipulators' redundancy resolution. We recall some fundamental techniques for such a unification and then specify it in full details based on quadratic programming (QP) and its online solutions. Such a QP formulation is general in the sense that it incorporates equality, inequality, and bound constraints, simultaneously. This QP formulation covers the online avoidance of joint physical limits and environmental obstacles, as well as the optimization of various performance indices. Every term is endowed with clear physical meaning and utility. Computer-simulation results based on various robotic models show the effectiveness of the presented methods and schemes. For substantiating the physical realizability, some of these methods and schemes are applied to an actual six-DOF planar robot manipulator.

The idea for this book was conceived during the research discussion in the laboratory and at international scientific meetings. Most of the materials of this book are derived from the authors' papers published in journals and proceedings of international conferences. In fact, in recent decades, the field of robotics has undergone the phases of exponential growth, generating many new theoretical concepts and applications (including the authors' ones). Our first priority is thus to cover each central topic in enough details to make the material clear and coherent; in other words, each part (and even each chapter) is written in a relatively self-contained manner.

This book contains 15 chapters that are classified into the following seven parts.

Part I: Pseudoinverse-Based ZD Approach (Chapter 1);
Part II: Inverse-Free Simple Approach (Chapter 2 through Chapter 4);
Part III: QP Approach and Unification (Chapter 5);
Part IV: Illustrative Velocity-Level QP Schemes and Performances (Chapter 6 through Chapter 8);
Part V: Self-Motion Planning (Chapter 9 through Chapter 11);
Part VI: Manipulability Maximization (Chapter 12 and Chapter 13);
Part VII: Encoder Feedback and Joystick Control (Chapter 14 and Chapter 15).

Chapter 1 – This chapter presents and investigates the applications of discrete-time ZD models to the kinematic control of redundant robot manipulators via time-varying matrix pseudoinversion. That is, by computing the time-varying pseudoinverse of the Jacobian matrix (of the robot manipulator), the resultant ZD models are applied to the redundant-manipulator kinematic control. Computer-simulation results based on two robot manipulators further illustrate the effectiveness of the presented ZD models for time-varying matrix pseudoinversion applied to the redundancy resolution of robot manipulators.

Chapter 2 – In this chapter, base on GD method, an inverse-free scheme is presented at the joint-velocity level to avoid calculating the inverse of Jacobian matrix. The scheme is called a G1 type as it uses GD once. In addition, two path tracking simulations based on five-link and six-DOF redundant robot manipulators illustrate the efficiency and the accuracy of the presented scheme. What is more, the physical realizability of G1 type scheme is also verified by the physical experiments based on the six-DOF planar redundant robot manipulator hardware system.

Chapter 3 – In this chapter, another inverse-free simple solution based on GD method, named the D1G1 scheme, is presented at the joint-acceleration level for solving the inverse kinematics problem of redundant robot manipulators. Furthermore, simulation results based on a three-link redundant robot manipulator substantiate the effectiveness and accuracy of the presented inverse-free D1G1 scheme.

Chapter 4 – In this chapter, different from the Z0G1 type scheme and D1G1 type scheme presented in previous chapters, the third inverse-free solution, named the Z1G1 type scheme, is presented and investigated at the joint-acceleration level. In addition, we conduct the path-tracking simulations performed on three-, four-, and five-link planar robot manipulators to substantiate the effectiveness and accuracy of such an inverse-free solution. Moreover, the experiments based on a six-DOF redundant robot manipulator hardware system illustrates the physical realizability of the presented novel solution for handling the time-varying inverse kinematics problem of the redundant robot manipulator(s) in an inverse-free manner.

Chapter 5 – In this chapter, we present the results regarding the unification of redundancy resolution of robot manipulators via QP. The presented QP formulation covers the online avoidance of joint physical limits and environmental obstacles, as well as the optimization of various performance indices. Every term is endowed with clear physical meaning and utility. Motivated by the real-time solution to such robotic problems, some QP online solutions are briefly reviewed. The QP-based unification of robots' redundancy resolution is substantiated by a large number of computer simulation results based on PUMA560, PA10, and planar robot manipulators.

Chapter 6 – In this chapter, a varying joint-velocity limits (VJVL) constrained minimum-velocity-norm (MVN) scheme (termed VJVL-constrained MVN scheme) is presented and investigated for redundant manipulators; that is the joint-velocity limits change with the end-effector and joints movement. The scheme is then formulated as a QP, which is subject to an equality constraint and a bound constraint, and such a QP problem is solved by the numerical algorithm 94LVI. In addition, experimental results performed on a six DOF planar robot manipulator substantiate the physical realizability and efficacy of such a VJVL-constrained MVN scheme and the corresponding discrete QP solver. Furthermore, the position-error analyses verify the accuracy of the presented scheme on redundant manipulators.

Chapter 7 – In this chapter, a feedback-aided minimum joint motion (FAMJM) scheme is presented and investigated to keep the minimum movement of the joints of redundant manipulators in the task duration. The optimal index of the FAMJM scheme is to minimize the norm of the drift between the instant joint state and the initial joint state, which is subject to the robot forward kinematics equation, joint-angle limits, and the varying joint-velocity limits. Such an FAMJM scheme is then reformulated as a QP subject to an equality constraint and a bound constraint, which is solved by the numerical algorithm 94LVI. Furthermore, this FAMJM scheme is implemented on the six-DOF planar robot manipulator to write two English letters, that is "M"- and "P"-shaped paths. Experimental results verify the efficacy and accuracy of the presented FAMJM scheme and the corresponding numerical algorithm 94LVI.

Chapter 8 – In this chapter, for achieving a desired configuration, a scheme for state adjustment of a redundant robot manipulator (termed a state-adjustment scheme) is presented, with no end-effector task explicitly assigned. In addition, the presented state-adjustment scheme can finally be formulated as a QP and resolved at the joint-velocity level. Based on the conversion technique of the QP to linear variational inequalities, a numerical algorithm is presented to solve the robot state-adjustment scheme. By employing such a state-adjustment scheme, the robot manipulator can automatically move to the desired configuration from any initial configuration with the movement kept within its physical limits. Computer-simulation and experiment results using a practical six DOF planar robot manipulator (which has variable joint-velocity limits) further verify the realizability, effectiveness, accuracy, and flexibility of the presented state-adjustment scheme.

Chapter 9 – In this chapter, a criterion is presented in the form of a quadratic function for the purpose of self-motion planning of functionally redundant PUMA560 and kinematically redundant PA10 robot manipulators. The presented self-motion scheme with joint physical limits considered can be formulated as a QP problem subject to equality, (inequality) and bound constraints. The LVIAPDNN is developed as the real-time solver for the resultant QP. Computer simulations performed based on PUMA560 and PA10 robot manipulators substantiate the efficacy of the QP-based self-motion-planning scheme.

Chapter 10 – In this chapter, for comparison purposes, the classical pseudoinverse-based method is presented for self-motion planning of redundant robot manipulators. Computer-simulation results based on a three-link planar manipulator, PUMA560 and PA10 manipulators further substantiate the efficacy and superiority of the LVIAPDNN solver and the QP-based self-motion planning scheme. Besides, the effect of design parameters λ and γ is investigated, and the singularities of the self-motion for redundant manipulators are discussed.

Chapter 11 – This chapter presents the design and implementation of a zero-initial-velocity (ZIV) self-motion scheme on the six-DOF planar robot manipulator. In view of the existence of physical limits in an actual robot manipulator, both joint-angle limits and joint-velocity limits are initially incorporated into the presented self-motion scheme for practical purposes. The presented self-motion scheme is then reformulated as a QP and resolved at the joint-velocity level. By combining the ZIV constraint, the resultant QP can prevent the occurrence of a large initial joint velocity. The presented ZIV self-motion scheme eliminates the phenomenon of the abrupt and drastic increase in joint velocity at the beginning of the self-motion task execution. Simulative and experimental results based on the practical six-DOF planar robot manipulator further verify the realizability, effectiveness, and accuracy of the presented self-motion scheme.

Chapter 12 – In this chapter, to achieve optimal maneuverability, ag performance index in the form of quadratic function is presented and analyzed for the manipulability-maximizing self-motion planning (MMSMP) of redundant manipulators. The corresponding MMSMP scheme can automatically select the desirable configuration so that the manipulator is most flexible and best maneuverable. For practical and protective purposes, the zero-initial-velocity constraint is also incorporated into the MMSMP scheme to eliminate the large-initial-velocity weakness. The MMSMP scheme can further be converted and unified into a QP. Furthermore, the experiments are conducted on the actual six-DOF planar robot manipulator, which substantiates the effectiveness and physical realization of the presented MMSMP scheme.

Chapter 13 – In this chapter, a time-varying coefficient aided manipulability-maximizing (TVCMM) scheme is presented and investigated for the optimal motion control of redundant robot manipulators. In order to improve the manipulability during the end-effector task execution, a manipulability-maximizing index is considered into the scheme formulation. Besides, for the remedy of the nonzero initial/final joint-velocity problem, a time-varying coefficient is introduced and incorporated in the scheme, which is further reformulated as a QP subject to equality and bound constraints. For guaranteeing the physical realizability of such a scheme, the numerical algorithm 94LVI is employed to solve such a QP. Simulative and experimental results validate the physical realization, effectiveness and accuracy of the presented QP-based manipulability-maximizing scheme.

Chapter 14 – This chapter presents and investigates an online motion planning and feedback control (OMPFC) scheme for redundant manipulators via techniques of QP and rotary encoder. The presented OMPFC scheme is performed on the six DOF planar manipulator presented in previous chapters, which incorporates the feedback of task-space position error. Then the joint state is obtained in real time via rotary encoders equipped on the physical manipulator. The original scheme is reformulated as a unified QP. In addition, the QP is solved online during the joint motion by employing the numerical algorithm 94LVI. Finally, simulation and experimental results validate the physical realizability, online property, and efficacy of the presented OMPFC scheme.

Chapter 15 – In this chapter, to achieve the real-time joystick control of the redundant manipulator, two subschemes are presented: one subscheme is a cosine-based position-to-velocity mapping, which maps the position of the joystick in the motion range to the velocity of the robot end-effector for the real-time end-effector velocity generation; and the other is a real-time joystick-controlled motion planning (JCMP) subscheme, which is based on minimum-velocity-norm (MVN) and used to resolve the velocity-specified inverse kinematic problem of the redundant manipulator. In addition, the JCMP subscheme is formulated as a standard QP subject to an equality and a bound constraint. For the purpose of experimentation, the numerical algorithm 94LVI is adopted for the solution of the resultant QP problem. Furthermore, the subschemes are implemented on the six-DOF planar robot manipulator presented in the previous chapters and controlled by a joystick. Computer-simulation and experiment results validate the realization and effectiveness of such subschemes and the corresponding numerical algorithm 94LVI.

This book is written for academic and industrial researchers as well as graduate students studying in the developing fields of robotics, numerical algorithms, and neural networks. It provides a comprehensive view of the combined research of these fields, in addition to its accomplishments, potentials, and perspectives. We do hope that this book will generate curiosity and also happiness to its readers for learning more in the fields and the research, and that it will provide new challenges to seek new theoretical tools and practical applications. Without doubt, this book can be extended. Any comments or suggestions are welcome. The authors can be contacted via e-mail at zhynong@mail.sysu.edu.cn and jinlongsysu@foxmail.com.

March 1, 2016

Yunong Zhang and Long Jin
Sun Yat-sen University,
Guangzhou, China

Acknowledgments

This book basically comprises the results of many original research papers of the authors' research group, in which many authors of these original papers have done a great deal of detailed and creative research work. Therefore, we are much obliged to our contributing authors for their high-equality work. During the work on this book, we have had the pleasure of discussing its various aspects and results with many cooperators and students. We highly appreciate their contributions, which particularly allowed us to improve our manuscript. Especially valuable help was provided by Liangyu He, Jian Li, Sitong Ding, Huinan Xiao, Jinjin Wang, Dechao Chen, Binbin Qiu, Wan Li, Yang Shi, Zhengli Xiao, Yaqiong Ding, Yinyan Zhang, and Xiaogang Yan. We are grateful to them for their help and suggestions.

The continuous support of our research by the National Natural Science Foundation of China (number 61473323), by the Foundation of Key Laboratory of Autonomous Systems and Networked Control, Ministry of Education, China (number 2013A07), and also by the Science and Technology Program of Guangzhou, China (number 2014J4100057) is gratefully acknowledged. Besides, we would like to thank the editors sincerely for their time and efforts spent in handling this book, as well as the constructive comments and suggestions provided. We are very grateful to the nice people at Wiley and the IEEE for their strong support during the preparation and publishing this book.

In addition, the second author, Long Jin, would like to express his special gratitude to the Guohua Memorial Middle School for the financial and moral support since 2004.

To all these wonderful people, we owe a deep sense of gratitude; especially now when the research projects and the book have been completed.

Yunong Zhang and Long Jin

Acronyms

DOF	Degrees of freedom
FAMJM	Feedback-aided minimum joint movement
GD	Gradient dynamics
IIWT	Inertia inverse weighted torque
JAL	Joint-acceleration level
JCMP	Joystick-controlled motion planning
JVL	Joint-velocity level
LVI	Linear variational inequality
LVIAPDNN	LVI-aided primal-dual neural network
MAN	Minimum acceleration norm
MKE	Minimum kinetic energy
MM	Manipulability-maximizing, manipulability-maximization
MMSMP	Manipulability-maximizing self-motion planning
MTN	Minimum torque norm
MVN	Minimum velocity norm
NN	Neural networks
OMPFC	Online motion planning and feedback control
PCI	Peripheral component interconnect
PLE	Piecewise-linear equation
PPS	Pulses per second
PR	Push-rod
QP	Quadratic program, quadratic programming
RMP	Repetitive motion planning
SMMVA	Self-motion with middle-value approached
SMP	Self-motion planning
TVCMM	Time-varying coefficient aided manipulability-maximizing
VJVL	Varying joint-velocity limits
ZD	Zeroing dynamics
ZG	Zeroing-gradient
ZIV	Zero initial velocity

Part I

Pseudoinverse-Based ZD Approach

1

Redundancy Resolution via Pseudoinverse and ZD Models

1.1 Introduction

Recently, robotics has played a more and more important role in scientific research and engineering applications [1–4]. Being an essential topic, the problem of redundancy resolution (or mostly say, the problem of inverse kinematics, which is related to the kinematic control of some redundant robot manipulator) has attracted the extensive attention of many researchers [5–9]. The general description of such a problem is that, given the desired Cartesian path $\mathbf{r}_\mathrm{d}(t) \in \mathbb{R}^m$ of the end-effector versus time $t \in [t_0, t_\mathrm{f}] \subseteq [0, \infty)$, the corresponding trajectories of joint-variable vector $\theta(t) \in \mathbb{R}^n$ need to be obtained online (or offline and in advance). In mathematics, to find $\theta(t)$ such that

$$\mathbf{f}(\theta(t)) \to \mathbf{r}_\mathrm{d}(t),$$

where forward-kinematics mapping $\mathbf{f}(\cdot)$ is nonlinear and differentiable with a known structure and parameters for a given robot manipulator [1–3, 9]. Note that, for a redundant robot manipulator, it possesses more degrees of freedom (DOF) than necessary to perform a user-specified end-effector primary task (in mathematics, $m < n$) [1–3]. The extra DOF allow the existence of an infinite number of feasible solutions to the redundancy-resolution problem. This can be utilized to determine the best joint-variable, joint-velocity, and/or joint-acceleration vectors in some sense (for a specified end-effector pose, velocity, or acceleration vector), which corresponds to an optimality criterion [9]. Thus, many studies have reported on the kinematic control of redundant robot manipulators [5–11].

The pseudoinverse-based approach represents the conventional solution to the redundancy-resolution problem, which can be analytical. In general, it contains one minimum-norm particular solution plus a homogeneous solution. This simple characteristic has made the research and application of such a pseudoinverse-based approach popular in recent decades [3, 5–7, 9–11]. For example, in [7], a multi-objective approach is investigated for the motion planning of redundant robot manipulators, which combines the closed-loop pseudoinverse method and a multi-objective genetic algorithm. Among the pseudoinverse-based techniques for robotic redundancy resolution, the minimum velocity norm (MVN) scheme, which aims at minimizing the sum of squares of joint velocities, has been widely adopted by researchers for the kinematic control of redundant robot manipulators [1–3, 9]. In addition, to guarantee the high

Robot Manipulator Redundancy Resolution, First Edition. Yunong Zhang and Long Jin.

precision of the end-effector positioning (i.e., the small or tiny position error of the end-effector), the corresponding feedback can be added to such a pseudoinverse-based MVN scheme. Thus, with time argument t omitted here (and sometimes afterward) for presentation convenience, the pseudoinverse-based MVN scheme with feedback (being a closed-loop scheme) is formulated as

$$\dot{\theta} = J^{\dagger}(\dot{\mathbf{r}}_{\mathrm{d}} + \kappa_{\mathrm{p}}(\mathbf{r}_{\mathrm{d}} - \mathbf{f}(\theta))), \tag{1.1}$$

where $\dot{\theta}$ is the joint-velocity vector; $J^{\dagger} \in \mathbb{R}^{n\times m}$ denotes the pseudoinverse of the Jacobian matrix $J = \partial\mathbf{f}(\theta)/\partial\theta \in \mathbb{R}^{m\times n}$; and $\dot{\mathbf{r}}_{\mathrm{d}}$ is the desired Cartesian velocity vector, that is, the time-derivative vector of the desired end-effector path $\mathbf{r}_{\mathrm{d}}$. Besides, $\kappa_{\mathrm{p}} > 0 \in \mathbb{R}$ is the feedback gain of position error $\epsilon = \mathbf{r}_{\mathrm{d}} - \mathbf{f}(\theta)$. Note that pseudoinverse-based MVN scheme (1.1) can actually be obtained by defining the aforementioned error function ϵ and exploiting the neural-dynamics method presented in [12, 13] (i.e., the method of zeroing dynamics, ZD, presented in this chapter as well). Compared with the widely-used constraint (i.e., $J\dot{\theta} = \dot{\mathbf{r}}_{\mathrm{d}}$) in the existing literature [1, 2, 9], a prominent advantage of the feedback-added constraint is that it guarantees the error ϵ with no drifting/diverging happened. This also means that the error drift/divergence phenomenon in pseudoinverse-based MVN scheme (1.1) does not exist. Besides, one of the main reasons for the popularity of the two-norm as an optimality criterion is the fact that the related optimization problems yield closed-form analytical expressions. Thus, in many robotic applications, the two-norm optimality criterion is utilized, more because of its mathematical tractability than physical desirability [9]. On the basis of these considerations, pseudoinverse-based MVN scheme (1.1) for redundant-manipulator kinematic control is the focus of this chapter.

In recent years, as a branch of artificial intelligence, artificial neural networks (ANNs) have attracted considerable attention as candidates for novel computational systems [1, 2, 12, 13]. Being a special type of ANNs, recurrent neural networks (RNNs), which originate from the research of Hopfield neural network [14], have been developed and investigated for solving a wide variety of mathematical problems arising in numerous fields of science and engineering [1, 2, 12]. Note that, compared with conventional numerical algorithms, the dynamics approach based on RNNs has several potential advantages in real-time applications, for example, high-speed parallel-processing, distributed-storage, and adaptive self-learning natures. Therefore, such an approach is generally taken into account as a powerful alternative to online computation and optimization. Especially for robotic redundancy resolution, various RNN models have been developed, exploited, and investigated [1, 2, 15–17].

As for pseudoinverse-based MVN scheme (1.1), the problem of time-varying matrix pseudoinversion is involved in the scheme formulation. It is worth pointing out here that, owing to its fundamental roles, much effort has been devoted to the fast solution of matrix pseudoinversion problem, and many models have been presented by researchers [18–22]. However, the investigations and revisit on those models are not the emphases in this book and thus will not be covered.

Gradient dynamics (GD) and zeroing dynamics (ZD) are two powerful dynamics approaches based on RNNs, which are regarded as two alternatives for online computation, and have widely arisen in scientific and engineering fields, drawing extensive interests and investigation of researchers [18, 23]. In particular especially, GD models are proposed and investigated in [18], which use the Frobenius norm of the error matrix as the performance criterion and evolve along the negative gradient-descent direction to make the error norm decrease to zero with time in the time-invariant case. Recent researches have shown that such a GD approach can also be developed and generalized for time-varying problems solving [12, 13]. Besides, the ZD approach, which is based on an indefinite matrix/vector-valued error function, has been presented for time-varying problems solving, such as the time-varying matrix pseudoinversion [12, 23]. It is worth providing the conceptions of ZD and GD here to lay a basis for further investigation.

Concept 1.1 Zeroing dynamics (ZD), where the state dimension can be one or more, has been derived from the zeroing neural network. It is viewed as a systematic approach to the online solution of time-varying problems; it differs from the conventional GD in terms of the error function, design formula, dynamic equation, and the utilization of time derivatives [12, 23, 24].

Concept 1.2 Gradient dynamics (GD), which uses the Frobenius norm of the error matrix as the performance criterion and evolves along the negative gradient-descent direction to make the error norm decrease to zero with time, is another type of dynamic approach. It is intrinsically feasible and efficient to solve time-invariant problems and can be developed and generalized to solve time-varying problems. For more details about the differences between ZD and GD, please refer to [13].

Compared with GD, the ZD design is based on a matrix-valued indefinite error function and an exponent-type formula (i.e., the ZD design formula), which makes every element of the error function exponentially converge to zero. By making good use of the time-derivative information of the time-varying coefficient, the resultant ZD models can effectively avoid the lagging errors and can exponentially converge to the theoretical pseudoinverse of time-varying matrix (e.g., the Jacobian matrix J involved in this chapter) [23].

On the basis of the successful work [12, 23, 24], this chapter presents and investigates the application of discrete-time ZD models to kinematic control of redundant robot manipulators via time-varying Jacobian matrix pseudoinversion. That is, by computing the time-varying pseudoinverse of the Jacobian matrix, the resultant ZD models are applied to redundant-manipulator kinematic control. Simulation results based on a five-link robot manipulator and a three-link robot manipulator are illustrated to show the effectiveness of the presented ZD models for time-varying matrix pseudoinversion applied to the redundancy resolution of robot manipulators.

1.2 Problem Formulation and ZD Models

In this section, based on the previous work [12, 23, 24], the discrete-time ZD models are presented and investigated for time-varying Jacobian matrix pseudoinversion.

1.2.1 Problem Formulation

In order to lay a basis for further investigation, the preliminaries and problem formulation of time-varying matrix pseudoinversion are presented next.

Definition 1.1 [19, 25, 26] For a given time-varying matrix $J(t) \in \mathbb{R}^{m\times n}$, if $X(t) \in \mathbb{R}^{n\times m}$ satisfies at least one of the following four Penrose equations:

$$J(t)X(t)J(t) = J(t),\ \ X(t)J(t)X(t) = X(t),$$
$$(J(t)X(t))^{\mathrm{T}} = J(t)X(t),\ \ (X(t)J(t))^{\mathrm{T}} = X(t)J(t),$$

$X(t)$ is called the generalized inverse of $J(t)$. If matrix $X(t)$ satisfies all of the Penrose equations, then matrix $X(t)$ is called the pseudoinverse of matrix $J(t)$, which is often denoted by $J^{\dagger}(t)$.

Note that the time-varying Jacobian pseudoinverse $J^{\dagger}(t)$ always exists and is unique [25]. In this chapter, we only consider the situation of $m < n$. In particular, if Jacobian matrix $J(t)$ is of full rank at any time instant t, that is, $\mathrm{rank}(J(t)) = m, \forall t \in [t_0, t_f] \subseteq [0, \infty)$, we have the following lemma about the time-varying pseudoinverse of $J(t)$.

Lemma 1.1 [25] For time-varying matrix $J(t) \in \mathbb{R}^{m\times n}$ with $m < n$, if $\mathrm{rank}(J(t)) = m$, $\forall t \in [t_0, t_f] \subseteq [0, \infty)$, then the unique time-varying pseudoinverse $J^{\dagger}(t)$ can be formulated as

$$J^{\dagger}(t) = J^{\mathrm{T}}(t)(J(t)J^{\mathrm{T}}(t))^{-1}. \tag{1.2}$$

Besides, for time-varying matrix $X(t) \in \mathbb{R}^{n\times m}$ with full rank, its unique time-varying pseudoinverse $X^{\dagger}(t) = (X^{\mathrm{T}}(t)X(t))^{-1}X^{\mathrm{T}}(t)$.

As the time-varying pseudoinverse $J^{\dagger}(t) \in \mathbb{R}^{n\times m}$ satisfies the corresponding matrix equation $J^{\dagger}(t)J(t)J^{\mathrm{T}}(t) = J^{\mathrm{T}}(t)$, the following problem formulation of continuous-time varying Jacobian matrix pseudoinversion can be considered and/or checked for the solution correctness:

$$X(t)J(t)J^{\mathrm{T}}(t) - J^{\mathrm{T}}(t) = 0 \in \mathbb{R}^{n\times m}, \tag{1.3}$$

where $X(t) \in \mathbb{R}^{n\times m}$ is the time-varying unknown matrix to be obtained. To lay a basis for further discussion, $J(t)$ is assumed to be of full rank at any time instant $t \in [t_0, t_f] \subseteq [0, \infty)$. In other words, the configuration singularity for the redundant robot manipulator does not exist during the motion-task execution.

1.2.2 Continuous-Time ZD Model

To design and control the solving process of the problem (1.3), the following matrix-valued indefinite error function is defined:

$$E(t) = J(t) - X^{\dagger}(t) \in \mathbb{R}^{m\times n}. \tag{1.4}$$

Then, based on the ZD design methodology [12], specifically, ZD design formula $\dot{E}(t) = -\gamma E(t)$ with $\gamma > 0$, we can obtain $\dot{J}(t) - \dot{X}^{\dagger}(t) = -\gamma(J(t) - X^{\dagger}(t))$. From $X^{\dagger}(t)X(t) = I$, with I being the identity matrix, we have an underdetermined system: $\dot{X}^{\dagger}(t)X(t) = -X^{\dagger}(t)\dot{X}(t)$, of which the minimum norm solution is $\dot{X}^{\dagger}(t) = -X^{\dagger}(t)\dot{X}(t)X^{\dagger}(t)$. Thus, we have the following dynamic equation:

$$X^{\dagger}(t)\dot{X}(t)X^{\dagger}(t) = -\dot{J}(t) - \gamma(J(t) - X^{\dagger}(t)). \tag{1.5}$$

From (1.5), postmultiplying $X(t)$ and knowing $X^{\dagger}(t)X(t) = I$, we have

$$X^{\dagger}(t)\dot{X}(t) = X^{\dagger}(t)\dot{X}(t)X^{\dagger}(t)X(t) = -\dot{J}(t)X(t) - \gamma(J(t)X(t) - I). \tag{1.6}$$

Evidently, Equation (1.6) is underdetermined and for this the minimum norm solution exists. That is, we have the minimum norm solution to it:

$$\dot{X}(t) = -X(t)\dot{J}(t)X(t) - \gamma(X(t)J(t)X(t) - X(t)), \tag{1.7}$$

which is termed the continuous-time ZD model. Note that continuous-time ZD model (1.7) is also the Getz–Marsden (G-M) dynamic system for time-varying pseudoinversion [23].

1.2.3 Discrete-Time ZD Models

For the purposes of possible hardware implementation, for example, the digital computer or the digital circuit, the discrete-time ZD (DTZD) models are presented and developed in this section for time-varying pseudoinversion (1.3).

1.2.3.1 Euler-Type DTZD Model with $\dot{J}(t)$ Known

In order to discretize the presented continuous-time ZD model (1.7) for solving time-varying pseudoinverse, we refer to the following Euler forward-difference rule [26, 27, 29]:

$$\dot{X}(t = k\tau) \approx \frac{X((k+1)\tau) - X(k\tau)}{\tau}, \tag{1.8}$$

where $\tau > 0$ denotes the sampling period, and update index $k = 0, 1, 2, \cdots$. In general, we denote $X_k = X(t = k\tau)$ for presentation convenience. In addition, $J(t)$ and $\dot{J}(t)$(which is assumed to be known) are discretized by the standard sampling method, of which the sampling period is also $\tau = t_{k+1} - t_k$. For convenience and also for consistency with X_k, we use J_k standing for $J(t = k\tau)$ and $\dot{J}_k$ standing for $\dot{J}(t = k\tau)$. Thus, we discretize continuous-time ZD model (1.7) as

$$X_{k+1} = X_k - \tau X_k\dot{J}_kX_k - h(X_kJ_kX_k - X_k), \tag{1.9}$$

where step-size $h = \tau\gamma > 0$. For presentation convenience, in this chapter, the discrete-time model (1.9) is called the Euler-type DTZD model with $\dot{J}(t)$ known, that is, the EDTZD-K model.

1.2.3.2 Euler-Type DTZD Model with $\dot{J}(t)$ Unknown

As probably we know, in real-world applications, the analytical form or numerical value of $\dot{J}(t)$ may be difficult to know. Thus, it is worth investigating the discrete-time ZD model when $\dot{J}(t)$ is unknown. In this situation, $\dot{J}(t)$ is generally estimated from the existing data of $J(t)$ by employing the following Euler backward-difference rule [27]:

$$\dot{J}(t = k\tau) \approx \frac{J(k\tau) - J((k-1)\tau)}{\tau},$$

where $\tau > 0$ is defined the same as before. Similarly, we define $J_k = J(t = k\tau)$ and $J_{k-1} = J(t = (k-1)\tau)$ for presentation convenience. From EDTZD-K model (1.9), we derive the Euler-type DTZD model with $\dot{J}(t)$ unknown (i.e., EDTZD-U model) for time-varying pseudoinversion as follows:

$$X_{k+1} = X_k - X_k(J_k - J_{k-1})X_k - h(X_kJ_kX_k - X_k). \tag{1.10}$$

Note that, from the Euler backward-difference rule, we can not estimate $\dot{J}(0)$ since t starts from 0 s and J_{-1} is undefined. In this situation, we choose $\dot{J}(0) = 0 \in \mathbb{R}^{m\times n}$ (i.e., $J_0 = J_{-1}$) to start the update (1.10). That is, at the first update, the EDTZD-U model is

$$X_1 = X_0 - h(X_0 J_0 X_0 - X_0),$$

which initiates the iterative computation of EDTZD-U model (1.10).

1.2.3.3 Taylor-Type DTZD Models

In order to achieve a higher computational accuracy in the ZD discretization for solving time-varying pseudoinverse, we refer to the following Taylor forward-difference rule [26]:

$$\dot{X}_k \approx \frac{2X_{k+1} - 3X_k + 2X_{k-1} - X_{k-2}}{2\tau}. \tag{1.11}$$

Thus, we discretize continuous-time ZD model (1.7) as

$$X_{k+1} = -\tau X_k \dot{J}_k X_k - h(X_k J_k X_k - X_k) + \frac{3}{2}X_k - X_{k-1} + \frac{1}{2}X_{k-2}. \tag{1.12}$$

For presentation convenience, Equation (1.12) is called the Taylor-type discrete-time ZD model with $\dot{J}(t)$ known, that is, the TDTZD-K model.

As we know, it may be difficult to know or obtain the value of $\dot{J}(t)$ directly in certain real-world applications. In this situation, $\dot{J}(t)$ can be estimated from $J(t)$ by employing the backward-difference rule of the first-order derivative with a third-order accuracy [27]:

$$\dot{J}_k \approx \frac{11J_k - 18J_{k-1} + 9J_{k-2} - 2J_{k-3}}{6\tau}. \tag{1.13}$$

Thus, the Taylor-type discrete-time ZD model with $\dot{J}(t)$ unknown (i.e., the TDTZD-U model) can be formulated as

$$\begin{aligned} X_{k+1} = &-X_k\left(\frac{11}{6}J_k - 3J_{k-1} + \frac{3}{2}J_{k-2} - \frac{1}{3}J_{k-3}\right)X_k \\ &-h(X_k J_k X_k - X_k) + \frac{3}{2}X_k - X_{k-1} + \frac{1}{2}X_{k-2}. \end{aligned} \tag{1.14}$$

The classical Newton iteration is generalized and developed to solve (1.3), which is formulated as [28]:

$$X_{k+1} = X_k - (X_k J_k X_k - X_k). \tag{1.15}$$

It is noted that three initial states X_0, X_1 and X_2 are needed for the initialization of TDTZD-U model (1.14). Besides, from (1.13), we can not obtain $\dot{J}_0$ since t starts from 0 s and thus J_{-1} is undefined. Thus, in the ensuing applications to the redundancy resolution of robot manipulators, in view of the fact that the Newton iteration has the simplest structure, we exploit Newton iteration (1.15) for the initializations of the TDTZD-K (1.12) and TDTZD-U models (1.14).

Remark 1.2 As for pseudoinverse-based MVN scheme (1.1) and ZD models, design parameters κ and $\gamma = h/\tau$ play important roles. Specifically, κ is the gain of the outer loop to solve for θ while γ decides the convergence rate of the inner loop for solving the ZD for the pseudoinverse of Jacobian matrix. This is a cascaded system (see also Figure 1.1), and it is thus necessary to choose the value of γ much larger than the value of κ such that the inner loop converges much faster.

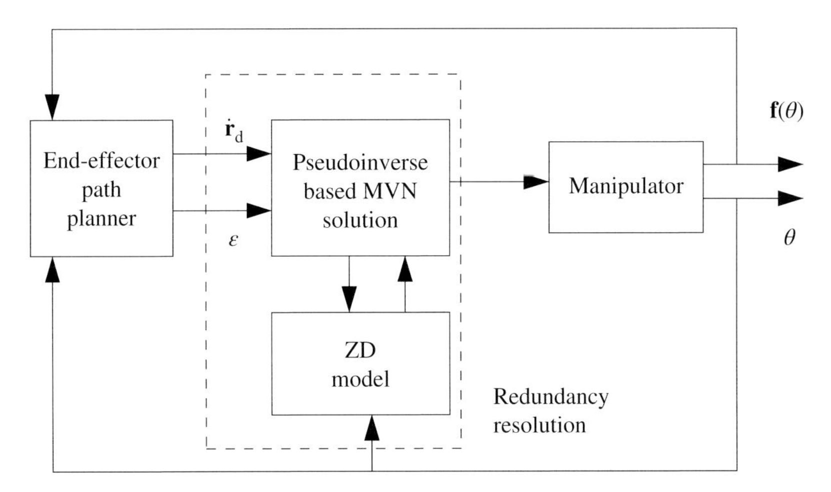

Figure 1.1 Block diagram of a kinematic-control system for a redundant robot manipulator by combining the MVN scheme (1.1) and ZD model, where $\epsilon = \mathbf{r}_\mathrm{d} - \mathbf{f}(\theta)$.

1.3 ZD Applications to Different-Type Robot Manipulators

In this section, based on a five-link planar robot manipulator and a three-link planar robot manipulator, computer simulations are conducted to illustrate the effectiveness of the presented discrete-time ZD models for redundant-manipulator kinematic control.

For a better understanding, the block diagram of the kinematic-control system that incorporates pseudoinverse-based MVN scheme (1.1) and the presented ZD models is illustrated in Figure 1.1.

1.3.1 Application to a Five-Link Planar Robot Manipulator

In this subsection, the presented discrete-time ZD models and Newton iteration are applied to the kinematic control of a five-link planar robot manipulator with its geometry illustrated in Figure 1.2. For such a robot manipulator, each element of Jacobian matrix $J \in \mathbb{R}^{2\times 5}$ is presented as follows:

$$\begin{aligned}
&J_{11} = -l_1 s_1 - l_2 s_2 - l_3 s_3 - l_4 s_4 - l_5 s_5,\ J_{12} = -l_2 s_2 - l_3 s_3 - l_4 s_4 - l_5 s_5,\\
&J_{13} - l_3 s_3 - l_4 s_4 - l_5 s_5,\ J_{14} = -l_4 s_4 - l_5 s_5,\ J_{15} = -l_5 s_5,\\
&J_{21} = l_1 c_1 + l_2 c_2 + l_3 c_3 + l_4 c_4 + l_5 c_5,\ J_{22} = l_2 c_2 + l_3 c_3 + l_4 c_4 + l_5 c_5,\\
&J_{23} = l_3 c_3 + l_4 c_4 + l_5 c_5,\ J_{24} = l_4 c_4 + l_5 c_5,\ J_{25} = l_5 c_5,
\end{aligned}$$

where l_i (with $i = 1, 2, 3, 4, 5$) denotes the length of the ith link. In addition, $s_i = \sin\left(\sum_{j=1}^{i} \theta_j\right)$ and $c_i = \cos\left(\sum_{j=1}^{i} \theta_j\right)$. For simplicity and illustration, with each link length being 1 m, the five-link planar robot manipulator is investigated to track a square path with the side length being 2.4 m, where $T = 20$ s and initial joint state $\theta(0) = [\pi/4, \pi/12, \pi/4, \pi/12, \pi/4]^\mathrm{T}$ rad. Besides, feedback gain is set as $\kappa = 0$, the sampling period is chosen as $\tau = 1$ ms and step-size is set as $h = 0.3$. The end-effector position error ϵ is defined as $\epsilon = \mathbf{r}_\mathrm{d} - \mathbf{f}(\theta) \in \mathbb{R}^2$, where $\mathbf{r}_\mathrm{d}$ denotes the desired end-effector square path.

Numerical results synthesized by pseudoinverse-based MVN scheme (1.1) aided with TDTZD-U model (1.14) are shown in Figure 1.3 and Figure 1.4. From Figure 1.3(a) and

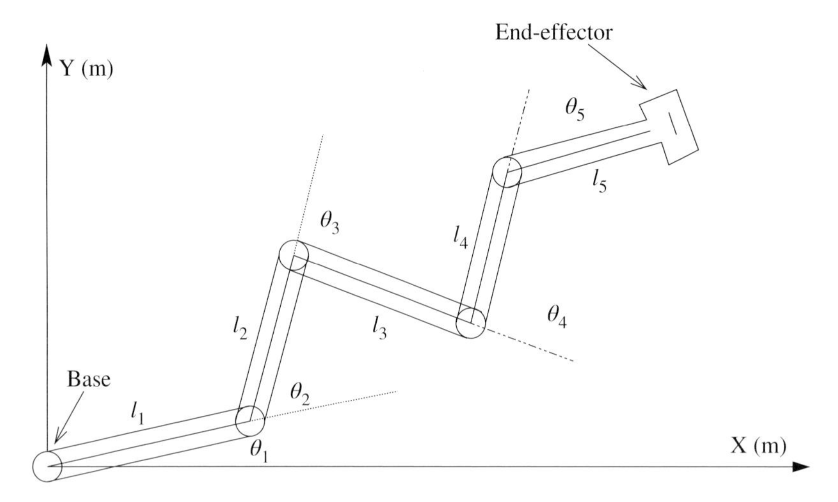

Figure 1.2 Geometry of a five-link planar robot manipulator used in simulations.

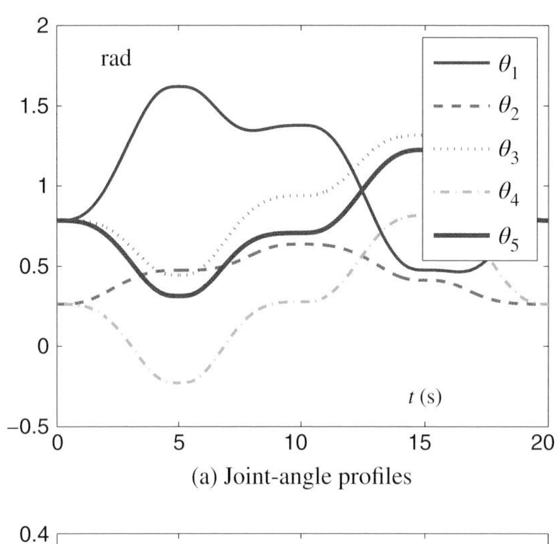

(a) Joint-angle profiles

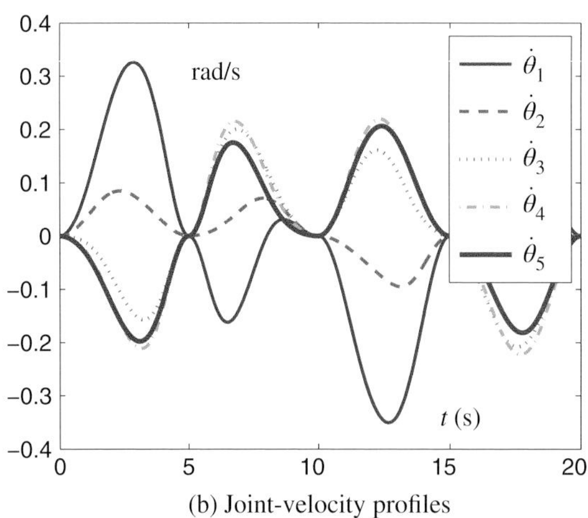

(b) Joint-velocity profiles

Figure 1.3 Joint-angle and joint-velocity profiles of a five-link planar robot manipulator synthesized by pseudoinverse-based MVN scheme (1.1) aided with TDTZD-U model (1.14).

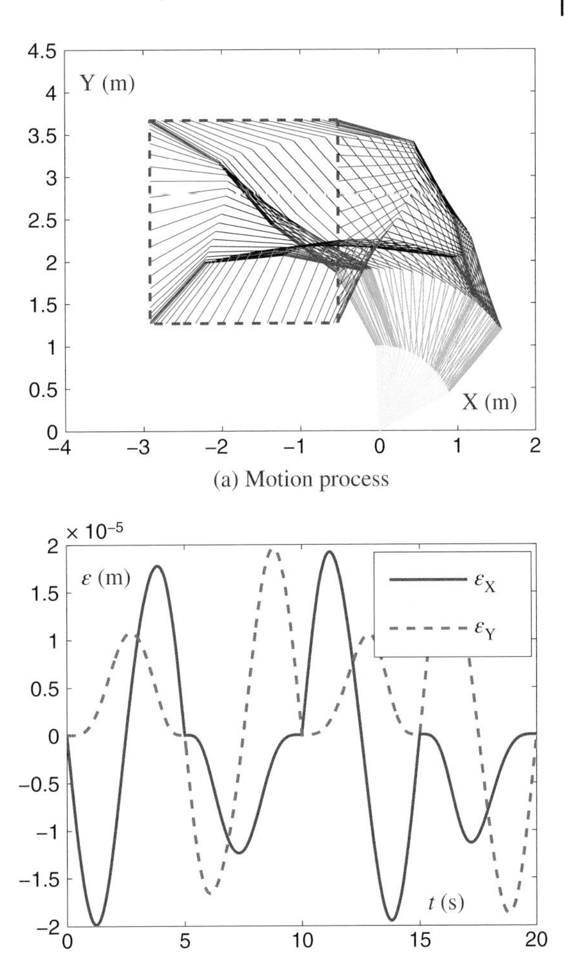

Figure 1.4 (a) Motion process and (b) position error of a five-link planar robot manipulator synthesized by the pseudoinverse-based MVN scheme (1.1) and aided by the TDTZD-U model (1.14).

(b), we can see that the joint variables (i.e., the joint angle θ and the joint velocity $\dot{\theta}$) are smooth and have not undergone abrupt changes, which is suitable for engineering applications. In addition, it can be seen from Figure 1.4 that the simulated end-effector trajectory of the robot manipulator is very close to the desired path, with the maximum end-effector position error being less than 2.0×10^{-5} m. This illustrates the effectiveness of the presented TDTZD-U model (1.14). These results substantiate well that pseudoinverse-based MVN scheme (1.1) aided with TDTZD-U model (1.14) is effective on redundant-manipulator kinematic control.

For further comparison, numerical results synthesized by the pseudoinverse-based MVN scheme (1.1) aided with a EDTZD-U model (1.10) and Newton iteration (1.15) are shown in Figure 1.5. Note that, similar to Figure 1.3 and Figure 1.4 synthesized by TDTZD-U model (1.14), the joint-angle, joint-velocity, and motion process generated by Newton iteration (1.15) are omitted due to space limitation.

It can be seen from Figure 1.5(a) that the maximum position error synthesized by EDTZD-U model (1.10) is 6×10^{-5} m, which is roughly three times larger than that

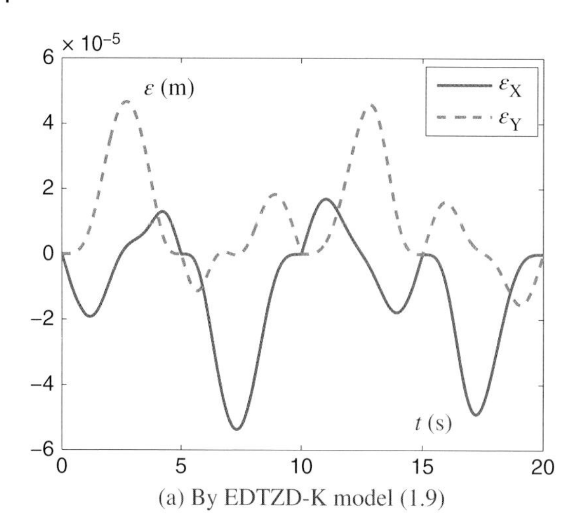

(a) By EDTZD-K model (1.9)

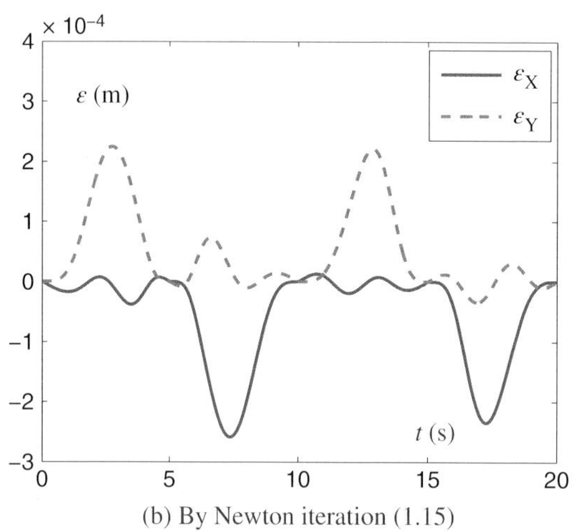

(b) By Newton iteration (1.15)

Figure 1.5 Position error of a five-link planar robot manipulator synthesized by a pseudoinverse-based MVN scheme (1.1) aided with the EDTZD-K model (1.9) or Newton iteration (1.15).

TDTZD-U model (1.14) in Figure 1.4. Besides, the maximum position error synthesized by Newton iteration (1.15) is 3×10^{-4} m, which is roughly 10 times larger than that TDTZD-U model (1.14) in Figure 1.4.

This application to the motion generation of the five-link planar robot manipulator further illustrates the superiority of the presented discrete-time ZD models for time-varying matrix pseudoinversion applied to the redundancy resolution of robot manipulators.

1.3.2 Application to a Three-Link Planar Robot Manipulator

In this section, the presented EDTZD-K model (1.9) and EDTZD-U model (1.10) are applied to the kinematic control of a three-link planar robot manipulator via online solution of time-varying pseudoinverse. In this application, the lengths of the links are set as $l_1 = l_2 = l_3 = 1.0$ m, and the initial joint state $\theta(0) = [\pi/6, \pi/6, \pi/6]^{\mathrm{T}}$ rad. In addition, the end-effector of the three-link planar robot manipulator is expected to track a square path with the side length being 0.5 m, and the motion-task duration is 40 s. Also, feedback gain is set as $\kappa = 0$.

It is worth pointing out that the motion process and the profiles of joint-angle and joint-velocity are omitted due to space limitation and we only present the simulation results on position error. Besides, the figure on position errors generated by EDTZD-U model (1.10) is very similar to that generated by EDTZD-K model (1.9) and thus omitted here.

EDTZD-K model (1.9) is applied to the robot's kinematic control, with the results shown in Figure 1.6. As seen from the figure, better control precision can also be achieved by using EDTZD-K model (1.9) with appropriate values of step size h and sampling period τ. That is, as synthesized by EDTZD-K model (1.9), the end-effector trajectories of the three-link planar robot manipulator are both sufficiently close to the desired square path with small position errors (i.e., of orders 10^{-4} and 10^{-5} m). These substantiate the effectiveness of the discrete-time models on robots' kinematic control.

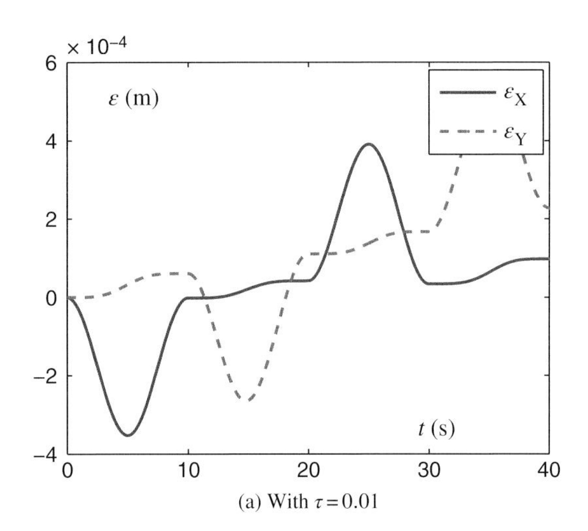

(a) With $\tau=0.01$

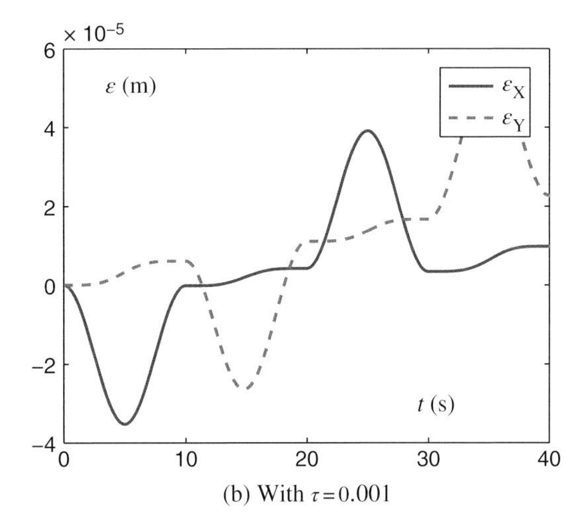

(b) With $\tau=0.001$

Figure 1.6 Position errors of a three-link planar robot manipulator synthesized by a pseudoinverse-based MVN scheme (1.1) aided with EDTZD-K model (1.9) with $h = 1.0$.

1.4 Chapter Summary

In this chapter, discrete-time ZD models have been presented and investigated for application to kinematic control of redundant robot manipulators. Specifically, to obtain the theoretical pseudoinverse of time-varying Jacobian matrix involved in robotic redundancy resolution, the continuous-time ZD model (1.7) and different discrete-time ZD models have been developed and investigated. Theoretical results have also been given showing the ZD models' feasibility for redundant-manipulator kinematic control. Then, the presented discrete-time ZD models have been applied to different types of redundant robot manipulators (i.e., the five-link planar robot manipulator and the three-link planar robot manipulator). Simulation results have illustrated the effectiveness of the presented discrete-time ZD models for time-varying matrix pseudoinversion applied to the redundancy resolution of robot manipulators.

Part II

Inverse-Free Simple Approach

2

G1 Type Scheme to JVL Inverse Kinematics

2.1 Introduction

In recent years, there have been numerous investigations in to robot manipulators [30–33], especially the redundant robot manipulators. Compared with the ordinary robot manipulators, the redundant one has more degrees of freedom (DOF) than necessary for position and orientation. It is worth noting that this characteristic improves the kinematic and the dynamic performance of the robot manipulator, such as increasing dexterity, avoiding obstacles and singularities, and optimizing joint velocity, which makes redundant the manipulators widely applied in the field of robotic manipulator control [34]. As one of the central issues in robot control, path tracking refers to making the end-effector move as expected by controlling the joints of robot manipulators. When the end-effector moves at a desired speed, it is often called path tracking control at the velocity level (JVL), which can fit in with the needs of the operators well with two advantages [35, 36]. One is that the high velocity can economize the execution time while the other one is that the low velocity can enhance the objective precision [37]. So the control at the velocity level is suited particularly well to tasks for machining operations, such as cutting and milling [38].

Robot kinematics, which studies the relationships between joint space and cartesian space, is a very effective way to control the robot manipulator and attracts numberous researchers [39–41]. The velocities of all joints play a decisive role in realizing the desired speed of the end-effector. So the most fundamental problem is how to use the kinematic equations to calculate the homologous joint velocities, also called inverse kinematics. However, because most kinematic equations involve complex (inverse) trigonometric functions, the inverse kinematics mapping has no closed-form solutions for most manipulators and animation figures [42]. Conventionally, most researchers exploit pseudoinverse approaches to obtain a simple general-form solution [11, 43]. However, these approaches do not only need a lot of time in calculating the inverse of Jacobian matrix, but also require the Jacobian matrix to be of full rank, which may be far from the reality. Thus various approaches have been presented, investigated, and developed to avoid the calculation of the inverse of Jacobian matrix, such as the quadratic programming (QP) method [44].

Gradient dynamics (GD) [45, 46] is a significant dynamic method that attracts many researchers to investigate and develop it. Now, the GD method is proving to be useful

Robot Manipulator Redundancy Resolution, First Edition. Yunong Zhang and Long Jin.

and effective and is widely acknowledged in scientific and engineering fields, thus generalizing such a GD method has become the primary work [44, 47–50]. In this chapter, based on GD method's advantage that it can help find a minimum of a nonnegative objective function effectively, we present and investigate a scheme named G1 type for path tracking in the redundant robot manipulator at the joint-velocity level. Also, in the framework of zeroing-gradient (ZG) method, the G1 type scheme is a special Z0G1 situation that only uses the GD method once. The ZG method is an effective method built by combining ZD presented in Chapter 1 and GD to solve the tracking-control and singularity problems [51, 52].

By exploiting the GD method, this chapter presents and investigates a G1 type scheme in an inverse-free manner at the joint-velocity level. The presented G1 type scheme can solve the inverse kinematics problem effectively but avoids calculating the inverse of Jacobian matrix. Besides, the path-tracking simulations and the physical experiments are conducted to further illustrate the effectiveness, the high accuracy, and the physical realizability of the G1 type scheme.

2.2 Preliminaries and Related Work

For a redundant robot manipulator with n joints, the end-effector pose (or position in this chapter) vector $\mathbf{r}_{\mathrm{e}} \in \mathbb{R}^m$ can be described by the following equation:

$$\mathbf{r}_{\mathrm{e}} = \mathbf{f}(\boldsymbol{\theta}), \tag{2.1}$$

where $\boldsymbol{\theta} \in \mathbb{R}^n$ refers to the variables (or angles in this chapter) of the n joints, and $\mathbf{f}(\cdot)$ is a differentiable nonlinear function with a known structure and parameters for a given manipulator. Then by differentiating Equation (2.1), the end-effector velocity is

$$\dot{\mathbf{r}}_{\mathrm{e}} = J\dot{\boldsymbol{\theta}}, \tag{2.2}$$

where $\dot{\mathbf{r}}_{\mathrm{e}} \in \mathbb{R}^m$ refers to the end-effector velocity, and $\dot{\boldsymbol{\theta}} \in \mathbb{R}^n$ refers to the velocities of all joints. Note that $J \in \mathbb{R}^{m\times n}$ is the Jacobian matrix defined as $J = \partial\mathbf{f}(\boldsymbol{\theta})/\partial\boldsymbol{\theta}$. According to Equation (2.2), for tracking the desired path $\mathbf{r}_{\mathrm{d}} \in \mathbb{R}^m$ via the desired speed $\dot{\mathbf{r}}_{\mathrm{d}} \in \mathbb{R}^m$, the velocities of all joints can be obtained by the following equation if J is a square matrix and of full rank:

$$\dot{\boldsymbol{\theta}} = J^{-1}\dot{\mathbf{r}}_{\mathrm{d}}, \tag{2.3}$$

If J is rectangular, the velocities of joints may be computed by the following equation of generalized inverse [35]:

$$\dot{\boldsymbol{\theta}} = J^{\dagger}\dot{\mathbf{r}}_{\mathrm{d}}, \tag{2.4}$$

where $J^{\dagger}$ denotes the pseudoinverse of Jacobian matrix J. However, the inverse scheme theoretically requires the Jacobian matrix to be of full rank, which, in real world application, may be unavailable sometimes in practice.

2.3 Scheme Formulation

In this section, we generalize the GD method to obtain an inverse-free scheme at the joint-velocity level. By following the GD method, the design procedure of such an inverse-free scheme can be presented detailedly via the following steps.

Firstly, to monitor and control the process of solving the time-varying inverse kinematics problem of redundant manipulators, we define a scalar-valued norm-based energy function according to Equation (2.1):

$$\rho = \|\mathbf{r}_{\mathrm{d}} - \mathbf{f}(\boldsymbol{\theta})\|_2^2/2, \tag{2.5}$$

where $\|\cdot\|_2$ denotes the two-norm of a vector.

Secondly, a computational rule is designed to evolve along a descent direction of this energy function until the minimum point is reached. The typical descent direction is the negative gradient of ρ, that is,

$$-\partial\rho/\partial\boldsymbol{\theta} = J^{\mathrm{T}}(\mathbf{r}_{\mathrm{d}} - \mathbf{f}(\boldsymbol{\theta})). \tag{2.6}$$

Then we combine the aforementioned negative gradient (2.6) and the following GD design formula [45, 46]:

$$\dot{\boldsymbol{\theta}} = -\lambda\partial\rho/\partial\boldsymbol{\theta}, \tag{2.7}$$

where the design parameter $\lambda > 0$ is used to scale the convergence rate of the GD method.

Finally, we have the following generalized G1 type scheme for solving the time-varying inverse kinematics problem of redundant robot manipulators:

$$\dot{\boldsymbol{\theta}} = \lambda J^{\mathrm{T}}(\mathbf{r}_{\mathrm{d}} - \mathbf{f}(\boldsymbol{\theta})). \tag{2.8}$$

Evidently, scheme (2.8) does not require the Jacobian inversion appearing in (2.3) and (2.4). Since scheme (2.8) is obtained by applying the GD method only once, it is named a G1 type (or Z0G1 type in the ZG framework).

2.4 Computer Simulations

In this section, the corresponding path-tracking simulations (square path tracking and "Z"-shaped-path tracking) are performed on five-link and six-DOF redundant robot manipulators, respectively, to illustrate the effectiveness and the accuracy of the presented G1 type scheme. Note that the design parameter $\lambda = 10^5$ is used throughout this section.

2.4.1 Square-Path Tracking Task

In the first example, G1 type scheme (2.8) is applied to tracking a square path via a desired velocity. The side length of the square path is 0.8 m, and the path tracking is simulated on a five-link redundant robot manipulator, with an initial state $\boldsymbol{\theta}(0)$ being $[\pi/3, \pi/3, \pi/2, -\pi/4, \pi/4]^{\mathrm{T}}$ rad. The corresponding simulation results are shown in Figures 2.1 through Figure 2.3, which illustrate the effectiveness and high accuracy of the presented G1 type scheme (2.8) for solving the time-varying inverse kinematics of robot manipulators.

Specifically, the results can be seen from Figure 2.1(a) and Figure 2.1(b), from which we can see how the manipulator tracks the desired path with five joints. It is easily found that the actual end-effector trajectory coincides with the desired square path. Figure 2.2(a) and Figure 2.2(b) show us more details about the trajectory. That is, Figure 2.2(a) presents the desired path and the actual trajectory, which illustrates

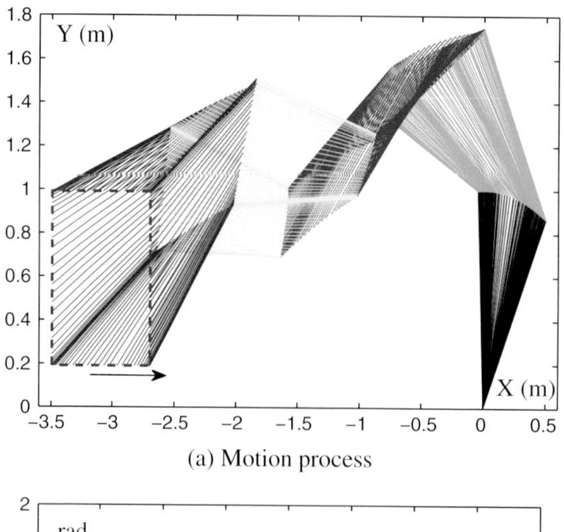

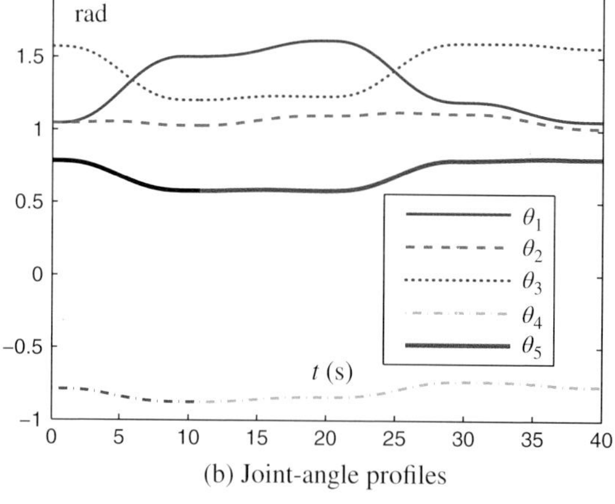

Figure 2.1 (a) Motion process and (b) joint-angle profiles of a five-link redundant robot manipulator tracking the desired square path synthesized by the G1 type scheme (2.8).

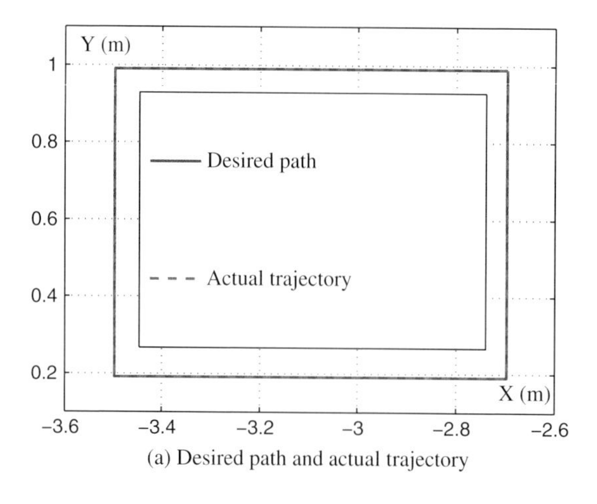

Figure 2.2 (a) Desired path, actual trajectory, and (b) position error of a five-link redundant robot manipulator tracking square path synthesized by a G1 type scheme (2.8).

Figure 2.2 (*Continued*)

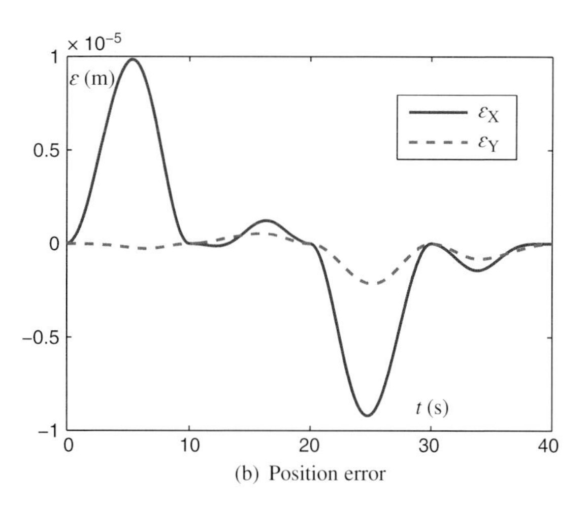

(b) Position error

Figure 2.3 (a) Desired velocity and (b) velocity error of a five-link redundant robot manipulator tracking a square path synthesized by a G1 type scheme (2.8).

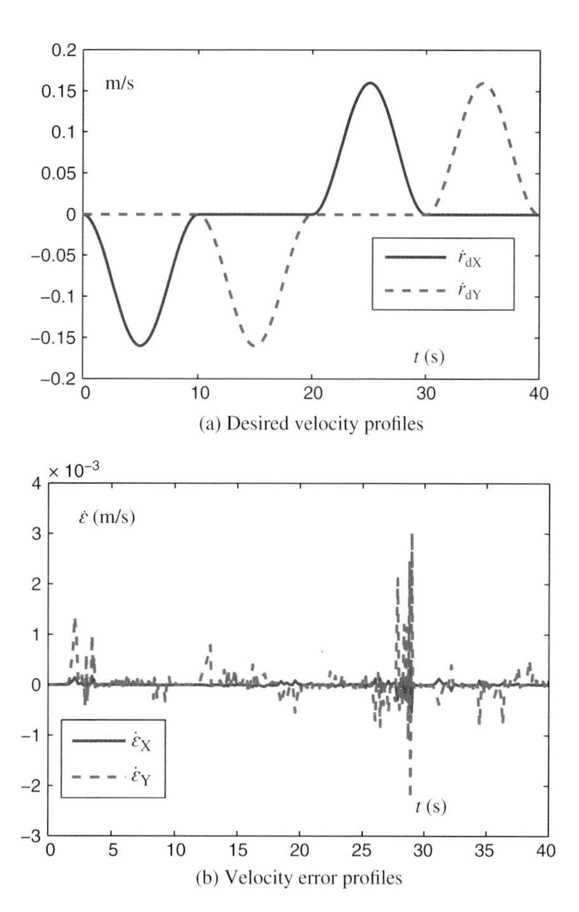

(a) Desired velocity profiles

(b) Velocity error profiles

the effectiveness and the accuracy intuitively. The corresponding errors shown in Figure 2.2(b) are all less than 1×10^{-5} m. This implies that the five-link robot manipulator completes the given square-path tracking task well. In addition, the desired velocities and the corresponding velocity errors are shown in Figure 2.3(a) and Figure 2.3(b). We see that the velocity errors are less than 4×10^{-3} m/s, validating the high accuracy of such an inverse-free type scheme.

2.4.2 "Z"-Shaped Path Tracking Task

In the second example, G1 type scheme (2.8) is applied to tracking a "Z"-shaped-path task with the side length being 0.8 m. The corresponding simulation results based on a six-DOF redundant robot manipulator are shown in Figure 2.4 through Figure 2.6, where an initial state $\boldsymbol{\theta}(0)$ is selected as $[\pi/4, \pi/4, \pi/4, \pi/4, \pi/4, \pi/4]^{\mathrm{T}}$ rad. These results illustrate once more the effectiveness and the high accuracy of the presented G1 type scheme (2.8) for solving the time-varying inverse kinematics of robot manipulators.

Specifically, the results can be seen from Figure 2.4(a) and Figure 2.4(b), from which we can see how the manipulator tracks the desired path with six joints. It is easily found that the actual end-effector trajectory coincides well with the desired "Z"-shaped path. In particular, Figure 2.5(a) and Figure 2.5(b) show us more details about the actual

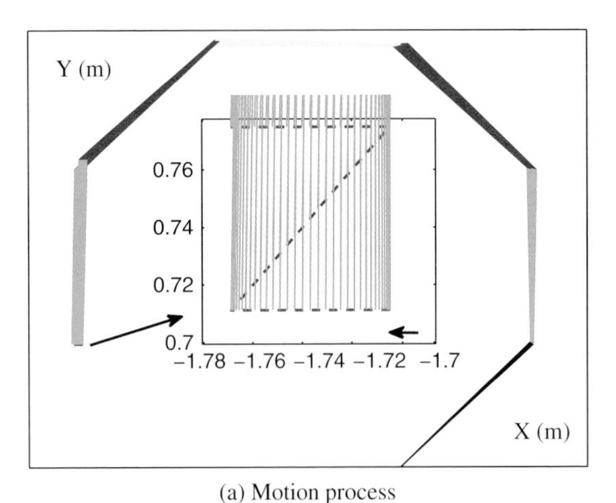

(a) Motion process

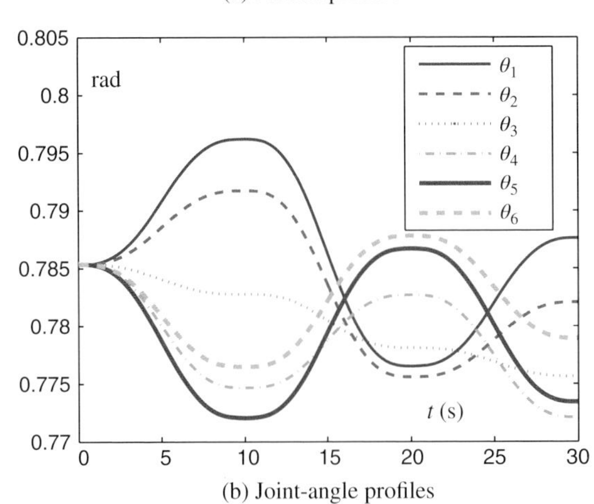

(b) Joint-angle profiles

Figure 2.4 Motion process (a) and joint-angle profiles (b) of six-DOF redundant robot manipulator tracking the desired "Z"-shaped path synthesized by a G1 type scheme (2.8).

Figure 2.5 (a) Desired path, actual trajectory, and (b) position error of a six-DOF redundant robot manipulator tracking a "Z"-shaped path synthesized by a G1 type scheme (2.8).

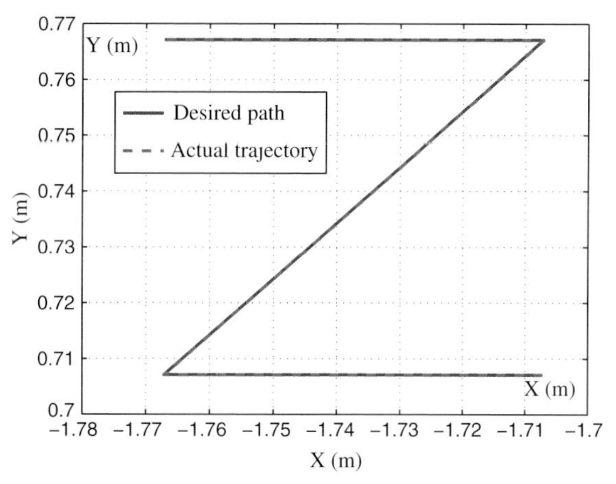

(a) Desired path and actual trajectory

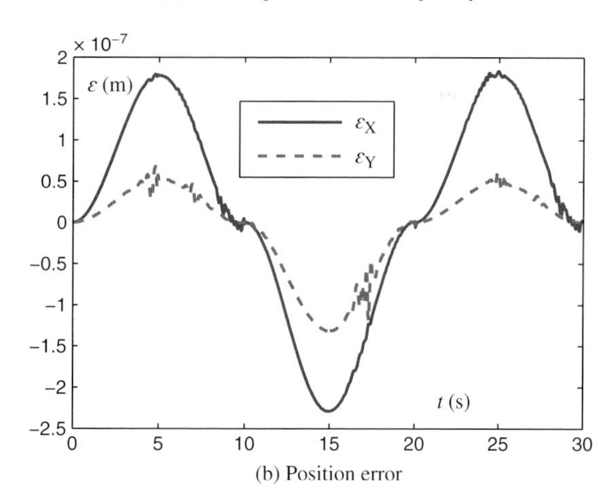

(b) Position error

Figure 2.6 (a) Desired velocity and (b) velocity error of a six-DOF redundant robot manipulator tracking a "Z"-shaped path synthesized by a G1 type scheme (2.8).

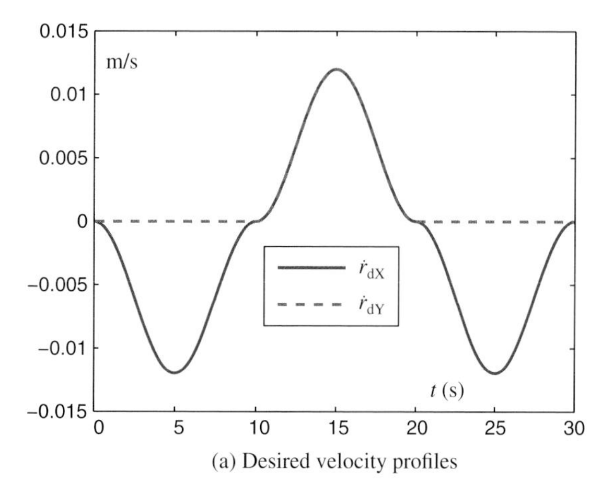

(a) Desired velocity profiles

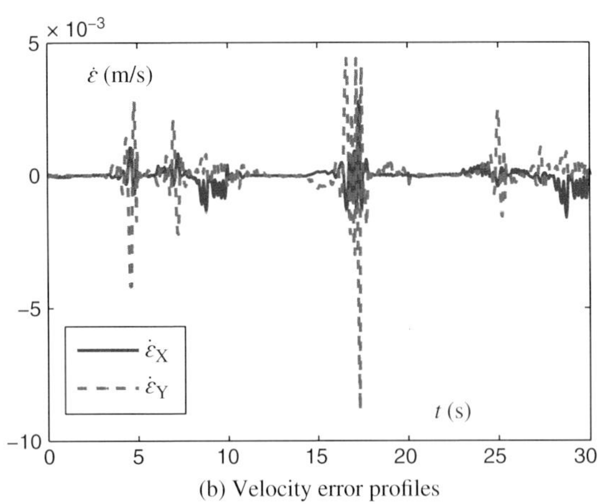

(b) Velocity error profiles

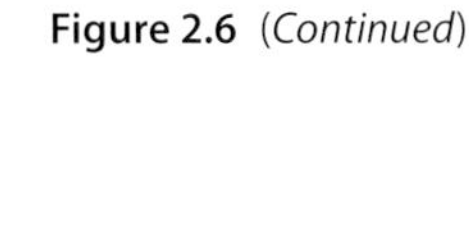

Figure 2.6 (*Continued*)

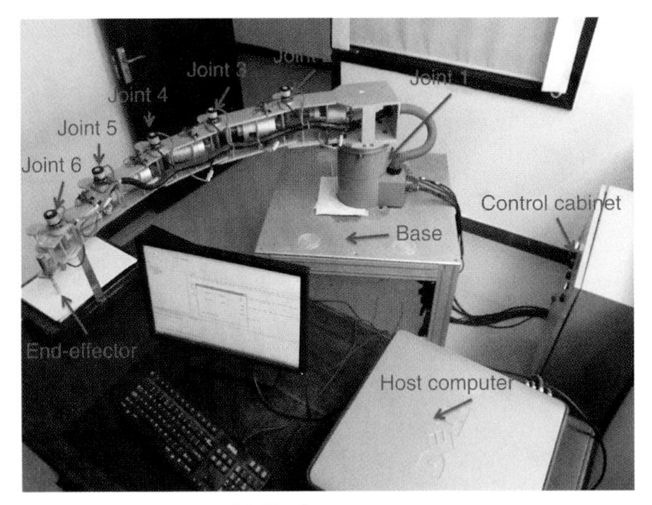

(a) Hardware system

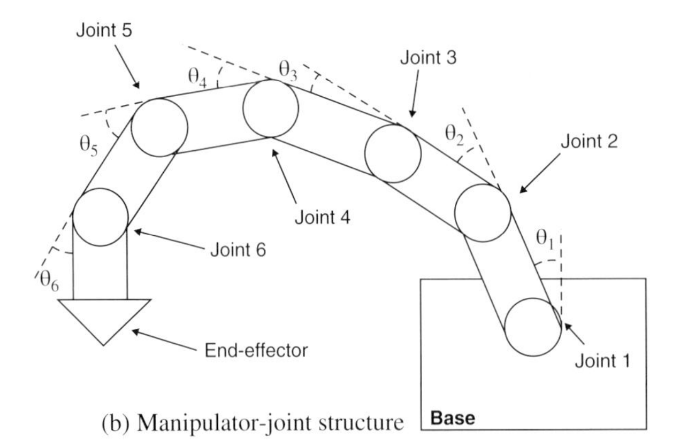

(b) Manipulator-joint structure

Figure 2.7 Hardware system (a) of six-DOF planar redundant robot manipulator with its structure platform (b). *Source:* Zhang et al. 2015. Reproduced from Y. Zhang, L. He, J. Ma et al., Inverse-free scheme of G1 type to velocity-level inverse kinematics of redundant robot manipulators, Figure 3, Proceedings of the Twelfth International Symposium on Neural Networks, pp. 99-108, 2015. ©Springer-Verlag 2015. With kind permission of Springer-Verlag (License Number 3978560065761).

trajectory and corresponding errors. These imply that the six-DOF robot manipulator completes the given "Z"-shaped-path tracking task well. Besides, the desired velocities and the corresponding velocity errors are shown in Figure 2.6(a) and Figure 2.6(b). We can find that any directional velocity error is less than 5×10^{-3} m/s, which validates the high accuracy of such an inverse-free type scheme.

2.5 Physical Experiments

To verify the physical realizability of the presented G1 type scheme (2.8), a six-DOF planar redundant robot manipulator hardware system is developed, investigated, and shown. The whole manipulator system is mainly composed of a robot manipulator, a control cabinet, and a host computer. Specifically, Figure 2.7(a) shows this planar robot hardware system, and Figure 2.7(b) depicts its manipulator-joint structure including a base and an end-effector. Note that the detailed description on the six-DOF planar redundant robot manipulator hardware system will be given in Chapter 6 and thus omitted here.

For the experiments, in order to verify the presented G1 type scheme (2.8) at the joint-velocity level, the end-effector is expected to move along a "Z"-shaped path with an initial state $\theta(0) = [\pi/12, \pi/12, \pi/12, \pi/12, \pi/12, \pi/12]^{\mathrm{T}}$ rad and the length of 4.5 cm, and the design parameter λ is also set as 10^5. The task execution can be seen from Figure 2.8, that is, the end-effector of the manipulator moves smoothly and draws a "Z"-shaped path precisely. Besides, the video of the process takes 24 seconds. Thus, the experiments illustrate well that the presented G1 type scheme (2.8) is effective on the redundant robot manipulator's inverse-free redundancy resolution (or, say, motion planning and control).

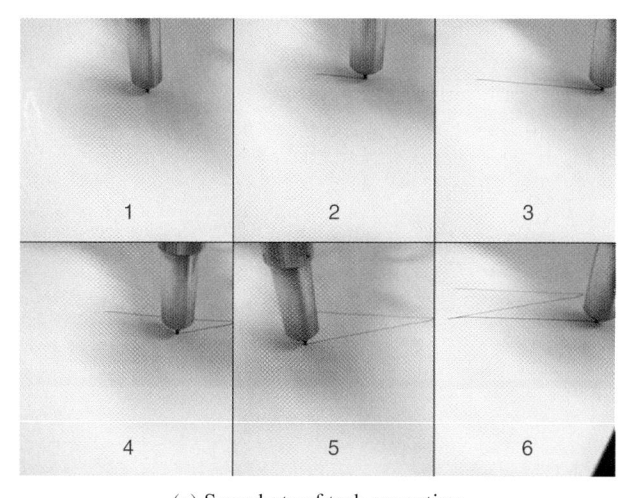

(a) Snapshots of task execution

Figure 2.8 "Z"-shaped-path-tracking experiment of six-DOF redundant robot manipulator synthesized by G1 type scheme (2.8) at joint-velocity level. Reproduced from Y. Zhang, L. He, J. Ma et al, Inverse-free scheme of G1 type to velocity-level inverse kinematics of redundant robot manipulators, Figure 4, Proceedings of the Twelfth International Symposium on Neural Networks, pp. 99-108, 2015. ©Springer-Verlag 2015. With kind permission of Springer-Verlag (License Number 3978560065761).

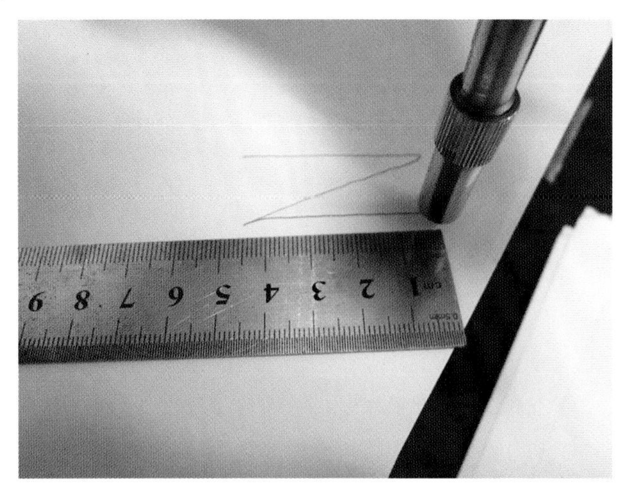

(b) Measurement of experimental result

Figure 2.8 *(Continued)*

2.6 Chapter Summary

In this chapter, to solve the time-varying inverse kinematics problem for redundant robot manipulators with high efficiency and accuracy, a special type of inverse-free scheme named G1 type scheme has been presented and investigated at the joint-velocity level. This scheme can avoid the Jacobian inversion in traditional pseudoinverse methods that not only costs expensive calculating time but also encounters many difficulties in practice. Besides, the corresponding path-tracking simulations have been performed on five-link and six-DOF redundant robot manipulators using such an inverse-free scheme. The simulation results have illustrated the effectiveness and the accuracy of the aforementioned scheme for solving the time-varying inverse kinematics problem of redundant robot manipulators in an inverse-free manner. In addition, the physical realizability of G1 type scheme has been verified further based on a six-DOF planar redundant robot manipulator hardware system.

3

D1G1 Type Scheme to JAL Inverse Kinematics

3.1 Introduction

In recent decades, robotic systems have been playing an increasingly important role in numerous fields of engineering applications. They have been applied in industrial automation to perform repetitive and high-intensitive work, such as painting, welding, ironing, and palletizing. Redundant robot manipulators are robotic devices whose available DOF are more than required for executing a specific end-effector task [2, 12, 53, 54], which leads to greater dexterity and flexibility for robot manipulators. One fundamental issue in operating the redundant robot manipulator is the redundant resolution problem (i.e., the inverse kinematics problem [2, 53]). In recent years, numerous redundancy-resolution schemes have been presented, developed, and investigated at different levels (e.g., the joint-velocity level and joint-acceleration level) for solving the inverse kinematics problem of redundant robot manipulators [12, 54–59]. Note that the Z0G1 type scheme and the inverse scheme have already been presented in Chapter 2.

To obtain the inverse-free and accurate solution of the inverse kinematics problem for redundant robot manipulators, an important branch of dynamics methods, that is, the gradient dynamics (GD) method [2, 12, 53, 54, 58], is thus exploited to present and investigate an inverse-free D1G1 scheme at the joint-acceleration level rather than the joint-velocity level. It is worth pointing out that many inverse kinematics approaches are resolved only at the joint-velocity level, which may not be practicable for the manipulators controlled at the joint-acceleration level or the joint-torque level. The presented joint-acceleration level inverse-free D1G1 scheme can effectively avoid the complex Jacobian pseudoinversion problem and obtain the accurate solution for the time-varying inverse kinematics problem.

In this chapter, an inverse-free D1G1 scheme is presented and investigated at the joint-acceleration level for redundancy resolution of redundant robot manipulators. Besides, the design process of such an inverse-free solution is presented in detail. The D1G1 scheme can effectively avoid the pseudoinversion of the Jacobian matrix and obtain an accurate solution to the complicated time-varying inverse kinematics problem for redundant robot manipulators. Compared with the traditional pseudoinverse-based solution, the simulations performed on a three-link redundant robot manipulator substantiate the effectiveness and superiority of the presented D1G1 scheme.

Robot Manipulator Redundancy Resolution, First Edition. Yunong Zhang and Long Jin.

3.2 Preliminaries and Related Work

The inverse scheme at the joint-velocity level has been discussed in Section 2.2, and thus this section focuses on the inverse scheme at the joint-acceleration level (JAL). The following two equations are presented for better understanding and have already been presented in Section 2.2:

$$\dot{\mathbf{r}}_{\mathrm{e}} = J(\theta)\dot{\theta}, \tag{3.1}$$

$$\dot{\theta} = J^{\dagger}(\theta)\dot{\mathbf{r}}_{\mathrm{d}}(t). \tag{3.2}$$

By differentiating (3.1), the relation between the end-effector's Cartesian acceleration $\ddot{\mathbf{r}}_{\mathrm{e}} \in \mathbb{R}^m$ and the joint- acceleration $\ddot{\theta} \in \mathbb{R}^n$ can be obtained as

$$\ddot{\mathbf{r}}_{\mathrm{e}} = \dot{J}(\theta)\dot{\theta} + J(\theta)\ddot{\theta}, \tag{3.3}$$

which is expected to track the desired path's acceleration $\ddot{\mathbf{r}}_{\mathrm{d}}(t)$, that is, $\ddot{\mathbf{r}}_{\mathrm{e}} = \dot{J}(\theta)\dot{\theta} + J(\theta)\ddot{\theta} \rightarrow \ddot{\mathbf{r}}_{\mathrm{d}}(t)$. Similarly, the acceleration-level pseudoinverse solution to (3.3) is generally formulated as

$$\ddot{\theta} = J^{\dagger}(\theta)(\ddot{\mathbf{r}}_{\mathrm{d}}(t) - \dot{J}(\theta)\dot{\theta}). \tag{3.4}$$

However, the pseudoinverse solutions (3.2) and (3.4) both require the Jacobian pseudoinversion during the solving process, which are computationally expensive and may not satisfy the real-time requirements.

Note that the traditional pseudoinverse-based solution may encounter some limitations [2, 12, 53–58]. For example, it may not solve the inverse kinematics problem effectively when the Jacobian matrix is not of full rank at some time instants. The pseudoinversion of a rank-deficient Jacobian matrix may result in kinematic singularities that are undesirable in many practical applications.

3.3 Scheme Formulation

To effectively solve the complicated inverse kinematics problem and obtain its accurate solution, the gradient dynamics (GD) method is applied to proposing and investigating an inverse-free D1G1 scheme at the joint-acceleration level. Note that the traditional joint-velocity level solution may not be applicable directly for the manipulators that are controlled at the joint-acceleration and/or joint-torque levels. Motivated by this, it makes sense to propose and investigate the inverse-free joint-acceleration level solution as follows.

Firstly, based on (3.1), we define a scalar-valued norm-based energy function:

$$\varrho(\dot{\theta}, \theta, t) = \|\dot{\mathbf{r}}_{\mathrm{d}}(t) - J(\theta)\dot{\theta}\|_2^2/2. \tag{3.5}$$

Secondly, a GD scheme can be designed to evolve along the negative gradient of $\varrho(\dot{\theta}, \theta, t)$, that is, $-\partial\varrho/\partial\dot{\theta}$, until the minimum point is reached. In view of $\partial J(\theta)/\partial\dot{\theta} = 0$, we obtain

$$-\partial\varrho/\partial\dot{\theta} = J^{\mathrm{T}}(\theta)(\dot{\mathbf{r}}_{\mathrm{d}}(t) - J(\theta)\dot{\theta}). \tag{3.6}$$

Finally, according to GD design formula $\ddot{\theta} = -\lambda \partial \varrho / \partial \dot{\theta}$, the dynamics equation of such a D1G1 scheme for solving the time-varying inverse kinematics problem of redundant robot manipulators is thus shown as follows.

$$\ddot{\theta} = \lambda J^{\mathrm{T}}(\theta)(\dot{\mathbf{r}}_{\mathrm{d}}(t) - J(\theta)\dot{\theta}), \tag{3.7}$$

where design parameter $\lambda > 0$ is used to adjust the convergence performance, and therefore should be selected to be appropriately large for simulative and experimental requirements.

Evidently, comparing acceleration-level pseudoinverse solution (3.4) with D1G1 scheme (3.7), we just need to calculate the transpose of $J(\theta)$ rather than the pseudoinverse of $J(\theta)$, which effectively reduces the computational complexity. In addition, the traditional pseudoinversion of a rank-deficient Jacobian matrix may result in the intolerable torque or force on links and the breakdown of control algorithms.

In fact, by exploiting the GD method and another powerful dynamics method, that is, the zeroing dynamics (ZD) method [53], two more types of inverse-free solutions, namely, the G1 type and the Z1G1 type, are presented and investigated respectively at the joint-velocity level and the joint-acceleration level. Note that the G1 type scheme has been presented in Chapter 2 and the Z1G1 type scheme will be presented in Chapter 4.

3.4 Computer Simulations

In this section, to verify the effectiveness, accuracy, and superiority of the presented inverse-free D1G1 scheme (3.7), computer simulations are performed based on a three-link robot manipulator with different path-tracking tasks. Specifically, the end-effector of such a robot manipulator is expected to track a rhombus path and a triangle path. For comparative purposes, the acceleration-level pseudoinverse solution (3.4) is also applied to resolving the inverse kinematics problem. Besides, we conduct the simulations of inverse-free D1G1 scheme (3.7) with different values of λ, where, being the simulation conditions, the relative tolerance is 10^{-6} and the absolute tolerance is 10^{-8}.

3.4.1 Rhombus-Path Tracking Task

For better illustration of the effectiveness of the presented inverse-free D1G1 scheme (3.7), the end-effector of the three-link robot manipulator is required to track a rhombus path with initial state $\theta(0) = [\pi/7, \pi/2, \pi/4]^{\mathrm{T}}$ rad. The corresponding simulation results are presented in Figures 3.1 through 3.5 as well as Table 3.1.

3.4.1.1 Verifications

Firstly, we show the simulative results synthesized by inverse-free D1G1 scheme (3.7) with design parameter $\lambda = 10^4$. Figure 3.1(a) shows that the end-effector of the three-link robot manipulator moves from the initial position to the desired position successfully during completing such a rhombus-path-tracking task. As seen from Figure 3.1(b), the desired path coincides well with the actual trajectory, which

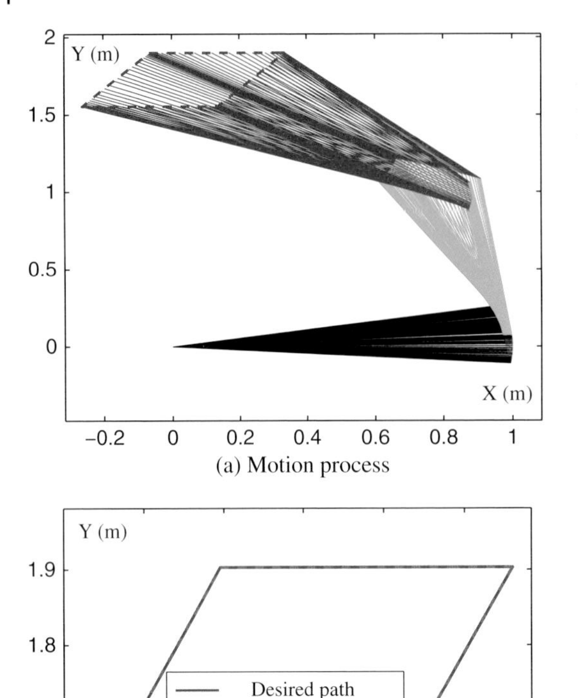

(a) Motion process

(b) Desired path and actual trajectory of rhombus

Figure 3.1 (a) Motion process, (b) desired path, and actual trajectory of a three-link redundant robot manipulator tracking a rhombus path synthesized by the D1G1 scheme (3.7) with $\lambda = 10^4$.

illustrates the effectiveness of such an inverse-free solution. Besides, Figure 3.2(a), Figure 3.3(a), and Figure 3.4(a) show that the resolved joint variables (i.e., the joint angle θ, joint velocity $\dot{\theta}$, and joint acceleration $\ddot{\theta}$) are continuous and smooth, which are suitable for engineering applications. In addition, as observed from Figure 3.2(b), Figure 3.3(b), and Figure 3.4(b), as well as Table 3.1, the maximal absolute value of the manipulator end-effector's position error ϵ is less than 7.9843×10^{-6} m, where $\epsilon = [\epsilon_{\mathrm{X}}, \epsilon_{\mathrm{Y}}]^{\mathrm{T}} = \mathbf{r}_{\mathrm{d}}(t) - \mathbf{f}(\theta)$ with ϵ_{X} and ϵ_{Y} denoting the elements of error ϵ along the X and Y axes, respectively. Furthermore, the velocity error $\dot{\epsilon} = [\dot{\epsilon}_{\mathrm{X}}, \dot{\epsilon}_{\mathrm{Y}}]^{\mathrm{T}}$ and acceleration error $\ddot{\epsilon} = [\ddot{\epsilon}_{\mathrm{X}}, \ddot{\epsilon}_{\mathrm{Y}}]^{\mathrm{T}}$ are both very tiny. These simulation results illustrate that the given rhombus-path-tracking task is fulfilled well using the presented D1G1 scheme (3.7).

3.4.1.2 Comparisons

For demonstrating the superiority of the presented inverse-free D1G1 scheme (3.7), the acceleration-level pseudoinverse solution (3.4) is applied to completing the same rhombus-path-tracking task for comparative purposes, where initial state $\theta(0)$ is set the same as before. In the pseudoinverse simulation, the path-tracking task can also

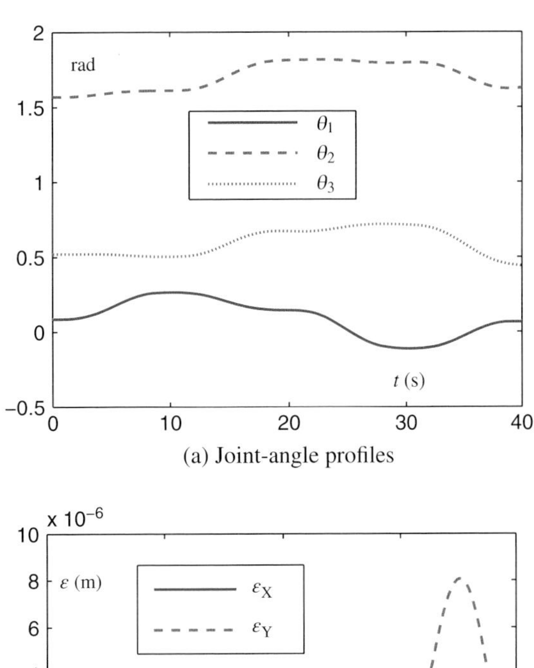

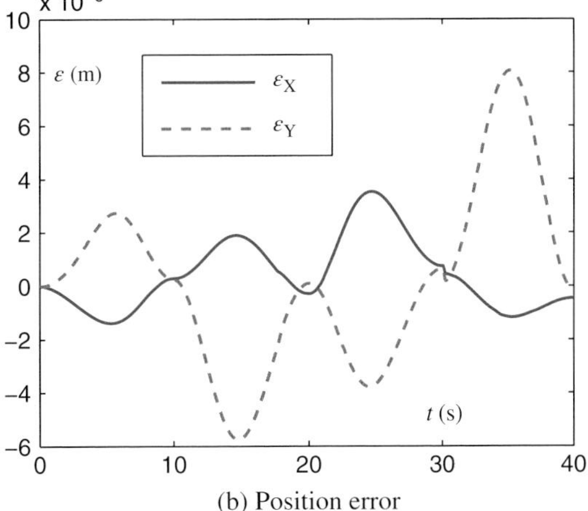

Figure 3.2 Joint-angle profiles (a) and position error (b) of a three-link redundant robot manipulator tracking a rhombus path synthesized by the D1G1 scheme (3.7) with $\lambda = 10^4$.

be achieved, and the related trajectories and joint profiles are very similar to simulation results of D1G1 scheme; thus, these results are omitted here. The maximal absolute value of synthesized position error shown in Figure 3.5(a) is less than 8.5615×10^{-6} m, which is larger than that of inverse-free D1G1 scheme (3.7) shown in Figure 3.5(b). Besides, the eventual position error ϵ of the acceleration-level pseudoinverse solution (3.4) is not convergent. Therefore, the presented inverse-free D1G1 scheme (3.7) has a superior performance compared to the traditional pseudoinverse solution (3.4) in terms of redundancy resolution of the redundant robot manipulator.

Furthermore, as seen from Figures 3.1 through 3.5, as well as Table 3.1, by adjusting the value of design parameter λ from 10^3 to 10^6, the corresponding maximal absolute value of position error reduces significantly from 8.3563×10^{-5} m to 1.2662×10^{-7} m, and all of the results with $\lambda \geqslant 10^4$ are better than those of the pseudoinverse solution. In summary, the presented inverse-free D1G1 scheme (3.7) is more effective than the traditional pseudoinverse solution (3.4), and design parameter λ plays an important role during the redundancy-resolution process.

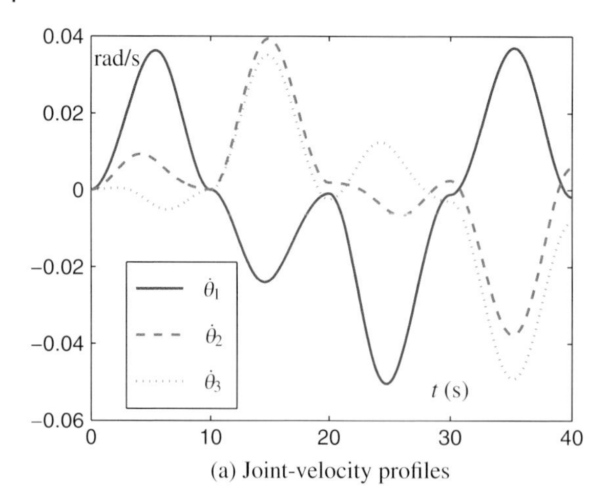

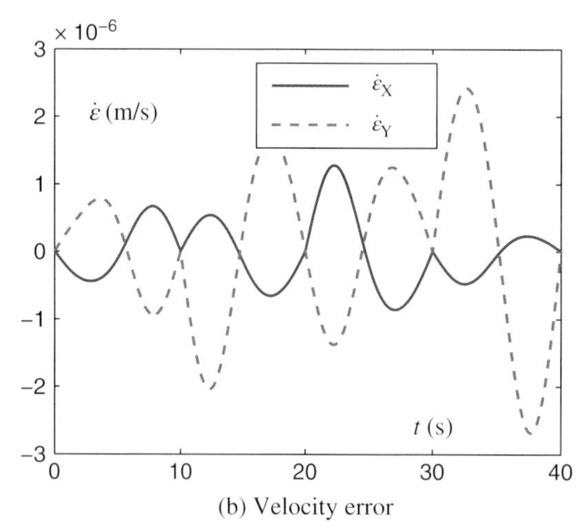

Figure 3.3 Joint-velocity profiles (a) and velocity error (b) of a three-link redundant robot manipulator tracking a rhombus path synthesized by the D1G1 scheme (3.7) with $\lambda = 10^4$.

3.4.2 Triangle-Path Tracking Task

For completeness and further demonstrating the effectiveness and superiority of the presented inverse-free D1G1 scheme (3.7), in this subsection, the three-link robot manipulator is expected to track a triangle path using the D1G1 and pseudoinverse solutions. Without loss of generality, we set the same initial state $\theta(0) = [\pi/3, \pi/3, \pi/3]^{\mathrm{T}}$ rad, and design parameter λ is adjusted with different values. The simulation results are presented in Figure 3.6 and Figure 3.7, in addition to Table 3.1.

Specifically, Figure 3.6(a) shows that the three-link robot manipulator moves from the initial configuration to the desired configuration successfully after completing a triangle-path-tracking task via the presented inverse-free D1G1 scheme (3.7). Besides, as shown in Figure 3.7(a) and Table 3.1, the maximal absolute value of the end-effector's position error is less than 1.3110×10^{-6} m, which is acceptable for practical applications. When the end-effector tracks the desired triangle path, the corresponding joint-angle

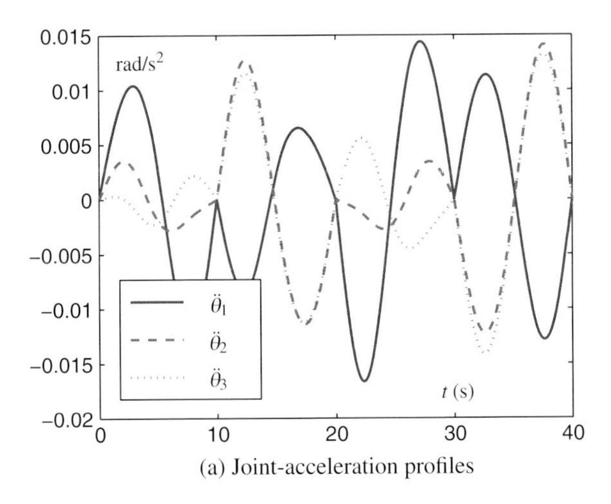

(a) Joint-acceleration profiles

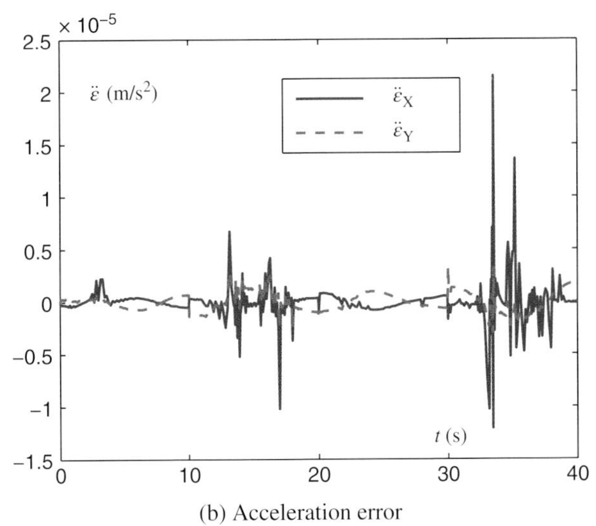

(b) Acceleration error

Figure 3.4 (a) Joint-acceleration profiles and (b) acceleration error of a three-link redundant robot manipulator tracking a rhombus path synthesized by the D1G1 scheme (3.7) with $\lambda = 10^4$.

profiles synthesized by the presented D1G1 scheme are shown in Figure 3.6(b). Similarly, the resolved joint-angle variables are also continuous and smooth, which ensures the feasibility of our presented solution. By comparison, when the robot manipulator tracks the desired triangle path using pseudoinverse solution (3.4), the corresponding position error is depicted in Figure 3.7(b). Evidently, we conclude that, compared to the error situation of the inverse-free D1G1 scheme, the position error ϵ of the pseudoinverse solution is eventually more divergent. It is worth pointing out that the synthesized joint-velocity and joint-acceleration errors using the D1G1 scheme are both within acceptable ranges (which can satisfy the practical application requirements) that, however, are omitted in this chapter due to space limitation.

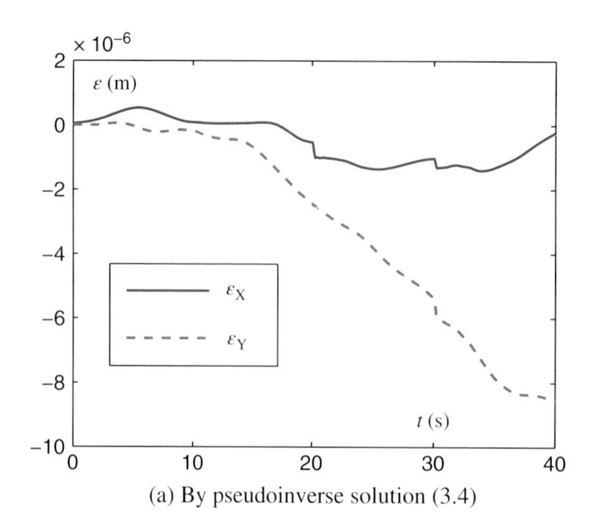

(a) By pseudoinverse solution (3.4)

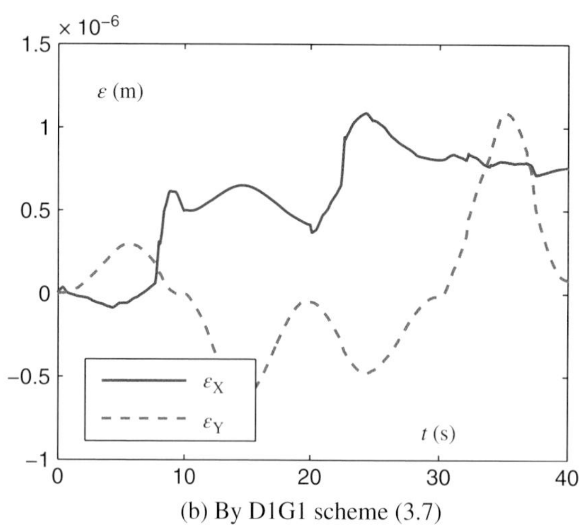

(b) By D1G1 scheme (3.7)

Figure 3.5 Position errors in rhombus-path-tracking task of a three-link robot manipulator synthesized by (a) pseudoinverse solution (3.4) and (b) inverse-free D1G1 scheme (3.7) with $\lambda = 10^5$.

Table 3.1 Position and velocity errors of a three-link robot manipulator's end-effector tracking rhombus and triangle paths synthesized by a inverse-free D1G1 scheme (3.7).

	Rhombus Path	
	Maximum position error (m)	**Maximum velocity error (m/s)**
$\lambda = 10^3$	8.3563×10^{-5}	2.6868×10^{-5}
$\lambda = 10^4$	7.9843×10^{-6}	2.6858×10^{-6}
$\lambda = 10^5$	1.0840×10^{-6}	2.6872×10^{-7}
$\lambda = 10^6$	1.2662×10^{-7}	2.6872×10^{-8}
$\lambda = 10^3$	1.3332×10^{-4}	4.4274×10^{-5}
$\lambda = 10^4$	1.3268×10^{-5}	4.4271×10^{-6}
$\lambda = 10^5$	1.3110×10^{-6}	4.4280×10^{-7}
$\lambda = 10^6$	1.1406×10^{-6}	4.4280×10^{-8}

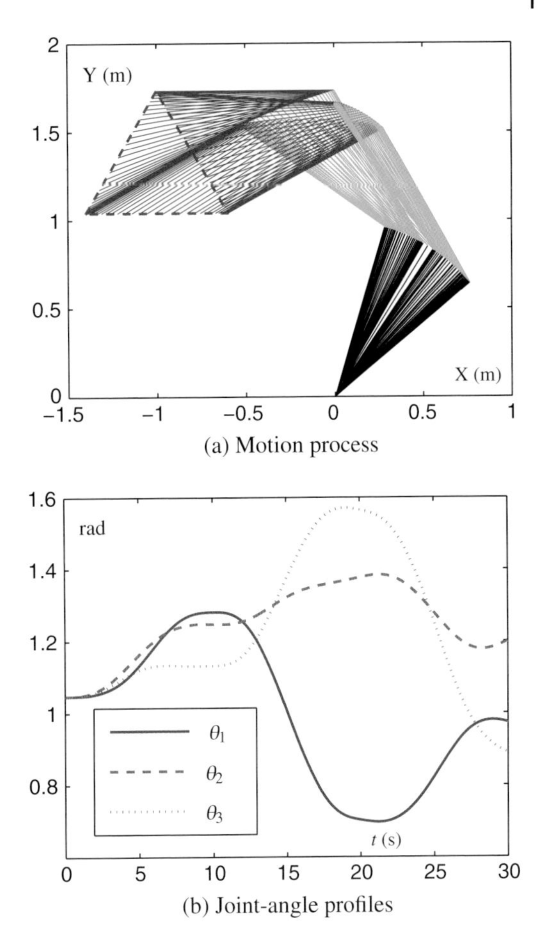

Figure 3.6 (a) Motion process and (b) joint-angle profiles of a three-link redundant robot manipulator tracking a desired triangle path synthesized by a D1G1 scheme (3.7) with $\lambda = 10^5$.

For further investigation, as seen from Table 3.1, using the inverse-free D1G1 scheme for the manipulator tracking the desired triangle path, when λ is adjusted from 10^3 to 10^6, the maximal absolute value of the position error ϵ reduces significantly from 1.3332×10^{-4} m to 1.1406×10^{-6} m. Similarly, we can draw the conclusion that the inverse-free D1G1 scheme is more effective than the conventional pseudoinverse-based solution, and design parameter λ is important in the inverse-free resolution-redundancy scheme, which should be selected to be appropriately large in order to satisfy different accuracy requirements in practical applications.

In summary, all of these simulation results based on a three-link robot manipulator further substantiate the effectiveness, accuracy, and superior performance of the presented inverse-free D1G1 scheme (3.7). Note that such an inverse-free redundancy-resolution scheme can also be applied to other robot manipulators with more DOF and more complicated path-tracking tasks (e.g., robot manipulators operating in three-dimensional space).

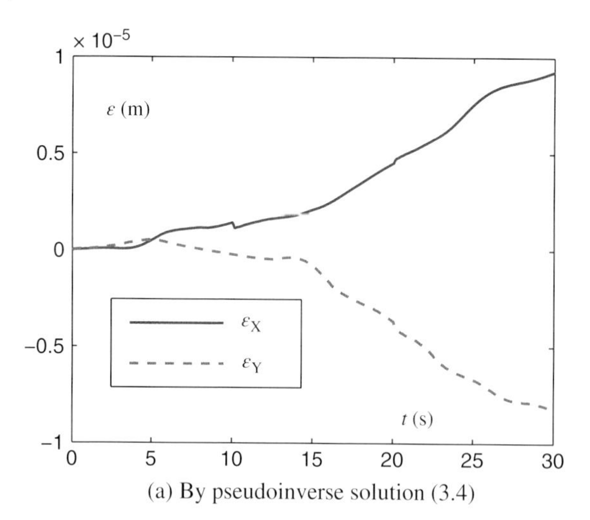

(a) By pseudoinverse solution (3.4)

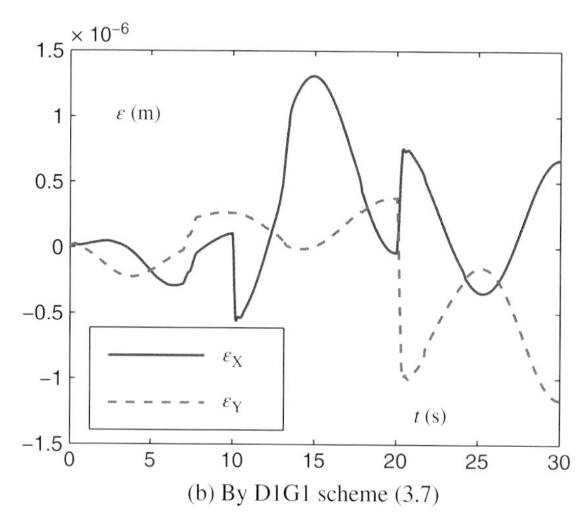

(b) By D1G1 scheme (3.7)

Figure 3.7 Position errors in triangle-path-tracking task of the three-link robot manipulator synthesized by (a) pseudoinverse solution (3.4) and (b) inverse-free D1G1 scheme (3.7) with $\lambda = 10^5$.

3.5 Chapter Summary

In this chapter, an acceleration-level inverse-free D1G1 scheme has been presented and investigated for the inverse kinematics of redundant robot manipulators, which is designed based on the powerful gradient dynamics (GD) method. For comparative purposes, the traditional pseudoinverse solution has also been presented. Computer-simulation results conducted on a three-link robot manipulator have further substantiated the effectiveness, accuracy, and superiority of the presented inverse-free D1G1 scheme. More importantly, it is an interesting and significant bridge, along which we can achieve many more inverse-free schemes to solve the inverse kinematics problem in the future. For example, the acceleration-level Z1G1 type solution scheme using a zeroing dynamics (ZD) method can be presented immediately as well.

4

Z1G1 Type Scheme to JAL Inverse Kinematics

4.1 Introduction

By exploiting the powerful gradient dynamics (GD) method and zeroing dynamics (ZD) method, this chapter presents and investigates the Z1G1 type scheme in an inverse-free manner at the joint-acceleration level for the redundant robot manipulator. The Z1G1 type scheme can effectively solve the inverse kinematics problem of redundant robot manipulators without calculating the computationally expensive inverse of Jacobian matrix. The path-tracking simulations and physical experiments are conducted to further substantiate the effectiveness, high accuracy, and physical realizability of the presented inverse-free solution.

4.2 Problem Formulation and Z1G1 Type Scheme

As mentioned in the Chapter 3, the end-effector position-and-orientation vector $\mathbf{r}_\mathrm{e} \in \mathbb{R}^m$ in Cartesian space is related to the joint space by a forward kinematics equation as depicted in [60]: $\mathbf{r}_\mathrm{e} = \mathbf{f}(\theta)$, which is expected to track the desired path $\mathbf{r}_\mathrm{d}(t)$, that is, $\mathbf{r}_\mathrm{e} = \mathbf{f}(\theta) \to \mathbf{r}_\mathrm{d}(t)$. Note that $\theta \in \mathbb{R}^n$ is the joint variable vector and $\mathbf{f}(\cdot)$ is a differentiable nonlinear function with a structure and parameters known for a given manipulator.

The conventional pseudoinverse solution with minimum acceleration norm to (3.3) is simply formulated as

$$\ddot{\theta} = J^{\dagger}(\ddot{\mathbf{r}}_\mathrm{d} - \dot{J}\dot{\theta}), \tag{4.1}$$

where $J \in \mathbb{R}^{m\times n}$ denotes the Jacobian matrix, $J^{\dagger} \in \mathbb{R}^{n\times m}$ is the pseudoinverse of J, and $\mathbf{r}_\mathrm{d} \in \mathbb{R}^m$ is the desired end-effector position-and-orientation vector [57, 60]. Note that the pseudoinverse method requires the inversion of Jacobian matrix during the solving process, which is computationally expensive and may not meet the real-time requirements.

To effectively avoid the Jacobian inversion, by exploiting two powerful dynamics methods (i.e., GD method and ZD method) the Z1G1 type scheme is presented for solving the inverse kinematics problem at the joint-acceleration level. The particular design steps are presented as follows.

Firstly, by following the ZD method presented in Chapter 1 [2, 60], we define a vector-valued error function (instead of the scalar-valued energy function usually used

Robot Manipulator Redundancy Resolution, First Edition. Yunong Zhang and Long Jin.

in GD-type solution design):

$$\epsilon = \mathbf{r}_\text{d} - \mathbf{f}(\theta). \tag{4.2}$$

Secondly, to make each element of error vector ϵ converge to zero, the following ZD design formula can be adopted:

$$\dot{\epsilon} = \frac{\mathrm{d}\epsilon}{\mathrm{d}t} = -\gamma\epsilon, \tag{4.3}$$

where design parameter $\gamma > 0$ is used to scale the exponential convergence rate of ϵ. Substituting (4.2) into equation (4.3) yields $\dot{\mathbf{r}}_\text{d} - J\dot{\theta} + \gamma(\mathbf{r}_\text{d} - \mathbf{f}(\theta)) = 0$.

Thirdly, by following the GD method, we define a scalar-valued energy function ϵ together with $\rho = \mathbf{r}_\text{d} - \mathbf{f}(\theta)$:

$$\rho = \|\dot{\mathbf{r}}_\text{d} - J\dot{\theta} + \gamma(\mathbf{r}_\text{d} - \mathbf{f}(\theta))\|_2^2/2 = \|\dot{\epsilon} + \gamma\epsilon\|_2^2/2. \tag{4.4}$$

With $\partial J/\partial\dot{\theta} = 0$ and $\partial\mathbf{f}(\theta)/\partial\dot{\theta} = 0$, we further obtain the negative gradient as $-\partial\rho/\partial\dot{\theta} = J^\text{T}(\dot{\mathbf{r}}_\text{d} - J\dot{\theta} + \gamma(\mathbf{r}_\text{d} - \mathbf{f}(\theta)))$.

Finally, according to GD design formula $\ddot{\theta} = -\lambda\partial\rho/\partial\dot{\theta}$ with $\lambda > 0$ being used to scale the convergence rate of the GD solution, the dynamic equation of such a solution for solving the time-varying inverse kinematics problem is obtained as

$$\ddot{\theta} = \lambda J^\text{T}(\dot{\mathbf{r}}_\text{d} - J\dot{\theta} + \gamma(\mathbf{r}_\text{d} - \mathbf{f}(\theta))). \tag{4.5}$$

Evidently, solution (4.5) does not require the inversion of Jacobian matrix appearing in (4.1). Besides, since the solution (4.5) is obtained by applying the ZD method once and applying the GD method once, it is termed a Z1G1 type scheme.

It is worth pointing out here that, if the initial end-effector pose (i.e., position and orientation) of the manipulator is not on the desired path, the presented Z1G1 type scheme can still work. In this situation, the feedback control term (or say, technique) inside (4.5) is actually introduced to guarantee that the end-effector can gradually move to the desired (initial) position and complete the user-specified path-tracking task. Besides, λ and $\lambda\gamma$ are, respectively, related to the velocity and position feedback gains, which are used to adjust the convergence rates. The related simulations are conducted in the ensuing section.

4.3 Computer Simulations

In this section, to validate the effectiveness of the presented Z1G1 type scheme (4.5), computer simulations are performed based on three-, four-, and five-link planar robot manipulators. The end-effectors of the three planar robot manipulators are required to track the specified isosceles-trapezoid path, isosceles-triangle path, and square path, respectively. Note that, when we apply Z1G1 type scheme (4.5) to solving the time-varying inverse kinematics problem in such an inverse-free manner, design parameters $\lambda = 10^5$ and $\gamma = 10$ are used throughout this chapter. The corresponding simulation results are illustrated in Figure 4.1 and Figure 4.4 as well as Table 4.1. Besides also, a simulative example, in which the robot's end-effector starts from a nondesired initial position, is presented in Figure 4.5.

4.3.1 Desired Initial Position

Now, let us first consider the situation that the initial end-effector positions of the robot manipulators are on the desired paths correspondingly via three simulation examples.

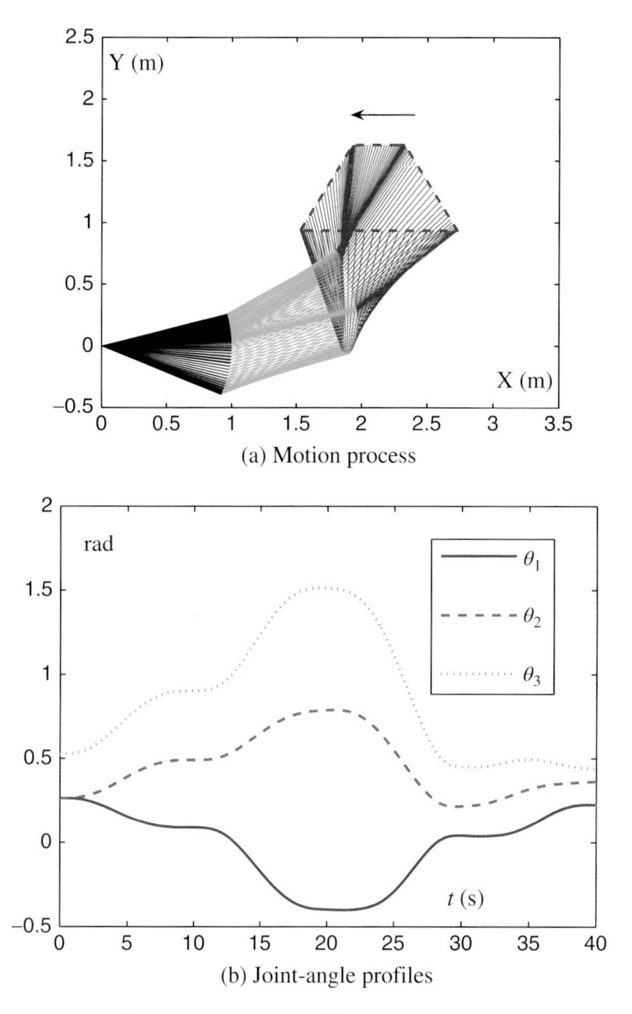

Figure 4.1 (a) Motion process and (b) joint-angle profiles of a three-link planar robot manipulator tracking a desired isosceles-trapezoid path synthesized by the Z1G1 type scheme (4.5).

Table 4.1 Maximum position, velocity, and acceleration errors when three planar robot manipulators track desired paths synthesized by Z1G1 type scheme (4.5).

Number of links	max$\{|\epsilon|\}$ (m)	max$\{|\dot{\epsilon}|\}$ (m/s)	max$\{|\ddot{\epsilon}|\}$ (m/s^2)
3	4.84×10^{-7}	9.25×10^{-7}	4.62×10^{-4}
4	1.24×10^{-7}	3.11×10^{-7}	2.49×10^{-4}
5	9.99×10^{-7}	1.52×10^{-6}	1.02×10^{-4}

4.3.1.1 Isosceles-Trapezoid Path Tracking

In the first example, the three-link planar robot manipulator's end-effector is expected to move along an isosceles trapezoid with the sloped angle being $\pi/3$ rad. During the task execution, the joints of the manipulator are expected to start from initial state $\theta(0) = [\pi/12, \pi/12, \pi/6]^{\mathrm{T}}$ rad and the task duration is 40 s. The simulation results are shown in Figure 4.1, Figure 4.2, and Table 4.1.

Specifically, Figure 4.1(a) shows the motion process of the three-link planar robot manipulator during the whole tracking process. As can be seen from it, the robot manipulator successfully achieves the purpose of tracking the isosceles-trapezoid path. Besides, what we are more interested in are some important variables, such as joint variables θ, $\dot{\theta}$, and $\ddot{\theta}$. This is because, as mentioned previously, the tasks to be performed by a robot manipulator are in Cartesian space, whereas the actuators directly relate to the joint space. So, Figure 4.1(b) through Figure 4.2(b), respectively, show the profiles of θ, $\dot{\theta}$, and $\ddot{\theta}$, when the end-effector tracks the specified isosceles-trapezoid path. From

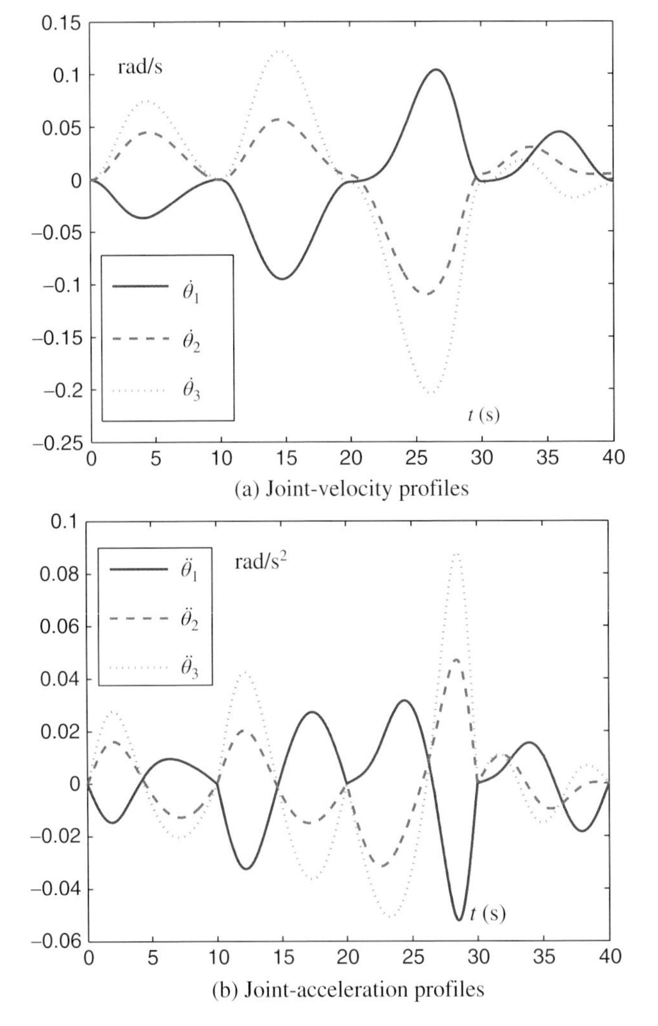

Figure 4.2 (a) Joint-velocity profiles and (b) joint-acceleration profiles of three-link planar robot manipulator tracking isosceles-trapezoid path synthesized by Z1G1 type scheme (4.5).

these three subfigures, we see that θ, $\dot{\theta}$, and $\ddot{\theta}$ are continuously time-varying during the task execution, and thus potentially suitable for practical applications.

In addition, Table 4.1 shows the corresponding tracking position, velocity, and acceleration errors. They are, respectively, less than 4.84×10^{-7} m, 9.25×10^{-7} m/s, and 4.62×10^{-4} m/s^2 in terms of absolute value, which are quite small and accurate for tracking control purposes.

4.3.1.2 Isosceles-Triangle Path Tracking

In the second example, the four-link planar robot manipulator's end-effector is expected to track an isosceles-triangle path with the right-angle side length being 0.8 m and the initial state $\theta(0)$ being $[3\pi/4, -\pi/2, -\pi/4, \pi/6]^{\mathrm{T}}$ rad. In order to investigate the effectiveness of Z1G1 type scheme (4.5) with respect to different task durations, we thus set 30 s in this example. The corresponding simulation results synthesized by Z1G1 type scheme (4.5) are shown in Figure 4.3 and Figure 4.4, as well as in Table 4.1.

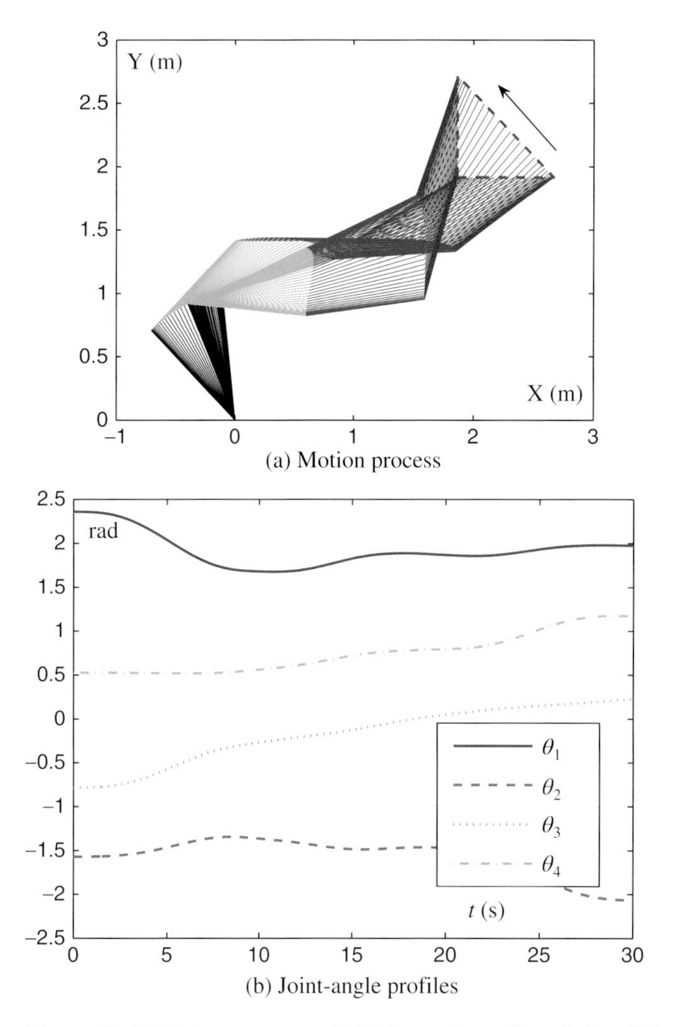

Figure 4.3 (a) Motion process and (b) joint-angle profiles of a four-link planar robot manipulator tracking a desired isosceles-triangle path synthesized by the Z1G1 type scheme (4.5).

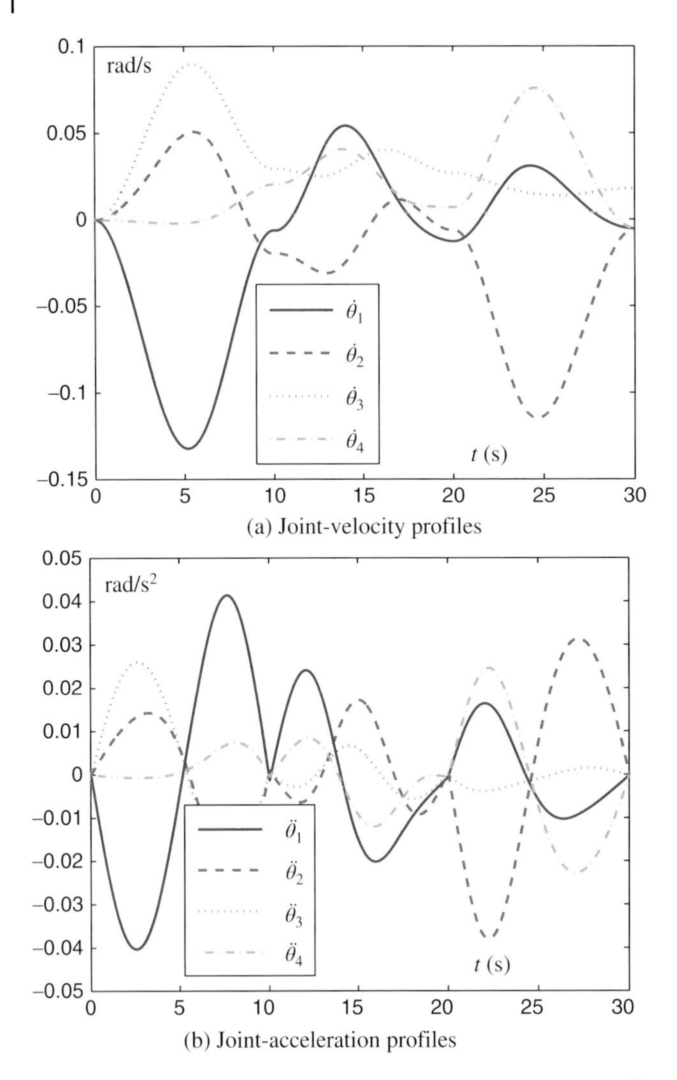

Figure 4.4 (a) Joint-velocity profiles and (b) joint-acceleration profiles of a four-link planar robot manipulator tracking an isosceles-triangle path synthesized by the Z1G1 type scheme (4.5).

As observed from Figure 4.3(a), we can conclude that the four-link planar robot manipulator also performs the specified path-tracking task well. Besides, as seen from Figure 4.3(b) and Figure 4.4, all of the variables vary with time t continuously during the task execution, and thus are acceptable for practical applications. In addition, it follows from Table 4.1 that the maximum absolute values of tracking position, velocity and acceleration errors are all very tiny. These results further illustrate the efficacy and high accuracy of Z1G1 type scheme (4.5) for the time-varying inverse kinematics problem solving of the four-link planar robot manipulator.

4.3.1.3 Square Path Tracking

In the third example, as a further investigation, Z1G1 type scheme (4.5) is applied to tracking another path task; that is, a square path with the side length being 0.8 m and initial state $\theta(0)$ being $[\pi/3, \pi/3, \pi/2, -\pi/4, \pi/4]^{\mathrm{T}}$ rad. The tiny error results presented in Table 4.1 verify once again the high accuracy of the presented Z1G1 type scheme (4.5). Note that the motion process and joint profiles of such a five-link planar robot manipulator are omitted due to space limitation. Similarly, we can find that the actual end-effector trajectory coincides with the desired square path well. This means that the five-link planar robot manipulator completes the specified square path task.

In summary, these computer-simulation results based on the three-, four-, and five-link planar robot manipulators have illustrated the feasibility, efficacy, and accuracy of the presented Z1G1 type scheme (4.5) at the joint-acceleration level in an inverse-free manner.

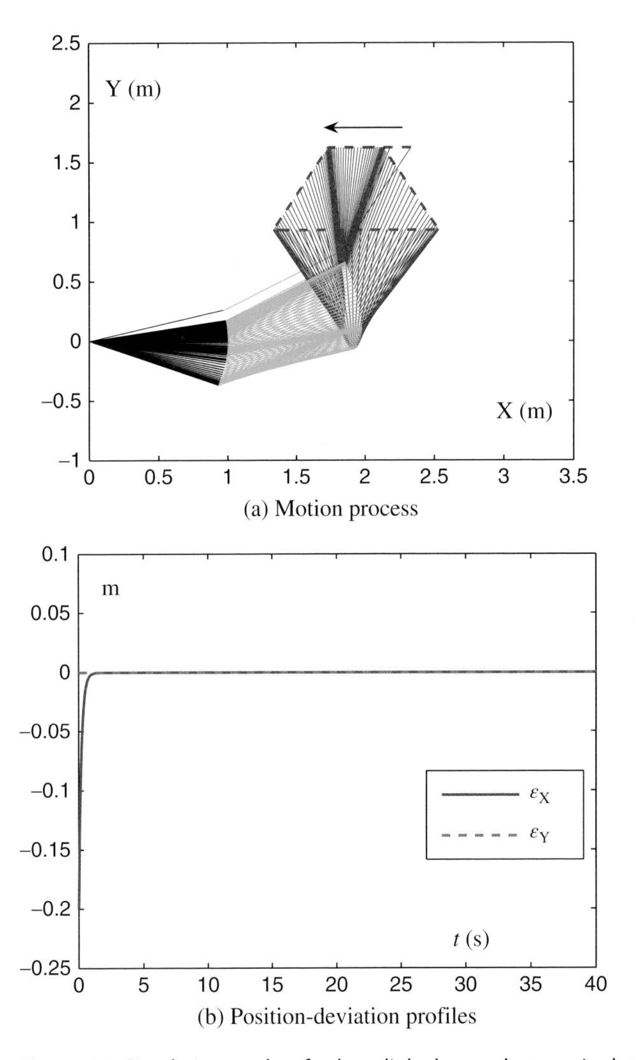

Figure 4.5 Simulation results of a three-link planar robot manipulator tracking an isosceles-trapezoid path synthesized by the Z1G1 type scheme (4.5) with an initial end-effector position not on the desired path.

4.3.2 Nondesired Initial Position

We conduct the simulation to test the efficacy of the presented Z1G1 type scheme in another situation where the initial end-effector position of the robot manipulator is not on the desired path. For example, the distance between the initial end-effector position of the three-link planar robot manipulator and the entry point of the desired path is 0.2 m along the x axis. The task duration, initial state, and parameters λ and γ are set the same as before. The motion process corresponding to such a task execution is presented in Figure 4.5(a), and the position deviation is shown in Figure 4.5(b), with ϵ_{X} converging to 0 (specifically, 2.12×10^{-5} m at 1.83 s).

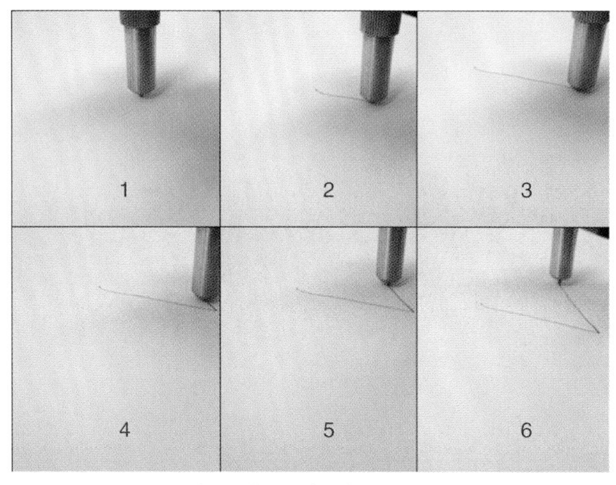

(a) Snapshots of task execution

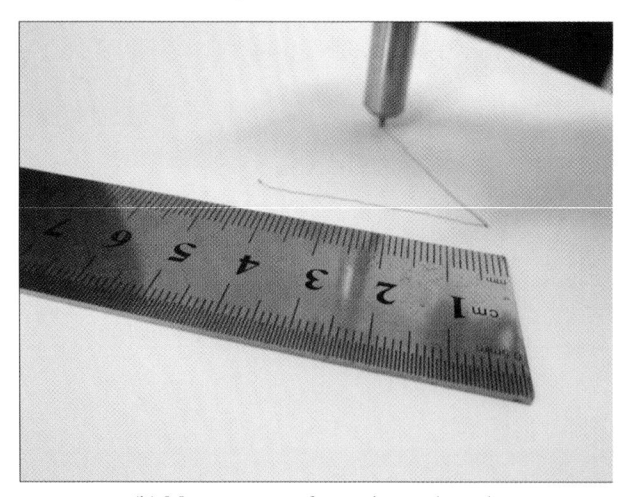

(b) Measurement of experimental result

Figure 4.6 "V"-shaped-path-tracking experiment of a six-DOF redundant robot manipulator synthesized by the Z1G1 type scheme (4.5) at joint-acceleration level. *Source*: Wang 2015. Reproduced with permission of IEEE.

In summary, these computer-simulation results have substantiated that, regardless of whether or not the initial end-effector positon of the robot manipulator is on the desired path, Z1G1 type scheme (4.5) at the joint-acceleration level is always effective for solving the inverse kinematics problem.

4.4 Physical Experiments

To verify the physical realizability of the acceleration-level (JAL) Z1G1 type scheme (4.5), a six-DOF redundant robot manipulator hardware system is developed, investigated, and shown. The whole manipulator system is mainly composed of a robot manipulator and a host computer, which is shown in Figure 2.7.

In order to substantiate the effectiveness of the presented Z1G1 type scheme (4.5) at the JAL, the end-effector is expected to move along a "V"-shaped path with length 4.5 cm, where initial state $\theta(0) = [\pi/12, \pi/12, \pi/12, \pi/12, \pi/12, \pi/12]^{\mathrm{T}}$ rad, and design parameters λ and γ are set as 10^5 and 10, respectively. The experimental results are presented in Figure 4.6; that is, the end-effector of the manipulator moves smoothly and draws a "V"-shaped path. Thus, the experiments substantiate well that the Z1G1 type scheme (4.5) is effective on the redundant robot manipulator's inverse-free redundancy resolution.

In summary, these experimental results based on the six-DOF planar robot manipulator have further shown the physical realizability of the acceleration-level Z1G1 type scheme (4.5).

4.5 Chapter Summary

In this chapter, to effectively avoid the Jacobian inversion existing in traditional pseudoinverse approaches and to obtain the accurate solution of the time-varying inverse kinematics problem for redundant robot manipulators, a special type of inverse-free solution, namely the Z1G1 type, has been presented and investigated at the joint-acceleration level. Also, the corresponding path-tracking simulations have been performed on three-, four-, and five-link planar robot manipulators by using such an inverse-free solution. The simulation results have illustrated the effectiveness and accuracy of these solutions for solving the time-varying inverse kinematics problem of redundant robot manipulators in such an inverse-free manner. The physical realizability of the Z1G1 type solution has also been further verified based on a six-DOF redundant robot manipulator hardware system.

Part III

QP Approach and Unification

5

Redundancy Resolution via QP Approach and Unification

5.1 Introduction

A redundant manipulator is defined when more degrees of freedom (DOF) are available than the minimum number of DOF required to execute a given end-effector primary task [61, 62]. Our human arm, elephant trunk, and snake are also such redundant systems [63, 64]. Compared to non-redundant manipulators, the redundant manipulator naturally has wider operational space and extra degrees to meet more functional constraints, such as the online avoidance of joint physical limits [36] and environmental obstacles [65, 66]. One of the most fundamental issues in operating the redundant manipulators is the redundancy-resolution problem. That is, given the Cartesian velocity/acceleration trajectories of the end-effector, we are required to generate the corresponding joint velocity, acceleration, and/or torque trajectories in real time [61].

Redundancy resolution for optimizing the joint torques is one most important part of redundancy-resolution problems, which is aimed at making a more effective utilization of input power from actuators by exploiting the extra DOF in redundant manipulators. Similar to the velocity-level redundancy resolution, most researchers use the pseudoinverse-type solution for the torque-minimizing redundancy resolution of manipulators. In particular, initial formulation for local torque minimization was presented in the mid 1980s by Hollerback and Suh [67]. They designed several schemes at the joint-acceleration level through the pseudoinverse and null-space method. However, the torque-minimizing redundancy resolution problem remains unsolved since the schemes exhibited instabilities/divergence, especially in long-range motions. Since then, many researchers have attempted to formulate other joint torque optimization schemes to eliminate the instability problem [68, 69]. Thorough analysis showed that almost all the pseudoinverse-type acceleration-level redundancy resolution schemes, including the remedy methods, may still exhibit local instabilities in the form of abrupt increases in joint velocities, accelerations, and torques.

Since the mid-1990s, quadratic programming (QP) approaches have been developed for the redundancy resolution of robot manipulators, for example, [36, 65–78]. The QP formulations corresponding to specific pseudoinverse-type solutions have the potential to contain directly the physical/mechanical constraints in the form of inequalities, like joint limits, joint velocity/acceleration limits, and environmental obstacles [36, 65]. To solve the QP, a compact QP solution method has also been established by using the

Robot Manipulator Redundancy Resolution, First Edition. Yunong Zhang and Long Jin.

serial-processing techniques of workspace decomposition, Gaussian elimination with partial pivoting, and matrix inversion [71, 74]. As a parallel alternative to solve the QP-based redundancy-resolution problem in real time and to provide explicit solutions, a dynamic system approach in the form of recurrent neural networks has been developed, for example, [76–78]. In view of the nature of parallel distributed computation and hardware implementation, the dynamic system approach is thought to have opened new avenues for online optimization.

In the last several decades, various dynamic system solutions in the form of recurrent neural networks have been developed for solving the constrained QP problems online. These include, among others, the Lagrange neural networks [79], the gradient and projected network [80], the usual primal-dual network [77], and the dual neural network [78]. For a survey of the aforementioned dynamic solutions, see [81]. Here, two types of dynamic neural systems for solving QP-based redundancy problems are reviewed. One is the usual primal-dual dynamic system in [77]. The primal-dual system was developed by minimizing the duality gap via gradient method, and thus the dynamic equations are usually complicated and with high-order nonlinear terms [77]. Another result is the dual dynamic system [78]. To reduce system complexity, the dual network was presented by only using dual decision variables and directly using Karush–Kuhn–Tucker conditions with the projection operator. However, this dual network was for strictly convex optimization problems only and entails the explicit matrix inversion [78].

This chapter unifies these local redundancy-resolution schemes into a general QP-based formulation and approach. The unification of these redundancy-resolution schemes may bring more insights into the wealth of existing solutions as well as a better understanding of future research. Specifically, in this chapter a general QP problem formulation is first established. This formulation can unify both the acceleration-level and the velocity-level redundancy-resolution schemes into a QP problem subject to equality and inequality/bound constraints. Second, motivated by the real-time solution to such robotic problems, some QP online solutions are briefly reviewed. That is, traditional QP optimization routines, compact QP method, dual neural network as a QP solution, linear variational inequality (LVI)-aided primal-dual neural network as a QP solution, and two numerical algorithms E47 and 94LVI. Note that, based on the LVI-reformulation of a QP problem, the presented solution has simple piecewise-linear dynamics and does not entail explicit matrix inversion. In addition, this QP solution can also provide accurate joint acceleration for the velocity-level redundancy-resolution schemes, which, in pseudoinverse-type or numerical QP solutions, is usually a problem that has to use numerical differentiation to approximate. Third, a number of comparison simulations are performed based on the PUMA560 robot arm to illustrate the characteristics and performance of the presented QP-aided primal-dual neural network approach. The efficiency and effectiveness are shown via a few illustrative simulations.

5.2 Robotic Formulation

The robotic issue of our interest here is that, given the trajectory $\mathbf{r}_\text{d}(t) \in \mathbb{R}^m$ of the end-effector, we want to generate online the joint trajectory $\theta(t) \in \mathbb{R}^n$ so as to command the manipulator motion. However, the redundancy (i.e., $n - m$ DOF) is generally resolved at the joint-velocity level $\dot{\theta}(t) \in \mathbb{R}^n$ or joint-acceleration level $\ddot{\theta}(t) \in \mathbb{R}^n$, due to the nonlinearity and redundancy of the forward-kinematic function from $\theta(t)$ to $\mathbf{r}_\mathrm{d}(t)$ [61, 62]. From Equation (2.1) in Chapter 2, we have

$$\dot{\mathbf{r}}_\mathrm{e} = J(\theta)\dot{\theta},$$

where $\dot{\mathbf{r}}_\mathrm{e}$ denotes the end-effector velocity. For tracking the desired path $\mathbf{r}_\mathrm{d}$ via the desired speed $\dot{\mathbf{r}}_\mathrm{d} \in \mathbb{R}^m$, the relation between the desired end-effector velocity $\dot{\mathbf{r}}_\mathrm{d}$ and the joint velocity $\dot{\theta}(t)$ via Jacobian matrix $J(\theta) \in \mathbb{R}^{m\times n}$ can be represented as

$$J(\theta)\dot{\theta} \to \dot{\mathbf{r}}_\mathrm{d} \tag{5.1}$$

Differentiating (5.1) yields the relation between joint acceleration $\ddot{\theta}(t)$ and end-effector acceleration $\ddot{\mathbf{r}}_\mathrm{d}(t)$:

$$J(\theta)\ddot{\theta} \to \ddot{\mathbf{r}}_\mathrm{a} \tag{5.2}$$

where $\ddot{\mathbf{r}}_\mathrm{a} = \ddot{\mathbf{r}}_\mathrm{d} - \dot{J}(\theta)\dot{\theta} \in \mathbb{R}^m$, and $\dot{J}(\theta)$ is the time derivative of $J(\theta)$. Because the manipulator system is redundant, $m < n$. Equation (5.1) and Equation (5.2) are thus under-determined, admitting an infinite number of solutions. The conventional solutions to Equation (5.1) and Equation (5.2) (i.e., to solve for $\dot{\theta}(t)$ and $\ddot{\theta}(t)$, respectively) were the pseudoinverse/nullspace-type solution [61, 62]. Research from the last 20 years [61, 70, 82] shows that the redundancy-resolution problem can be solved in a more favorable manner via online optimization techniques.

For example, we can start with the following pure robotic problem formulation for manipulators' redundancy resolution at the joint-velocity level:

$$\text{minimize} \quad \phi(\theta, \dot{\theta}) \tag{5.3}$$
$$\text{subject to} \quad J\dot{\theta} = \dot{\mathbf{r}}_\mathrm{d} \tag{5.4}$$
$$A\dot{\theta} \leqslant \mathbf{b} \tag{5.5}$$
$$\theta^- \leqslant \theta \leqslant \theta^+ \tag{5.6}$$
$$\dot{\theta}^- \leqslant \dot{\theta} \leqslant \dot{\theta}^+ \tag{5.7}$$

where

- Equation (5.4) is exactly Equation (5.1) used to perform the given end-effector primary tasks;
- joint physical limits $\theta^\pm$ and $\dot{\theta}^\pm$ are to be avoided via bound constraints (5.6) and (5.7) [36, 78];
- inequality constraint (5.5) is entailed for obstacle avoidance that will be examined in Section 5.4 [65, 83]; and,
- $\phi(\theta, \dot{\theta}) \in \mathbb{R}$ is any suitable performance index that will be examined in Section 5.5 [61].

Similar to the problem formulation in (5.3)–(5.7), if the redundancy is to be resolved at the joint-acceleration level, we have the following robotic problem formulation:

$$\text{minimize} \quad \phi(\theta, \dot{\theta}, \ddot{\theta}) \tag{5.8}$$
$$\text{subject to} \quad H(\theta)\ddot{\theta} + \mathbf{c}(\theta, \dot{\theta}) + \mathbf{g}(\theta) = \tau \tag{5.9}$$
$$J\ddot{\theta} = \ddot{\mathbf{r}}_\mathrm{a} \tag{5.10}$$
$$A\ddot{\theta} \leqslant \mathbf{b} \tag{5.11}$$
$$\theta^- \leqslant \theta \leqslant \theta^+ \tag{5.12}$$

$$\dot{\theta}^- \leqslant \dot{\theta} \leqslant \dot{\theta}^+ \tag{5.13}$$

$$\ddot{\theta}^- \leqslant \ddot{\theta} \leqslant \ddot{\theta}^+ \tag{5.14}$$

where

- Equation (5.10) is exactly Equation (5.2) used to perform the given end-effector primary tasks;
- joint physical limits $\theta^\pm$, $\dot{\theta}^\pm$ and $\ddot{\theta}^\pm$ are to be avoided via bound constraints (5.12)–(5.14) [36, 78];
- inequality constraint (5.11) plays the same role as inequality constraint (5.5) for obstacle avoidance [65, 83];
- $\phi(\theta, \dot{\theta}, \ddot{\theta})$ is a performance index similar to performance index (5.3) [61]; and
- Equation (5.9) is the manipulator's dynamic equation [62, 69] with $H(\theta) \in \mathbb{R}^{n\times n}$ denoting the positive-definite inertia matrix, $\mathbf{c}(\theta, \dot{\theta}) \in \mathbb{R}^n$ denoting the Coriolis/centrifugal force vector, $\mathbf{g}(\theta) \in \mathbb{R}^n$ denoting the gravitational force vector, and $\tau \in \mathbb{R}^n$ denoting the joint-torque vector.

5.3 Handling Joint Physical Limits

In this subsection, the avoidance of joint limits $\theta^\pm$ in bound constraints (5.6) is reformulated for handling joint physical limits at both joint-velocity level and joint-acceleration level.

5.3.1 Joint-Velocity Level

For velocity-level redundancy resolution, the performance index and constraints in (5.3)–(5.7) all have to be converted to the expressions based on joint velocity $\dot{\theta}$ [36]. For example, the avoidance of joint limits $\theta^\pm$ in bound constraints (5.6) can be reformulated as follows:

$$\mu_{\mathrm{p}}(\theta^- - \theta) \leqslant \dot{\theta} \leqslant \mu_{\mathrm{p}}(\theta^+ - \theta) \tag{5.15}$$

where $\mu_{\mathrm{p}} > 0$, termed the intensity coefficient, is used to scale the feasible region of $\dot{\theta}$ and determine the magnitude of a deceleration when a joint approaches its limit. The value of μ_{p} is chosen such that the feasible region of $\dot{\theta}$ made by joint limits conversion (5.15) is normally not smaller than the original one made by joint velocity limits [i.e., bound constraints (5.7)]. Note that larger values of μ_{p} cause joint deceleration more quickly. In general, μ_{p} is selected to be greater than $\max_{1\leqslant i\leqslant n}\{(\dot{\theta}_i^+ - \dot{\theta}_i^-)/(\theta_i^+ - \theta_i^-)\}$, while in a large number of computer simulations based on the PUMA560, PA10 and other manipulators, we choose $\mu_{\mathrm{p}} = 20$. By combining inequality (5.15), the velocity-level avoidance of joint physical limits [i.e., bound constraints (5.6) and (5.7)] becomes

$$\xi^- \leqslant \dot{\theta} \leqslant \xi^+ \tag{5.16}$$

where $\xi^- = \max(\mu_{\mathrm{p}}(\theta^- - \theta),\ \dot{\theta}^-)$, and $\xi^+ = \min(\mu_{\mathrm{p}}(\theta^+ - \theta),\ \dot{\theta}^+)$.

5.3.2 Joint-Acceleration Level

For acceleration-level redundancy resolution, the performance index and constraints in (5.8)–(5.14) have to be converted to the expressions based on joint acceleration $\ddot{\theta}$ [78]. In light of the inertia movement, by using the conversion techniques with $\theta^- < 0 < \theta^+$

generally considered [36, 61, 78], the avoidance of joint limits $\theta^{\pm}$ in bound constraint (5.12) is converted as

$$\mu_{\mathrm{p}}(\eta\theta^{-} - \theta) \leqslant \dot{\theta} \leqslant \mu_{\mathrm{p}}(\eta\theta^{+} - \theta) \tag{5.17}$$

where μ_{p} is defined the same as inequality (5.15), and $\eta \in (0, 1)$, termed the critical-area coefficient, is used to define two critical areas (i.e., $[\theta^{-}, \eta\theta^{-}]$ and $[\eta\theta^{+}, \theta^{+}]$) such that there will appear a deceleration when a joint enters such areas. The avoidance of joint velocity limits $\dot{\theta}^{\pm}$ in bound constraint (5.13) can be similarly converted as an $\ddot{\theta}$-based expression:

$$\mu_{\mathrm{v}}(\dot{\theta}^{-} - \dot{\theta}) \leqslant \ddot{\theta} \leqslant \mu_{\mathrm{v}}(\dot{\theta}^{+} - \dot{\theta}). \tag{5.18}$$

By using inequalities (5.17) and (5.18), the acceleration-level avoidance of joint physical limits (5.12)–(5.14) becomes $\xi^{-} \leqslant \ddot{\theta} \leqslant \xi^{+}$, where $\xi^{-} = \max(\mu_{\mathrm{p}}(\eta\theta^{-} - \theta),\ \mu_{\mathrm{v}}(\dot{\theta}^{-} - \dot{\theta}),\ \ddot{\theta}^{-})$, and $\xi^{+} = \min(\mu_{\mathrm{p}}(\eta\theta^{+} - \theta),\ \mu_{\mathrm{v}}(\dot{\theta}^{+} - \dot{\theta}),\ \ddot{\theta}^{+})$. In our computer simulations based on PUMA560, PA10, and other manipulators, $\mu_{\mathrm{p}} = \mu_{\mathrm{v}} = 20$ and $\eta = 0.9$.

5.4 Avoiding Obstacles

Avoiding collision with obstacles is a basic and important requirement for operating robot manipulators working in a crowded environment [65, 83, 84]. Like joint physical limits, various obstacles do exist in almost any manipulator's operational space. For example, even if there is no external obstacles, collisions may still occur between manipulator links and the pedestal. If a robot encounters a collision and then fails there, the desired end-effector trajectory $\mathbf{r}_{\mathrm{d}}(t)$ becomes impossible to realize, not to mention the physical damage possibly caused [84].

As compared to distance-maximizing methods [85, 86], some researches [87–89] treat the collision-free requirement as an equality constraint. That is, when a robot link L approaches an obstacle O within the influential range of radius, an escape velocity $\mathbf{b} \in \mathbb{R}^{3}$ is generated to push the link L to move away from the obstacle O. This escape velocity was imposed in an equality-constraint form [87, 89]; that is, $A\dot{\theta} = \mathbf{b}$ where $A \in \mathbb{R}^{3\times m}$ is the critical-point Jacobian matrix at the link L corresponding to the obstacle point O. The weakness of such an equality-based collision-free formulation is that it is always difficult (or sometimes impossible) to determine a suitable magnitude of escape velocity $\mathbf{b}$. For example, see Figure 5.1 for two contradicting situations [65, 83].

In view of this observation, an inequality-constraint based collision-free formulation is developed in [65, 66, 83] by only using the directions of escape velocities pointing from obstacles to links. Specifically speaking, the design procedure is as follows.

1) After calculating the Cartesian coordinates of obstacle point O and the critical point C of a most vulnerable link, the escape-velocity direction can be denoted by $\overrightarrow{OC} \in \mathbb{R}^{3}$.
2) After calculating the Jacobian matrix $J_C \in \mathbb{R}^{3\times n}$ at the critical point C of the link, the new matrix $A = -\mathrm{sgn}(\overrightarrow{OC}) \diamond J_C$ where $\diamond$ is a vector-matrix multiplication operator defined in [65, 66, 83].
3) The inequality-based criterion is thus $A\dot{\theta} \leqslant 0$.
4) To avoid suddenly imposing such a constraint, a smoothed version is used by modifying the critical-point speed gradually, in the form of $A\dot{\theta} \leqslant \mathbf{b}$ [65, 66, 83].

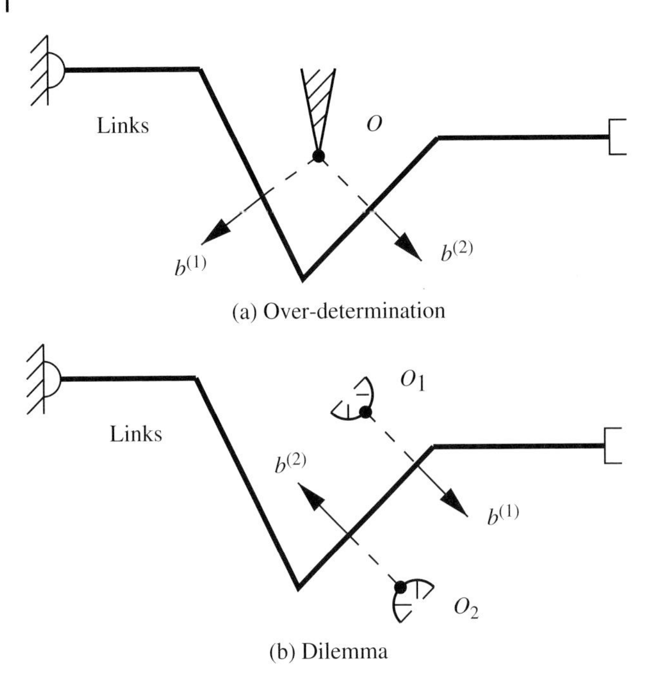

Figure 5.1 Contradicting situations in equality-based collision-free formulation.

The details and mathematical proof of such an inequality-constraint based formulation,

$$A\dot{\theta} \leqslant \mathbf{b}, \tag{5.19}$$

are given in [65, 66, 83], respectively, for point obstacles and window-shaped obstacle avoidance. Compared to distance-maximization and equality-based approaches, the inequality-based formulation (5.19) has the following advantages.

- Distance maximization as a performance index is only preferable, whereas constraint satisfaction such as (5.19) is more imperative.
- Equality constraints may unnecessarily reduce the solution space, whereas the inequality-based criterion (5.19) is proved to be able to generate a variable-magnitude escape velocity.
- Inequality-based criterion (5.19) is suitable for multiple obstacle avoidance with no gradient/derivative computation required, as compared to the distance-maximization method.

5.5 Various Performance Indices

After reformulating and interpreting the constraints in both velocity-level redundancy resolution (5.3)–(5.7) and acceleration-level redundancy resolution (5.8)–(5.14), we come to interpret and handle various typical performance indices $\phi(\theta, \dot{\theta}, \ddot{\theta})$.

5.5.1 Resolved at Joint-Velocity Level

In this subsection, different schemes (i.e., the MVN scheme, the RMP scheme and the MKE scheme) at joint-velocity level are presented for redundancy resolution.

5.5.1.1 MVN scheme

The simplest and most effective performance index at the velocity-level redundancy resolution is the so-called minimum velocity norm (MVN) scheme; that is,

$$\phi(\dot{\theta}) = \|\dot{\theta}\|_2^2/2 = \dot{\theta}^T\dot{\theta}/2.$$

If we rewrite it in a quadratic-minimization form such as $\phi(\dot{\theta}) = \dot{\theta}^{\mathrm{T}}W\dot{\theta}/2 + \mathbf{q}^{\mathrm{T}}\dot{\theta}$, then in the MVN redundancy-resolution scheme $W = I$ and $\mathbf{q} = 0$.

5.5.1.2 RMP scheme

A natural extension from the MVN scheme is the repetitive motion planning (RMP) by minimizing the joint displacements between current states and initial states [36, 90]. That is, to minimize the following performance index $\phi(\theta, \dot{\theta})$ so that the robot could perform cyclic tasks:

$$\phi(\theta, \dot{\theta}) = (\dot{\theta} + \mathbf{z})^{\mathrm{T}}(\dot{\theta} + \mathbf{z})/2 \quad \text{with} \quad \mathbf{z} = 4(\theta - \theta(0)).$$

If we rewrite it in a quadratic form such as $\phi(\dot{\theta}) = \dot{\theta}^{\mathrm{T}}W\dot{\theta}/2 + \mathbf{q}^{\mathrm{T}}\dot{\theta}$, then in the RMP redundancy-resolution scheme $W = I$ and $\mathbf{q} = \mathbf{z} = 4(\theta - \theta(0))$.

5.5.1.3 MKE scheme

Being a local counterpart of global kinetic-energy minimization [78, 91], the following inertia-weighted performance index is defined and resolved at the joint-velocity level:

$$\phi(\theta, \dot{\theta}) = \dot{\theta}^{\mathrm{T}}H\dot{\theta}/2.$$

This performance index minimizes the instantaneous kinetic energy, and hence the resulting scheme is termed the minimum kinetic energy (MKE) redundancy resolution. If we rewrite it in a quadratic form such as $\phi(\dot{\theta}) = \dot{\theta}^{\mathrm{T}}W\dot{\theta}/2 + \mathbf{q}^{\mathrm{T}}\dot{\theta}$, then in the MKE redundancy-resolution scheme $W = H(\theta)$ and $\mathbf{q} = 0$.

5.5.2 Resolved at Joint-Acceleration Level

In this subsection, different schemes (i.e., the MAN scheme, the MTN scheme and the IIWT scheme) at acceleration level are presented for redundancy resolution.

5.5.2.1 MAN scheme

Acceleration-level redundancy resolution can accommodate more joint physical limits and, with such physical limits considered, acceleration-level schemes can also be used for long trajectory movements. Based on a large number of computer simulations, the minimum acceleration norm (MAN) scheme is thought to be one of the most effective acceleration-level redundancy-resolution schemes for long trajectory movements [61, 92]. The performance index is $\phi(\ddot{\theta}) = \ddot{\theta}^{\mathrm{T}}\ddot{\theta}/2$ that defines $W = I$ and $\mathbf{q} = 0$, if we rewrite it in the quadratic form $\phi(\ddot{\theta}) = \ddot{\theta}^{\mathrm{T}}W\ddot{\theta}/2 + \mathbf{q}^{\mathrm{T}}\ddot{\theta}$.

5.5.2.2 MTN scheme

A natural extension from the MAN scheme to the joint torque minimization is the minimum torque norm (MTN) redundancy-resolution scheme. That is, to minimize the following two norm of instantaneous joint torques for a better distribution of actuator power [61, 69, 70]:

$$\phi(\theta,\dot{\theta},\ddot{\theta}) = \|\tau\|_2^2/2 = \tau^{\mathrm{T}}\tau/2. \tag{5.20}$$

As the redundancy is to be resolved at the joint-acceleration level, the performance index in (5.20) is converted to an expression based on $\ddot{\theta}$ via manipulator dynamic equation (5.9). Simple matrix manipulations yield the following minimization [61], $\phi(\theta,\dot{\theta},\ddot{\theta}) = \ddot{\theta}^{\mathrm{T}}H^2\ddot{\theta}/2 + (\mathbf{c}+\mathbf{g})^{\mathrm{T}}H\ddot{\theta}$, which defines $W = H^2$ and $q = H(\mathbf{c}+\mathbf{g})$, if we rewrite it in the quadratic form $\phi(\ddot{\theta}) = \ddot{\theta}^{\mathrm{T}}W\ddot{\theta}/2 + \mathbf{q}^{\mathrm{T}}\ddot{\theta}$.

5.5.2.3 IIWT scheme

Another acceleration-level redundancy-resolution scheme is the inertia inverse weighted torque (IIWT) minimization by minimizing the joint torque weighted by inertia inverse [78, 91]. As analyzed via calculus of variations, the IIWT scheme results in resolutions with global characteristics. The performance index to be minimized is

$$\phi(\theta,\dot{\theta},\ddot{\theta}) = \tau^{\mathrm{T}}H^{-1}\tau/2. \tag{5.21}$$

The redundancy is resolved at the joint-acceleration level (i.e., based on $\ddot{\theta}$). This performance index is thus converted to the following one by using the manipulator dynamic equation (5.9) [69, 78]:

$$\phi(\theta,\dot{\theta},\ddot{\theta}) = \ddot{\theta}^{\mathrm{T}}H\ddot{\theta}/2 + (\mathbf{c}+\mathbf{g})^{\mathrm{T}}\ddot{\theta}.$$

This defines $W = H$ and $\mathbf{q} = (\mathbf{c}+\mathbf{g})$ in the IIWT context, if we rewrite the performance index in the unified quadratic form $\phi(\ddot{\theta}) = \ddot{\theta}^{\mathrm{T}}W\ddot{\theta}/2 + \mathbf{q}^{\mathrm{T}}\ddot{\theta}$.

5.6 Unified QP Formulation

Following these analysis procedures (especially from Sections 5.3 to 5.5), we can see that both velocity-level redundancy resolution (5.3)–(5.7) and acceleration-level redundancy resolution (5.8)–(5.14) can be reformulated as a QP problem.

Theorem 5.1 Consider the avoidance of joint physical limits and environmental obstacles. The velocity-level redundancy resolution (i.e., the MVN, RMP, and MKE schemes) and acceleration-level redundancy resolution (i.e., the MAN, MTN, and IIWT schemes) all can be rewritten in the following QP form:

$$\text{minimize} \quad \mathbf{x}^{\mathrm{T}}W\mathbf{x}/2 + \mathbf{q}^{\mathrm{T}}\mathbf{x}, \tag{5.22}$$

$$\text{subject to} \quad J\mathbf{x} = \mathbf{d}, \tag{5.23}$$

$$A\mathbf{x} \leqslant \mathbf{b}, \tag{5.24}$$

$$\xi^- \leqslant x \leqslant \xi^+, \tag{5.25}$$

where decision vector $\mathbf{x}$ is defined, respectively, as $\dot{\theta}$ in velocity-level schemes or $\ddot{\theta}$ in acceleration-level schemes. Coefficients W, $\mathbf{q}$, A, $\mathbf{b}$, $\mathbf{d}$, and $\xi^{\pm}$ are defined accordingly for a specific redundancy-resolution scheme.

5.7 Online QP Solutions

In previous sections, we have reformulated the physically-constrained redundancy resolution problem into a time-varying QP subject to hybrid kinds of constraints. Each term has interpretably physical meaning and utility. This reformulation extracts major mathematic problems from an originally very complex robotic context, making the redundancy-resolution task much clearer and easier to understand. To solve QP (5.22)–(5.25), the following QP solutions are reviewed.

5.7.1 Traditional QP Routines

The first option without derivation could be the traditional QP optimization routines performed on digital computers. For example, among the MATLAB optimization routines [93], "QUADPROG" can be used with syntax being

$$x = \text{QUADPROG}(W, \mathbf{q}, A, \mathbf{b}, J, \mathbf{d}, \xi^-, \xi^+).$$

5.7.2 Compact QP Method

A compact QP method was developed to improve the computational efficiency of solving QP [71, 90]. The compact QP method entails Gaussian elimination with partial pivoting, which is possibly of $O(n^3)$ operations. In general, it is a more efficient numerical QP method, compared to general-purpose optimization algorithms.

5.7.3 Dual Neural Network

As a parallel-processing dynamic-system based alternative to continuous-time optimization, a dual-neural-network QP solution was developed in [81]. It has piecewise linear dynamics, global (exponential) convergence to optimal solutions, and capability of handling hybrid constraints simultaneously. However, because the dual-neural-network solution requires the inverse of coefficient matrix W, it is only able to solve strictly-convex QP problems (specifically, W being positive-definite and preferably constant).

5.7.4 LVI-Aided Primal-Dual Neural Network

Considering that W can be time-varying and positive semi-definite while $\dot{\mathbf{x}}$ is required for velocity-level redundancy resolution when applied to joint torque control [61], the LVIAPDNN has been developed. The LVIAPDNN has simple piecewise linear dynamics, global (exponential) convergence to optimal solutions, and capability to handle general QP problems in the same inverse-free manner [61, 94]. Given that the LVIAPDNN is one of the main QP solvers involved in this book, a relatively detailed description on LVIAPDNN is presented as follows for completeness.

We can first convert QP (5.22)–(5.25) to a set of LVI, and then to the following piecewise-linear equation (PLE):

$$\mathbf{P}_\Omega(\mathbf{y} - (Q\mathbf{y} + \mathbf{p})) - \mathbf{y} = 0, \tag{5.26}$$

where the primal-dual decision vector $\mathbf{y}$, together with its lower/upper bounds, is defined as

$$\mathbf{y} = \begin{bmatrix} \mathbf{x} \\ \mathbf{u} \\ \mathbf{v} \end{bmatrix}, \ \mathbf{y}^- = \begin{bmatrix} \xi^- \\ -\mathbf{1}_\mathrm{v}\varpi \\ 0 \end{bmatrix}, \ \mathbf{y}^+ = \begin{bmatrix} \xi^+ \\ \mathbf{1}_\mathrm{v}\varpi \\ \mathbf{1}_\mathrm{v}\varpi \end{bmatrix}, \tag{5.27}$$

with $\mathbf{1}_\mathrm{v} = [1, \cdots, 1]^\mathrm{T}$ denoting an appropriately dimensioned vector composed of ones, and $\varpi \gg 0$ being sufficiently large to represent $+\infty$. The augmented coefficients are defined as

$$Q = \begin{bmatrix} W & -J^\mathrm{T} & A^\mathrm{T} \\ J & 0 & 0 \\ -A & 0 & 0 \end{bmatrix}, \ \mathbf{p} = \begin{bmatrix} \mathbf{q} \\ -\mathbf{d} \\ \mathbf{b} \end{bmatrix}. \tag{5.28}$$

The set $\Omega = \{\mathbf{y} | \mathbf{y}^- \leqslant \mathbf{y} \leqslant \mathbf{y}^+\}$, and the piecewise linear projection operator $\mathbf{P}_\Omega(\cdot)$ is defined as $\mathbf{P}_\Omega(\mathbf{y}) = [P_\Omega(y_1), \cdots, P_\Omega(y_i), \cdots]^\mathrm{T}$ with the ith element being [81]

$$P_\Omega(y_i) = \begin{cases} y_i^- & \text{if} \quad y_i < y_i^- \\ y_i & \text{if} \quad y_i^- \leqslant y_i \leqslant y_i^+ \\ y_i^+ & \text{if} \quad y_i > y_i^+ \end{cases}, \ \forall i.$$

From our neural-network design experience ([61, 81, 94] and references therein), it follows that the LVIAPDNN, being the QP solution for (5.22)–(5.25), can use the following dynamics (see [61] and references therein):

$$\dot{\mathbf{y}} = \gamma(I + Q^\mathrm{T})(\mathbf{P}_\Omega(\mathbf{y} - (Q\mathbf{y} + \mathbf{p})) - \mathbf{y}) \tag{5.29}$$

where $\gamma > 0$ is the design parameter used to scale the network convergence.

Figure 5.2 shows the block diagram of the QP-based dynamic system approach to redundancy resolution and joint torque control of redundant manipulators. Based on the current robot state, the online path planner first generates the end-effector command, that is, $\mathbf{r}_\mathrm{d}$, $\dot{\mathbf{r}}_\mathrm{d}$, and $\ddot{\mathbf{r}}_\mathrm{d}$ if necessary. The redundancy resolution problem is then formulated as a QP in the form of (5.22)–(5.25). With simple matrix/vector augmentations and operations, the LVIAPDNN for online solution to the QP can generate optimal $\mathbf{x}$ (i.e., $\ddot{\theta}$ in acceleration-level schemes and $\dot{\theta}$ in velocity-level schemes). For velocity-level schemes, the signal $\dot{\mathbf{x}}$ is also brought into use from the LVIAPDNN for joint torque control. Note that the redundancy resolution generates desired joint motion trajectories based on accurate manipulator models. In a real system, the modelling errors exist and the feedback control should be applied [62, 95, 96].

It is also worth mentioning that an early model of (5.29) was presented in [97] in the form of recurrent neural networks to solve linear projection equations only, while this dynamic system is elaborated on here to solve general QP problems subject to hybrid constraints. As the primal and dual problems/variables are considered simultaneously, such a QP solution (5.29) is termed an LVIAPDNN or dynamic system. The network

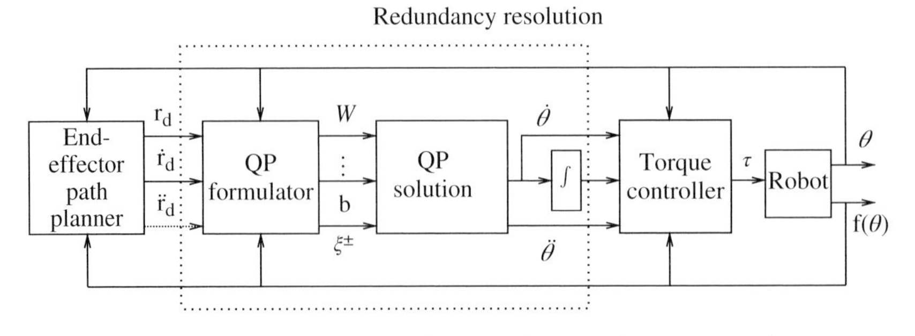

Figure 5.2 QP-based approach to redundancy resolution and torque control.

architecture or computational complexity is thus simpler than that other recurrent neural networks, also in light of the piecewise-linear dynamics. In addition, there is no online inversion of W in our approach, which is prohibitively of $O(n^3)$ numerical operations existing in others' approaches [73, 78]. Furthermore, we have the following convergence results of the LVI-aided primal-dual neural network QP solution [61].

Theorem 5.2 (solution Convergence) Starting from any initial state, the state vector $\mathbf{y}(t)$ of LVIAPDNN dynamic system (5.29) is convergent to an equilibrium point $\mathbf{y}^*$, of which the first n elements constitute the optimal solution $\mathbf{x}^*$ to the QP problem in (5.22)–(5.25). Moreover, the exponential convergence can be achieved, provided that there exists a constant $\rho > 0$ such that $\|\mathbf{y} - \mathbf{P}_\Omega(\mathbf{y} - (Q\mathbf{y} + \mathbf{p}))\|_2 \geqslant \rho\|\mathbf{y} - \mathbf{y}^*\|_2$.

5.7.5 Numerical Algorithms E47 and 94LVI

Given that numerical algorithms E47 and 94LVI are other two main QP solvers involved in this book, we also present the relatively detailed descriptions on these two numerical algorithms as follows for the completeness of this book. Define that the optimal solution set $\Omega_x^* = \{\mathbf{x}^* | \mathbf{x}^* \text{ is a solution of QP (5.22)–(5.25)}\}$ or equivalently, define that the solution set $\Omega_y^* = \{\mathbf{y}^* | \mathbf{y}^* \text{ is a solution of PLPE (5.26) with } \mathbf{e}(\mathbf{y}^*) = 0\}$. Given initial vector $\mathbf{y}^0 \in \mathbb{R}^{n+m}$, for iteration index $k = 0, 1, 2, 3, \cdots$ (which is used as a superscript), either of the following numerical algorithms E47 and 94LVI can be applied to solving the PLE (5.26) as well as the original QP problem (5.22)–(5.25).

5.7.5.1 Numerical Algorithm E47

Input: Coefficients W, $\mathbf{q}$, A, J, $\mathbf{b}$, $\mathbf{d}$, $\mathbf{x}^-$, and $\mathbf{x}^+$ as well as initial vector $\mathbf{x}^0$.
Output: $\mathbf{x}^*$.

- Step 1: Generate Q, $\mathbf{p}$, $\mathbf{y}^-$, and $\mathbf{y}^+$ via equalities (5.27) and (5.28).
- Step 2: Set $k = 0$, and generate $\mathbf{y}^0$ from $\mathbf{x}^0$ via equality (5.27).
- Step 3: Compute $\mathbf{e}(\mathbf{y}^k)$, where $\mathbf{e}(\mathbf{y}) = \mathbf{y} - \mathbf{P}_\Omega(\mathbf{y} - (Q\mathbf{y} + \mathbf{p}))$. If $\mathbf{y}^k \in \Omega_y^*$ (i.e., theoretically $\mathbf{e}(\mathbf{y}^k) = 0$, but practically and acceptably $\|\mathbf{e}(\mathbf{y}^k)\|_2 < 10^{-3}$ for example, with $\|\cdot\|_2$ denoting the two-norm), then go to Step 6. Otherwise, go to Step 4.
- Step 4: With I denoting an identity matrix, compute

$$\delta(\mathbf{y}^k) = Q^{\mathrm{T}}\mathbf{e}(\mathbf{y}^k) + Q\mathbf{y}^k + \mathbf{p}, \tag{5.30}$$

$$\rho(\mathbf{y}^k) = \frac{\|\mathbf{e}(\mathbf{y}^k)\|_2^2}{\|(Q^{\mathrm{T}} + I)\mathbf{e}(\mathbf{y}^k)\|_2^2}, \tag{5.31}$$

$$\mathbf{y}^{k+1} = \mathbf{P}_\Omega(\mathbf{y}^k - \rho(\mathbf{y}^k)\delta(\mathbf{y}^k)). \tag{5.32}$$

- Step 5: Increase k by one, go to Step 3.
- Step 6: Output $\mathbf{x}^*$ which is a vector made of the first n elements of $\mathbf{y}^k$.

5.7.5.2 Numerical Algorithm 94LVI

Input: Coefficients W, $\mathbf{q}$, A, J, $\mathbf{b}$, $\mathbf{d}$, $\mathbf{x}^-$, and $\mathbf{x}^+$ as well as initial vector $\mathbf{x}^0$.
Output: $\mathbf{x}^*$.

- Step 1: Generate Q, $\mathbf{p}$, $\mathbf{y}^-$, and $\mathbf{y}^+$ via equalities (5.27) and (5.28).
- Step 2: Set $k = 0$, and generate $\mathbf{y}^0$ from $\mathbf{x}^0$ via equality (5.27).

- Step 3: Compute $\mathbf{e}(\mathbf{y}^k)$. If $\mathbf{y}^k \in \Omega_y^*$ (i.e., acceptably with $\|\mathbf{e}(\mathbf{y}^k)\|_2 < 10^{-3}$ for example), then go to Step 6.
- Step 4: Compute

$$\delta(\mathbf{y}^k) = (Q^{\mathrm{T}} + I)\mathbf{e}(\mathbf{y}^k), \tag{5.33}$$

$$\rho(\mathbf{y}^k) = \frac{\|\mathbf{e}(\mathbf{y}^k)\|_2^2}{\|\delta(\mathbf{y}^k)\|_2^2}, \tag{5.34}$$

$$\mathbf{y}^{k+1} = \mathbf{y}^k - \rho(\mathbf{y}^k)\delta(\mathbf{y}^k). \tag{5.35}$$

- Step 5: Increase k by one, go to Step 3.
- Step 6: Output $\mathbf{x}^*$ which is a vector made of the first n elements of $\mathbf{y}^k$.

The global convergence of numerical algorithms E47 and 94LVI is presented in the following theorems. Theorem 5.3 is actually an intermediate result for proving the global linear convergence of numerical algorithms E47 and 94LVI (i.e., Theorem 5.4 and Theorem 5.5) [8, 17, 98].

Theorem 5.3 $\forall \mathbf{y}^* \in \Omega_y^*$, sequence $\{\mathbf{y}^k\}$ (with iteration index $k = 0, 1, 2, \cdots$) generated by either of numerical algorithms E47 and 94LVI satisfies

$$\|\mathbf{y}^{k+1} - \mathbf{y}^*\|_2^2 \leqslant \|\mathbf{y}^k - \mathbf{y}^*\|_2^2 - \rho(\mathbf{y}^k)\|\mathbf{e}(\mathbf{y}^k)\|_2^2. \tag{5.36}$$

Theorem 5.4 Sequence $\{\mathbf{y}^k\}$ generated by either of numerical algorithms E47 and 94LVI converges to a solution $\mathbf{y}^* \in \Omega_y^*$ of PLE (5.26) with its first n elements constituting the optimal solution $\mathbf{x}^* \in \Omega_x^*$ of QP (5.22)–(5.25).

Theorem 5.5 Sequence $\{\mathbf{x}^k\}$ generated by either of numerical algorithms E47 and 94LVI converges to the optimal solution $\mathbf{x}^* \in \Omega_x^*$ of QP (5.22)–(5.25) linearly.

Proof: We present the following two specific parts of the proof for the theorem.

Part 1 (Proof about numerical algorithm E47). According to the proof of Theorem 3 in [98] for numerical algorithm E47, we have

$$\|\mathbf{y}^{k+1} - \mathbf{y}^*\|_2^2 \leqslant \|\mathbf{y}^k - \mathbf{y}^*\|_2^2 - \rho(\mathbf{y}^k)\|\mathbf{e}(\mathbf{y}^k)\|_2^2. \tag{5.37}$$

Two situations of (5.37) are discussed next.

Situation 1. If $\|\mathbf{y}^k - \mathbf{y}^*\|_2^2 = 0$ (i.e., $\|\mathbf{e}(\mathbf{y}^k)\|_2^2 = 0$), then $\|\mathbf{y}^{k+1} - \mathbf{y}^*\|_2^2 = 0$. Therefore, $\mathbf{y}^k$ and $\mathbf{y}^{k+1}$ are the exact theoretical solution $\mathbf{y}^*$ for (5.26). Correspondingly, $\mathbf{x}^k$ is the exact solution for QP (5.22)–(5.25).

Situation 2. If $\|\mathbf{y}^k - \mathbf{y}^*\|_2^2 > 0$ (i.e., $\|\mathbf{e}(\mathbf{y}^k)\|_2^2 > 0$), then (5.37) yields

$$\begin{aligned}\|\mathbf{y}^{k+1} - &\mathbf{y}^*\|_2^2/\|\mathbf{y}^k - \mathbf{y}^*\|_2^2 \\ &\leqslant -1 - \rho(\mathbf{y}^k)\|\mathbf{e}(\mathbf{y}^k)\|_2^2/\|\mathbf{y}^k - \mathbf{y}^*\|_2^2 \leqslant 1 - \varsigma,\end{aligned} \tag{5.38}$$

where a sufficiently small constant $\varsigma > 0$ exists such that $\rho(\mathbf{y}^k)\|\mathbf{e}(\mathbf{y}^k)\|_2^2 \geqslant \varsigma\|\mathbf{y}^k - y^*\|_2^2$, that is,

$$\|\mathbf{y}^{k+1} - \mathbf{y}^*\|_2/\|\mathbf{y}^k - \mathbf{y}^*\|_2 \leqslant \sqrt{1 - \varsigma} < 1.$$

In addition, from (5.38), with iteration index $k = 0, 1, 2, 3, \cdots$, we obtain

$$\|\mathbf{y}^1 - \mathbf{y}^*\|_2^2/\|\mathbf{y}^0 - \mathbf{y}^*\|_2^2 \leqslant 1 - \varsigma$$
$$\|\mathbf{y}^2 - \mathbf{y}^*\|_2^2/\|\mathbf{y}^1 - \mathbf{y}^*\|_2^2 \leqslant 1 - \varsigma$$
$$\vdots$$
$$\|\mathbf{y}^{j+1} - \mathbf{y}^*\|_2^2/\|\mathbf{y}^j - \mathbf{y}^*\|_2^2 \leqslant 1 - \varsigma$$
$$\vdots$$

Multiplying all these inequalities, we have

$$\|\mathbf{y}^{k+1} - \mathbf{y}^*\|_2^2/\|\mathbf{y}^0 - \mathbf{y}^*\|_2^2 \leqslant (1 - \varsigma)^{k+1}.$$

With $\mathbf{x}^k$ being constituted by the first n elements of $\mathbf{y}^k$, according to the definition of the two-norm, we obtain

$$\|\mathbf{y}^k - \mathbf{y}^*\|_2^2 = \|\mathbf{x}^k - \mathbf{x}^*\|_2^2 + \|\mathbf{u}^k - \mathbf{u}^*\|_2^2.$$

As $\|\cdot\|_2^2 \geqslant 0$, we further obtain

$$\|\mathbf{x}^{k+1} - \mathbf{x}^*\|_2^2 \leqslant \|\mathbf{y}^0 - \mathbf{y}^*\|_2^2(1 - \varsigma)^{k+1}.$$

We thus prove that $\mathbf{x}^k$ constituted by the first n elements of $\mathbf{y}^k$ converges to the optimal solution $\mathbf{x}^* \in \Omega_x^*$ of QP (5.22)–(5.25) linearly.

The proof on the global linear convergence property of numerical algorithm E47 is thus complete.

Part 2 (Proof about numerical algorithm 94LVI). It can be seen and generalized from [99] and the previous Part 1 proof, thus omitted here.

So, the proof on the global linear convergence property of numerical algorithms E47 and 94LVI is complete. □

5.8 Computer Simulations

The research of the past 10 years has finally achieved a unification of various redundancy-resolution schemes through the QP formulation. In the 1990s, computer simulations were mainly based on planar or theoretical robot manipulators, for example, three- or four-link planar robots [86, 89, 90]. Recently, a large number of QP-based computer simulations have been performed more oriented towards PUMA560, PA10, or other industrial spatial robot manipulators [36, 61, 65]. Due to space limitation, only closely related simulation results are summarized next with illustrative observations presented.

- This QP formulation can keep the joints within their physical limits [36, 78]. One more example is in Figure 5.3.
- This QP formulation can avoid multiple obstacles in a crowded dynamic environment which includes point obstacles and window-shaped obstacle. See [65, 83]. In addition, this QP formulation can handle various performance indices in the same unified framework, such as, the aforementioned MVN, RMP, MKE, MAN, MTN, and IIWT schemes [36, 61, 78, 100].
- This QP formulation and its dynamic solution (5.29) can generate accurately joint acceleration information for velocity-level redundancy-resolution schemes if required for joint torque control. See Figure 5.4, which corresponds to MKE scheme.

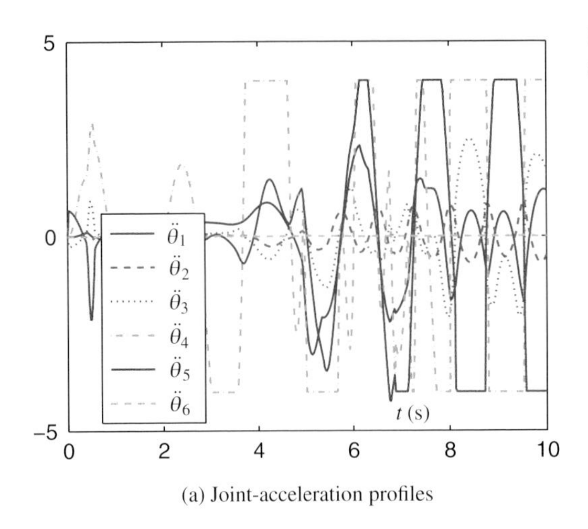

(a) Joint-acceleration profiles

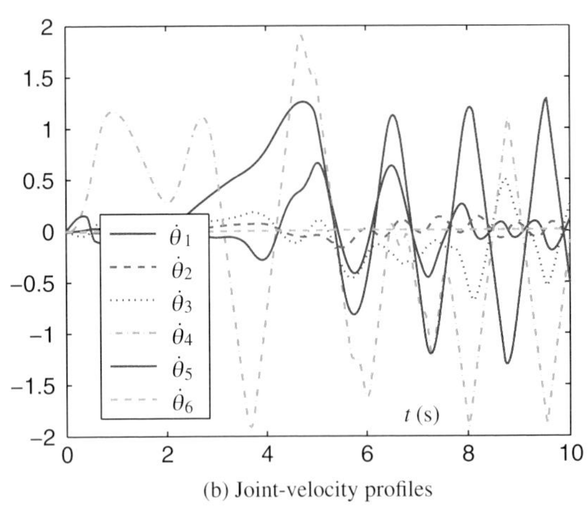

(b) Joint-velocity profiles

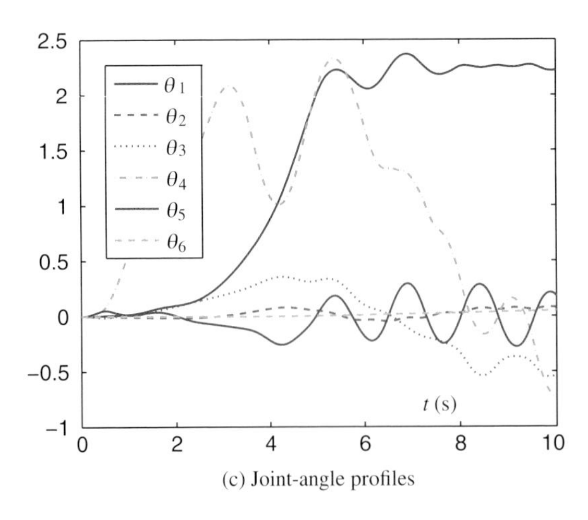

(c) Joint-angle profiles

Figure 5.3 PUMA560 transients synthesized by a QP-based MTN scheme.

Figure 5.4 PUMA560 transients synthesized by a QP-based MKE scheme.

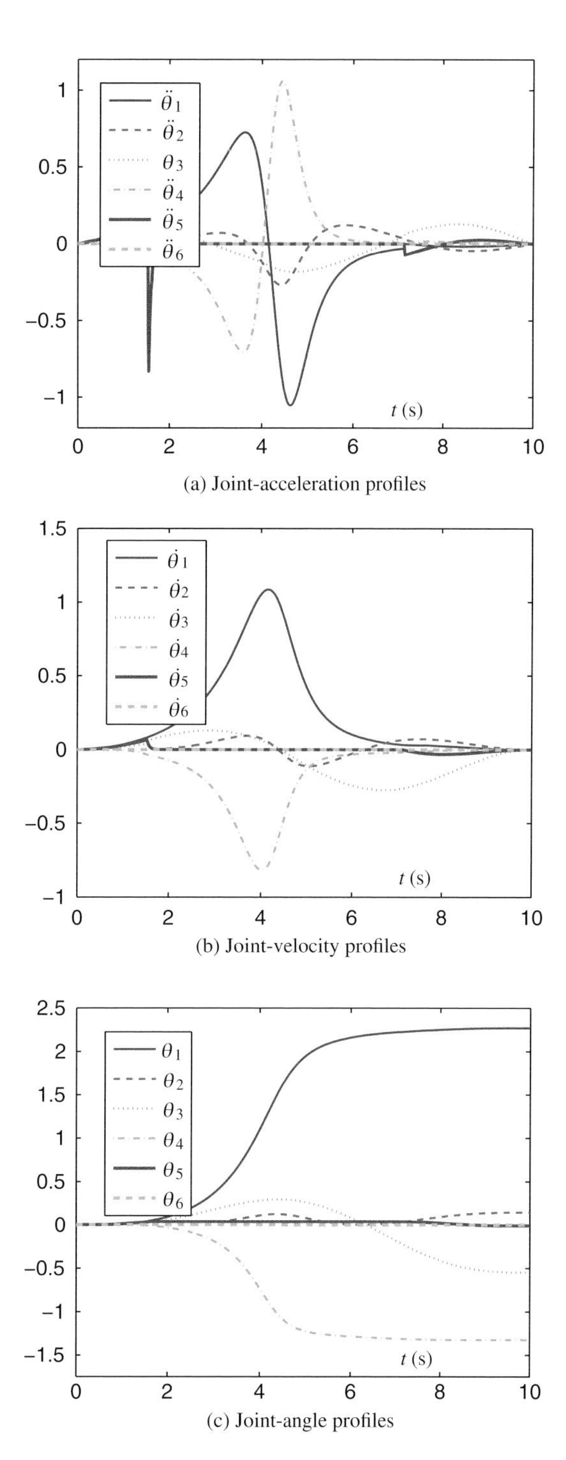

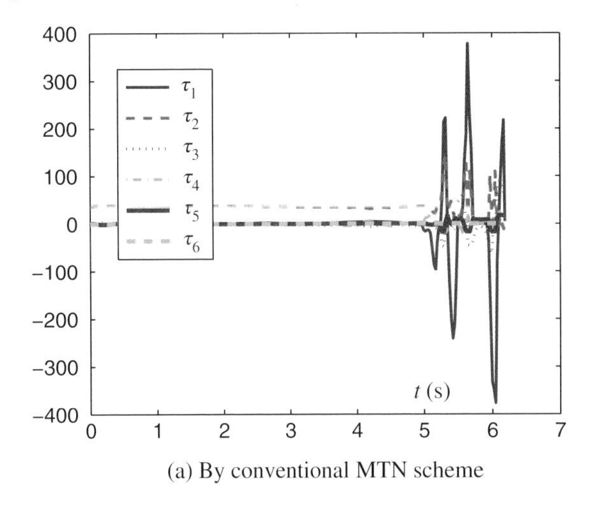

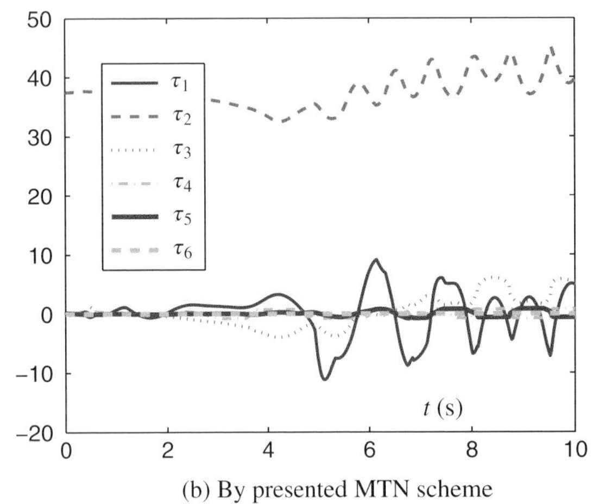

Figure 5.5 Comparison of (a) conventional and (b) presented MTN schemes.

- The velocity-level redundancy resolution in (5.3)–(5.7) is more desirable for long trajectory movements [61, 70, 101]. In contrast, the acceleration-level redundancy resolution in (5.8)–(5.14) can handle more joint physical limits simultaneously, which can naturally remedy the torque divergence problem as well [61, 92], where the MAN scheme is the preferred one.
- It is shown in Figure 5.5 through the MTN examples that taking into consideration joint physical limits can hinder the build-up of very large null-space joint velocities/accelerations and thus remedy the torque instability problem, whereas the conventional MTN fails at time $t = 6.2$ s without considering joint physical limits.
- In terms of torque profiles as in Figure 5.6 and Figure 5.7, the velocity-level MVN, MKE, and acceleration-level MAN schemes are superior to the MTN and IIWT schemes for long movements. On the other hand, MTN and IIWT schemes may exhibit natural motion characteristics because of considering the inertia matrix [78] and may encounter intensive computational burden/inaccuracy also because of this sometimes ill-conditioned H-weighting matrix.

Figure 5.6 Joint-torque profiles synthesized by other QP-based resolution schemes of (a) IIWT and (b) MAN.

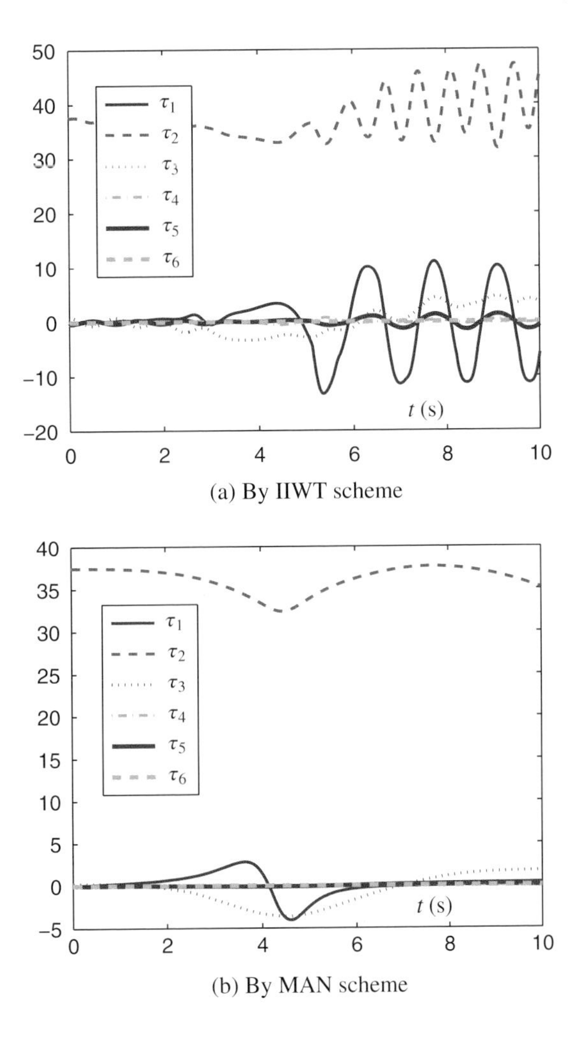

Figure 5.7 Joint-torque profiles synthesized by other QP-based resolution schemes of (a) MKE and (b) MVN.

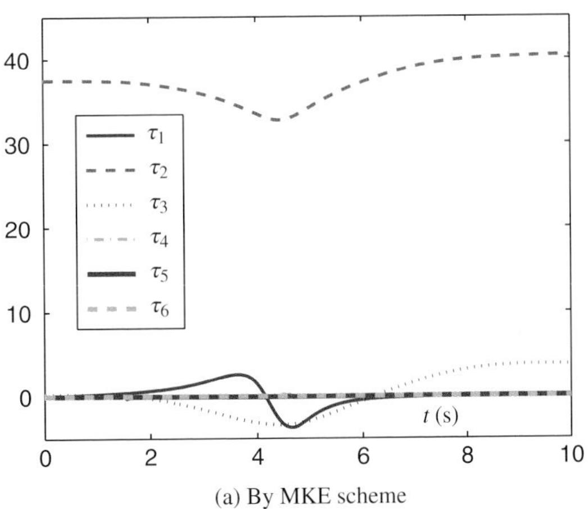

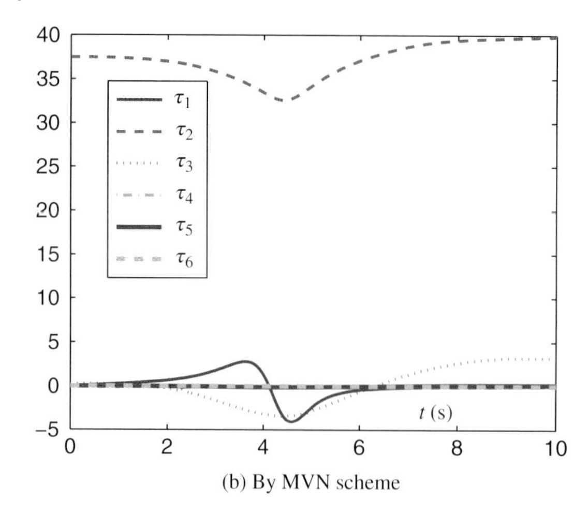

(b) By MVN scheme

Figure 5.7 (*Continued*)

- Joint physical variables have been kept within their mechanical limits by using the presented QP-based redundancy-resolution schemes. As seen from simulation data, the end-effector positioning error is usually of (or less than) the order 10^{-4} m.

5.9 Chapter Summary

This chapter has unified the redundancy resolution of robot manipulators into a general QP formulation; that is, in (5.22)–(5.25). This QP formulation can cover the online avoidance of joint physical limits and environmental obstacles, as well as optimizing various performance indices. Every term has clear physical meaning and utility. Some QP online solutions have also been reviewed for solving QP (5.22)–(5.25). Note that the LVIAPDNN and the two numerical algorithms E47 and 94LVI have been applied to the online redundancy resolution of redundant manipulators. In addition, a large number of simulation results have been summarized based on PUMA560, PA10, and other robot manipulators. This research from the last 10 years has basically answered the issue of manipulators' redundancy resolution via QP approaches.

Part IV

Illustrative JVL QP Schemes and Performances

6

Varying Joint-Velocity Limits Handled by QP

6.1 Introduction

Redundant manipulators can achieve subtasks readily and smartly such as generating cyclic motion [102–106], avoiding joint limits [36, 107], avoiding singularities [108–112], tolerating faults [113], avoiding obstacles [114–116], and optimization of multiple performance criteria [61, 92, 117–129] because they have more DOF than required to execute a desired primary task [61, 92]. Motion planning (or resolving the redundancy problem) of the manipulators is thus an appealing topic in the robotics area [36, 61, 92, 104]. The pseudoinverse-based approach is the conventional method for resolving the redundancy problem of the manipulators [130–133]. Research in the last decade [36, 61, 92, 104, 134, 135] shows that various online optimization strategies/techniques are preferred methods, and some of these optimization strategies are usually expressed as a quadratic program (QP), which is subject to equality and inequality constraints [61, 90]. Such a QP is then converted into linear variational inequality (LVI) [61, 128, 136], which may be solved approximately by many methods and techniques efficiently, such as numerical methods [136] and some types of neural networks (NNs) [36, 61, 92, 104, 137–139]. In [104], three recurrent NNs are applied to cyclic motion generation. In [61, 105, 106, 137–139], primal-dual NNs formulated in differential equations are used to solve various redundancy resolution problems. In [23], another NN, called zeroing dynamics, is used to solve the time-varying problem. In this chapter, for implementing the scheme on the digital computer and physical hardware system more readily, a numerical algorithm formulated in a difference equation is adopted to solve the QP.

Among those approaches and techniques, the mechanical limits are usually considered constant [104, 128]. This, however, may not be applicable to some kinds of manipulators, such as the push-rod (PR) manipulators, of which the joint-velocity limits change with end-effector and joints movement (i.e., the joint-velocity limits are functions of the joint angles). If these physical limits are assumed constant, the feasible-solution region of joint variables would decrease, which may mean some end-effector tasks cannot be finished. In this chapter, we present the design and analysis of a varying joint-velocity limits (VJVL) constrained minimum-velocity-norm (MVN) scheme (termed VJVL-constrained MVN scheme); furthermore, we implement and test such a technique on a six-DOF planar robot manipulator.

Robot Manipulator Redundancy Resolution, First Edition. Yunong Zhang and Long Jin.

6.2 Preliminaries and Problem Formulation

As shown in [1, 65, 105], in view of the robot forward-kinematic equation (i.e., $\mathbf{r}_\mathrm{e} = \mathbf{f}(\boldsymbol{\theta})$ with $\mathbf{r}_\mathrm{e} \in \mathbb{R}^m$ denoting the end-effector position and orientation vector, $\boldsymbol{\theta} \in \mathbb{R}^n$ denoting the joint variable vector, and $\mathbf{f}(\cdot)$ being a differentiable nonlinear function), its first-order differential equation (i.e., $J(\boldsymbol{\theta})\dot{\boldsymbol{\theta}} \rightarrow \dot{\mathbf{r}}_\mathrm{d}$ with $\dot{\mathbf{r}}_\mathrm{d} \in \mathbb{R}^m$ denoting the desired end-effector velocity, $\dot{\boldsymbol{\theta}} \in \mathbb{R}^n$ denoting the joint velocity, and $J(\boldsymbol{\theta}) \in \mathbb{R}^{m\times n}$ being the Jacobian matrix), and the feedback theory [36], it is realistic and useful to consider the following scheme formulation to determine joint velocity $\dot{\boldsymbol{\theta}}$ in an inverse-free manner, since nearly all robots are constrained by the mechanical limits:

$$\text{minimize} \quad \|\dot{\boldsymbol{\theta}}\|_2^2/2 \tag{6.1}$$

$$\text{subject to} \quad J(\boldsymbol{\theta})\dot{\boldsymbol{\theta}} = \dot{\mathbf{r}}_\mathrm{d} + \kappa_\mathrm{p}(\mathbf{r}_\mathrm{d} - \mathbf{f}(\boldsymbol{\theta})) \tag{6.2}$$

$$\boldsymbol{\theta}^- \leqslant \boldsymbol{\theta} \leqslant \boldsymbol{\theta}^+ \tag{6.3}$$

$$\dot{\boldsymbol{\theta}}^- \leqslant \dot{\boldsymbol{\theta}} \leqslant \dot{\boldsymbol{\theta}}^+ \tag{6.4}$$

where $\|\cdot\|_2$ denotes the two-norm of a vector, and $\|\dot{\boldsymbol{\theta}}\|_2^2/2 = \dot{\boldsymbol{\theta}}^\mathrm{T}\dot{\boldsymbol{\theta}}/2$ with a superscript $^\mathrm{T}$ denoting the transpose of a vector or matrix. Equation (6.2) expresses a linear relation between the desired Cartesian velocity $\dot{\mathbf{r}}_\mathrm{d} \in \mathbb{R}^m$ and the joint velocity $\dot{\boldsymbol{\theta}} \in \mathbb{R}^n$, with position-error feedback $\kappa_\mathrm{p}(\mathbf{r}_\mathrm{d} - \mathbf{f}(\boldsymbol{\theta}))$ included, where κ_p is the feedback gain of position error (in the ensuing simulations and experiments, $\kappa_\mathrm{p} = 8$. In (6.3) and afterwards, $\boldsymbol{\theta}^+$ and $\boldsymbol{\theta}^-$ denote, respectively, the upper and lower limits of joint-angle vector $\boldsymbol{\theta}$. In (6.4) and afterwards, $\dot{\boldsymbol{\theta}}^+$ and $\dot{\boldsymbol{\theta}}^-$ denote, respectively, the upper and lower limits of joint velocity $\dot{\boldsymbol{\theta}}$. Note that, due to redundancy, here $m < n$, so (6.2) is under-determined.

Remark 6.1 Note that feeding just the reference Cartesian velocity $\dot{\mathbf{r}}_\mathrm{d}$ in an inverse kinematic control scheme may not be sufficient to obtain a robust tracking behavior, since the model disturbance and computational round-off errors always exist. The feedback control should thus be applied. One way is to add a simple feedback of Cartesian position error; that is, $J(\boldsymbol{\theta})\dot{\boldsymbol{\theta}} = \dot{\mathbf{r}}_\mathrm{d} + \kappa_\mathrm{p}(\mathbf{r}_\mathrm{d} - \mathbf{f}(\boldsymbol{\theta}))$ in (6.2) instead of using $J(\boldsymbol{\theta})\dot{\boldsymbol{\theta}} = \dot{\mathbf{r}}_\mathrm{d}$.

Remark 6.2 Based on previous experience [36, 61, 65, 92, 104] as well as the previous chapters, Equation (6.1) through Equation (6.4) can be reformulated as a QP. Many previous works [36, 61, 65, 92, 104] illustrate the advantages of using such QPs. For instance, there is no explicit need to invert $J(\boldsymbol{\theta})$ in this method and another advantage is that additional constraints such as physical limits and obstacle-avoidance constraints [36, 65, 107, 114–116] can also be considered into the generalized QP-based scheme formulation.

6.2.1 Six-DOF Planar Robot System

In the computer-simulations about motion planning, there is usually no or less of a need to consider specifically how to design and realize the configuration of joints with hardware. It is generally thought that the design and motors are ideal, satisfying simulation requests. However, in practice, by considering various factors such as motor performance, the weight of the arm, the accuracy of motion, existing design experience, and available equipment, the planar manipulator system is developed,

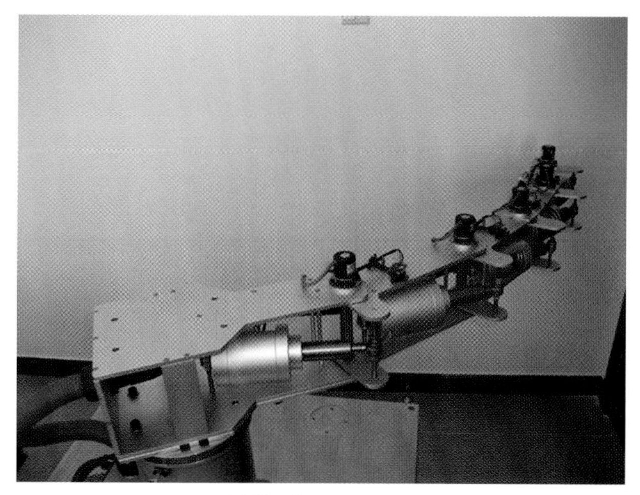

(a) Hardware system

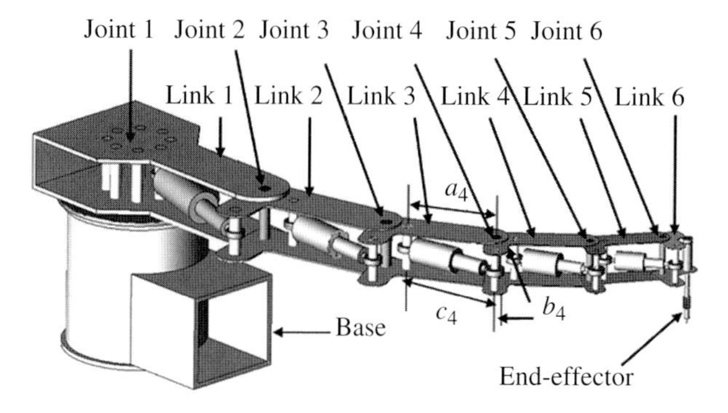

(b) Three-dimensional model

Figure 6.1 (a) Hardware system and (b) model of a six-DOF planar robot manipulator. *Source:* Zhang 2013. Reproduced with permission of IEEE.

which is shown in Figure 6.1. The second through the sixth joints of the robot are like triangles as shown in Figure 6.2(a), and the three edges are a_i, b_i, and c_i (with $i = 2, 3, \cdots, 6$) which is shown in Figure 6.2(b). The upper and lower limits of the joint angle used in this chapter can be, respectively, formulated as $\theta^- = [0, 0, 0, 0, 0, 0]$ rad, $\theta^+ = [+4.587, +0.785, +0.611, +0.576, +0.559, +0.445]$ rad. The other joint physical parameters are shown in Table 6.1, where l_i denotes the length of the ith link, with $i = 1, 2, \cdots, 6$.

This manipulator has six joints. The first joint, of which the mechanical material is steel, is driven by a big torque motor so as to meet the need of a wide range of adjustment and to bear the weight of the arm. The mechanical material of the second through to the sixth joints is aluminum so as to reduce the weight of the arm. The whole manipulator system is composed of a host computer and a manipulator. The host computer is a personal digital computer equipped with a Pentium E5300 2.6GHz CPU, 4GB DDR3 memory, and a Windows XP Professional operating system, which sends commands and

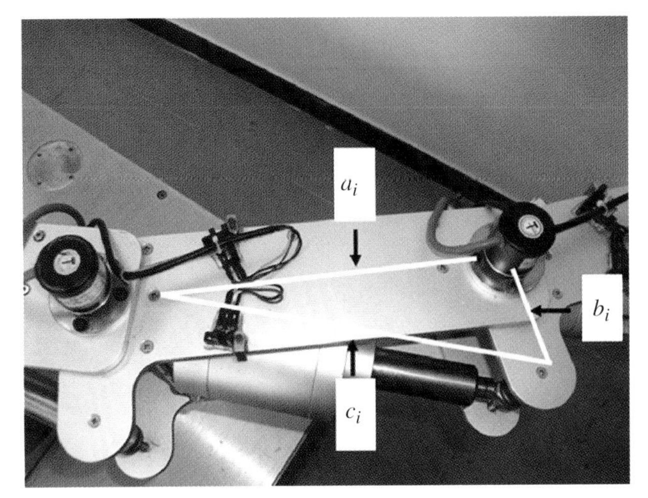

(a) Local configuration of hardware system

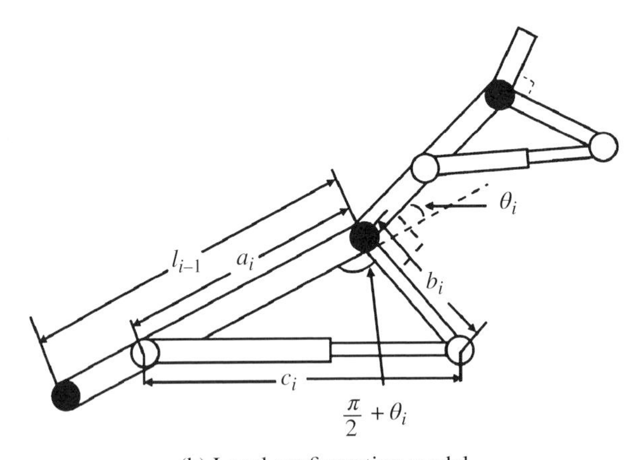

(b) Local configuration model

Figure 6.2 Local configuration of a six-DOF planar robot manipulator. *Source:* Zhang 2013. Reproduced with permission of IEEE.

Table 6.1 Parameters of a six-DOF planar robot manipulator hardware system.

♯	1	2	3	4	5	6
l (m)	0.301	0.290	0.230	0.225	0.214	0.103
a (m)	—	0.250	0.250	0.190	0.185	0.174
b (m)	—	0.080	0.080	0.080	0.080	0.080

signals to the robot motion-control module [e.g., a six-axis control card of peripheral component interconnect (PCI)]. Except for the first joint using a servomotor, stepping motors are employed in the other five joints.

6.2.2 Varying Joint-Velocity Limits

The side-lengths of the triangle shown in Figure 6.2 are a_i, b_i, and c_i, with $i = 2, 3, \cdots, 6$. By following the law of cosines, we have

$$c_i^2 = a_i^2 + b_i^2 - 2a_ib_i\cos(\pi/2 + \theta_i). \tag{6.5}$$

Since $c_i \geq 0$ and $\cos(\pi/2 + \theta_i) = -\sin\theta_i$, equation (6.5) becomes

$$c_i = \sqrt{a_i^2 + b_i^2 + 2a_ib_i\sin\theta_i} \tag{6.6}$$

and the time derivative of (6.5) is

$$2c_i\dot{c}_i = 2a_ib_i\sin(\pi/2 + \theta_i)\dot{\theta}_i \tag{6.7}$$

where $\dot{c}_i$ is the time derivative of c_i, $c_i = c_{0i} + \Delta c_i = c_{0i} + \int v_is_idt$, c_{0i} is the initial length of c_i, Δc_i is the elongation length of the ith push-rod, v_i is the rotation rate of the ith stepping motor (with its maximum value $|v_i|_{\max} = 10$ rot/s used in the equipment), and s_i is the elongation rate of the ith push-rod (i.e., the elongation length when the motor moves a full turn). For this equipment, $s_i = 2.5 \times 10^{-3}$ m/rot (with $i = 2, 3, \cdots, 6$). Thus $\dot{c}_i = v_is_i$, and substituting it into (6.7) yields

$$2c_i(v_is_i) = 2a_ib_i\cos\theta_i\dot{\theta}_i. \tag{6.8}$$

So, from (6.6) and (6.8), we have

$$\dot{\theta}_i = \frac{c_iv_is_i}{a_ib_i\cos\theta_i} = \frac{s_iv_i\sqrt{a_i^2 + b_i^2 + 2a_ib_i\sin\theta_i}}{a_ib_i\cos\theta_i}$$

which is evidently constrained by the following varying joint-velocity limits (VJVL):

$$\dot{\theta}_i^-(\theta_i) \leqslant \dot{\theta}_i \leqslant \dot{\theta}_i^+(\theta_i) \tag{6.9}$$

where

$$\dot{\theta}_i^-(\theta_i) = \frac{s_iv_i^-\sqrt{a_i^2 + b_i^2 + 2a_ib_i\sin\theta_i}}{a_ib_i\cos\theta_i} \tag{6.10}$$

$$\dot{\theta}_i^+(\theta_i) = \frac{s_iv_i^+\sqrt{a_i^2 + b_i^2 + 2a_ib_i\sin\theta_i}}{a_ib_i\cos\theta_i} \tag{6.11}$$

with v_i^- and v_i^+ being the negative and positive rotation rate limits of the stepping motor (in the equipment, $-v_i^- = v_i^+ = |v_i|_{\max} = 10$ rot/s).

This derivation shows that the joint-velocity limits $\dot{\theta}_i^\pm$ change with the joint angle θ_i (i.e., being functions of θ_i), with $i = 2, 3, \cdots, 6$, which are shown in Figure 6.3 through Figure 6.5.

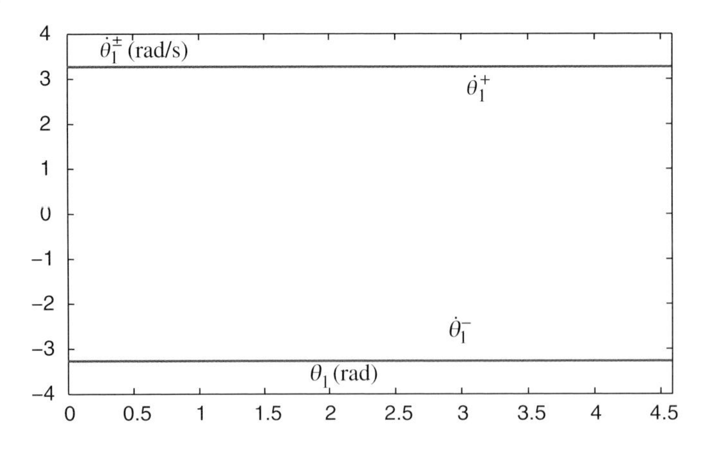

(a) Velocity limits $\dot{\theta}_1^{\pm}$ of Joint 1 being independent of joint angle θ_1

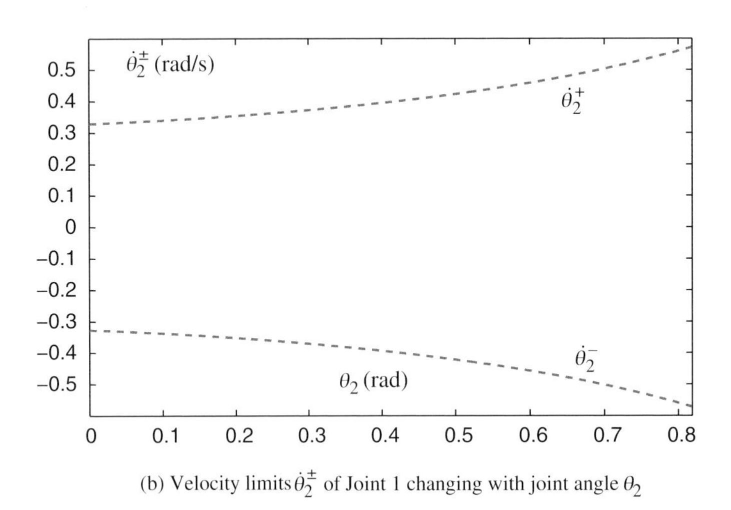

(b) Velocity limits $\dot{\theta}_2^{\pm}$ of Joint 1 changing with joint angle θ_2

Figure 6.3 Relationship between (a) $\dot{\theta}_1^{\pm}$ and θ_1 as well as (b) the relationship between $\dot{\theta}_2^{\pm}$ and θ_2.

In light of [1, 65, 128], by considering the VJVL (see the previous subsection), the above MVN scheme depicted in (6.1)–(6.4) for the six-DOF planar robot manipulator can be reformulated finally as the following QP:

$$\text{minimize} \quad \mathbf{x}^{\mathrm{T}} W \mathbf{x}/2 + \mathbf{c}^{\mathrm{T}} x \tag{6.12}$$

$$\text{subject to} \quad J\mathbf{x} = \dot{\mathbf{r}}_{\mathrm{d}} + \kappa_{\mathrm{p}}(\mathbf{r}_{\mathrm{d}} - \mathbf{f}(\boldsymbol{\theta})) \tag{6.13}$$

$$\xi^- \leqslant \mathbf{x} \leqslant \xi^+ \tag{6.14}$$

$$\text{with } W = I, \mathbf{c} = 0$$

where decision variable $\mathbf{x} \in \mathbb{R}^n$ denotes the joint-velocity $\dot{\boldsymbol{\theta}}$. Performance index (6.12) is obtained for the constrained MVN scheme and from the derivation of (6.1). To keep joint variables within their corresponding physical ranges, the ith elements of ξ^- and ξ^+ are defined, respectively, as $\xi_i^- = \max\{\dot{\theta}_i^-(\theta_i), \mu(\theta_i^- - \theta_i)\}$, $\xi_i^+ = \min\{\dot{\theta}_i^+(\theta_i), \mu(\theta_i^+ - \theta_i)\}$.

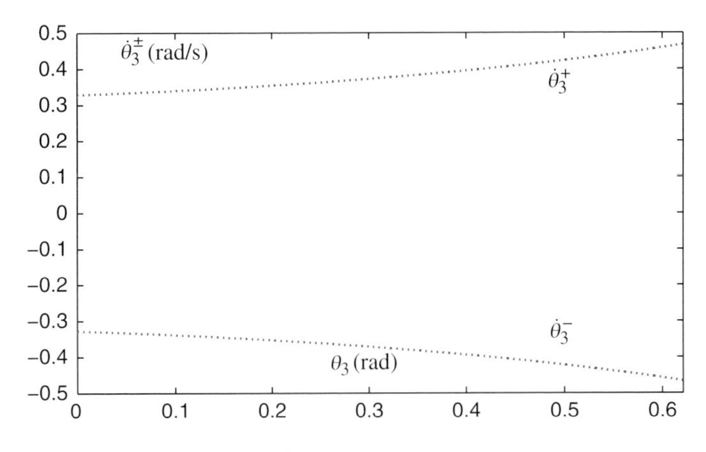

(a) Velocity limits $\dot{\theta}_3^{\pm}$ of Joint 3 changing with joint angle θ_3

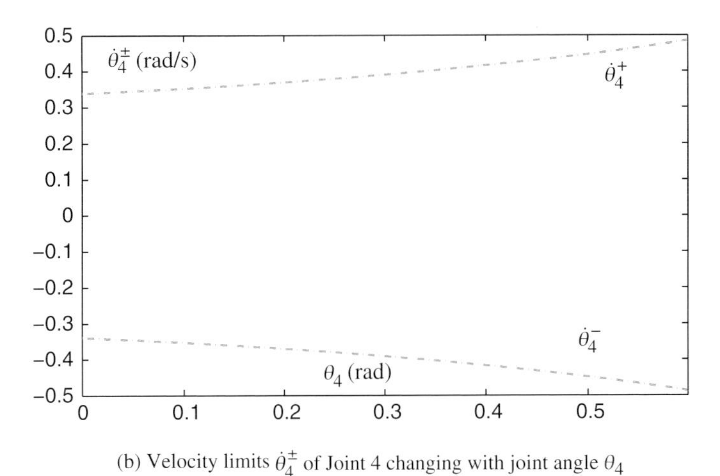

(b) Velocity limits $\dot{\theta}_4^{\pm}$ of Joint 4 changing with joint angle θ_4

Figure 6.4 Relationship between (a) $\dot{\theta}_3^{\pm}$ and θ_3 as well as (b) the relationship between $\dot{\theta}_4^{\pm}$ and θ_4.

The intensity coefficient $\mu > 0$ (e.g., being 4 in the ensuing simulations and experiments) is used to scale the feasible region of $\dot{\boldsymbol{\theta}}$ caused by this transformation.

Remark 6.3 When the Jacobian matrix $J(\boldsymbol{\theta})$ in Equation (6.13) is rank-deficient, the kinematic-singularity problem appears [1]. In addition, the singularity problem remains even if it does manifest itself via the inversion of $J(\boldsymbol{\theta})$, and the singularities indicate a branching of the solution. The latter exists independently of the numerical method used to solve the problem (possibly avoiding the inversion of $J(\boldsymbol{\theta})$). Note that the presented VJVL-constrained MVN scheme is an inverse-free method (i.e., avoiding the inversion of $J(\boldsymbol{\theta})$), which, in the situation of no feasible solution existing, can generate an approximate solution and make the robot movement continue. Besides, the additional constraints such as varying joint-velocity limits guarantee that the generated solution is within mechanical limits, agreeing well with actual situation and industrial applications.

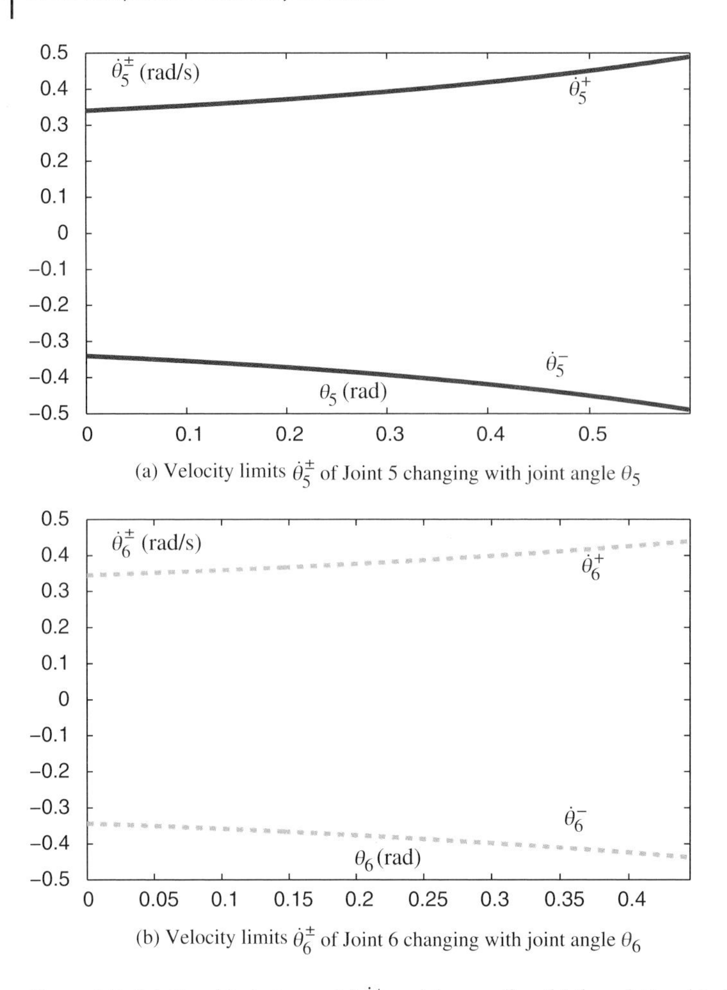

(a) Velocity limits $\dot{\theta}_5^{\pm}$ of Joint 5 changing with joint angle θ_5

(b) Velocity limits $\dot{\theta}_6^{\pm}$ of Joint 6 changing with joint angle θ_6

Figure 6.5 Relationship between (a) $\dot{\theta}_5^{\pm}$ and θ_5 as well as (b) the relationship between $\dot{\theta}_6^{\pm}$ and θ_6.

6.3 94LVI Assisted QP Solution

In view of duality theory [1] and the authors' previous work [61, 65, 97, 128, 136] as well as the results presented in Chapter 5, (6.12)–(6.14) can be converted to LVI, and the solution to the LVI can be obtained by solving the following equation:

$$P_{\Omega}(\mathbf{y} - (Q\mathbf{y} + \mathbf{p})) - \mathbf{y} = 0 \tag{6.15}$$

where $\Omega = \{\mathbf{y} \in \mathbb{R}^{n+m} | \mathbf{y}^- \leqslant \mathbf{y} \leqslant \mathbf{y}^+\} \subset \mathbb{R}^{n+m}$,

$$\mathbf{y} = \begin{bmatrix} \mathbf{x} \\ \mathbf{u} \end{bmatrix} \in \mathbb{R}^{n+m}, \mathbf{y}^+ = \begin{bmatrix} \xi^+ \\ \varpi \mathbf{1}_{\mathrm{v}} \end{bmatrix} \in \mathbb{R}^{n+m}, \quad \mathbf{y}^- = \begin{bmatrix} \xi^- \\ -\varpi \mathbf{1}_{\mathrm{v}} \end{bmatrix} \in \mathbb{R}^{n+m},$$

$$Q = \begin{bmatrix} W & -J^{\mathrm{T}} \\ J & 0 \end{bmatrix} \in \mathbb{R}^{(n+m)\times(n+m)}, \quad \mathbf{p} = \begin{bmatrix} \mathbf{c} \\ -\dot{\mathbf{r}}_{\mathrm{d}} - \kappa_{\mathrm{p}}(\mathbf{r}_{\mathrm{d}} - \mathbf{f}(\boldsymbol{\theta})) \end{bmatrix} \in \mathbb{R}^{n+m},$$

with $\mathbf{1}_v = (1, \cdots, 1)^T$. In addition, $\mathbf{P}_\Omega(\cdot) : \mathbb{R}^{n+m} \to \Omega$ is a projection operator, with the ith element of $\mathbf{P}_\Omega(\mathbf{y})$ defined as

$$\begin{cases} y_i^-, & \text{if } y_i < y_i^-, \\ y_i, & \text{if } y_i^- \leqslant y_i \leqslant y_i^+, \quad \forall i \in \{1, 2, \cdots, n+m\}. \\ y_i^+, & \text{if } y_i > y_i^+, \end{cases}$$

Here, $\mathbf{y} \in \mathbb{R}^m$ is the dual decision vector for (6.13), and $\varpi \gg 0$ is defined sufficiently large (e.g., $\varpi = 10^6$) to replace $+\infty$ for numerical-implementation purposes.

Let $\Omega^* = \{\mathbf{y}^* | \mathbf{y}^* \text{ is a solution of (6.15)}\}$. To solve QP (6.12)–(6.14), guided by [1, 61, 65, 136–139], solving (6.15) is equivalent to finding the equilibrium point of the following function:

$$\mathbf{e}(\mathbf{y}) = \mathbf{y} - \mathbf{P}_\Omega(\mathbf{y} - (Q\mathbf{y} + \mathbf{p})). \tag{6.16}$$

According to [136], given the initial augmented primal-dual decision variable vector $\mathbf{y}^0 \in \mathbb{R}^{n+m}$, for iteration index $k = 0, 1, 2, 3, \cdots$, if $\mathbf{y}^k \notin \Omega^*$, then

$$\mathbf{y}^{k+1} = \mathbf{y}^k - \frac{\|\mathbf{e}(\mathbf{y}^k)\|_2^2 \delta(\mathbf{y}^k)}{\|\delta(\mathbf{y}^k)\|_2^2}, \tag{6.17}$$

where $\mathbf{e}(\mathbf{y}^k) = \mathbf{y}^k - \mathbf{P}_\Omega(\mathbf{y}^k - (Q\mathbf{y}^k + \mathbf{p}))$ and $\delta(\mathbf{y}^k) = (Q^T + I)\mathbf{e}(\mathbf{y}^k)$. The convergence of numerical algorithm 94LVI (6.17) is guaranteed by Theorem 5.4 [136].

Remark 6.4 In terms of Equation (6.17), the presented numerical algorithm 94LVI for one iteration contains $((n+m)(2n+2m+5)-2)$ addition operations, $(n+m)(2n+2m+3)$ multiplication operations, $2(n+m)$ comparison operations (as for piecewise-linear activation function $\mathbf{P}_\Omega$), and one division operation. In general, the computational complexity of the method is of $O((n+m)^2)$ operations. Specifically, since $n = 6$ and $m = 2$, the method consists of 166 addition operations, 152 multiplication operations, 16 comparison operations, and one division operation per iteration.

6.4 Computer Simulations and Physical Experiments

In order to test the VJVL-constrained MVN scheme, computer simulations and experiments are performed by using the six-DOF planar robot manipulator to track a line-segment path and an elliptical path in the two-dimensional horizontal plane (i.e., the motion plane of the robot is parallel to the X-Y plane). The error precision $\mathbf{e}(\mathbf{y})_{\max}$ is set to be 1.0×10^{-6} rad/s for algorithm (6.17). The robot has six DOF, its maximum joint velocity of Joint 1 is 3.27 rad/s, and its other joints' parameters are given in Subsection 6.2.1 and Subsection 6.2.2.

6.4.1 Line-Segment Path-Tracking Task

In this subsection, the end-effector of the six-DOF planar robot manipulator with VJVL is expected to move forward and then backward along a line-segment that is shown

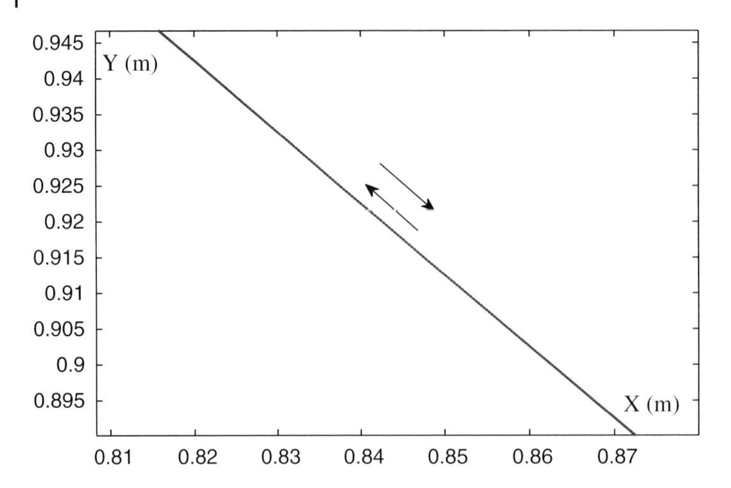

Figure 6.6 Desired line-segment path to be tracked by the end-effector of a six-DOF planar robot manipulator.

in Figure 6.6. The slope of the desired line-segment is −1. The x- and y-axis velocity functions of the line-segment path are

$$\dot{r}_{\mathrm{X}}(t)=\begin{cases}-l(1-\cos(2\pi t/T))/(\sqrt{2}T), & \forall t\in[0,T]\\ l(1-\cos(2\pi t/T))/(\sqrt{2}T), & \forall t\in[T,2T]\end{cases}$$

$$\dot{r}_{\mathrm{Y}}(t)=\begin{cases}l(1-\cos(2\pi t/T))/(\sqrt{2}T) & \forall t\in[0,T]\\ -l(1-\cos(2\pi t/T))/(\sqrt{2}T) & \forall t\in[T,2T]\end{cases}$$

where the task duration is $2T$, and parameter l should be set appropriately according to the desired length of the line-segment in a path-following task. Specifically, in the simulation and experiment of this subsection, $l = 0.08$ m and the task duration is 20.0 s (i.e., $T = 10.0$ s). In addition, such a path-tracking task starts with initial joint state $\theta(0) = [\pi/12, \pi/12, \pi/12, \pi/12, \pi/12, \pi/12]^{\mathrm{T}}$ rad. The schemes (6.12)–(6.14) involved in this experiment are solved by numerical QP solver (6.17), where the sampling period Δt is 0.01 s.

Firstly, we fix the camera, adjust the focus and take snapshots every 2.5 seconds. The actual task execution of the six-DOF planar robot manipulator hardware system synthesized by the VJVL-constrained MVN scheme is then shown in Figure 6.7. During the task execution, the end-effector of the manipulator moves smoothly, which illustrates that such a scheme is very effective for solving the redundancy of the manipulator. The experimental result, that is, the actual end-effector trajectory, is shown in Figure 6.8. Correspondingly, Figure 6.9 shows the simulated joint-motion trajectories and their proportions of the six-DOF planar robot manipulator. These illustrate that the line-segment path tracking experiment is performed well. In addition, it is worth noting that the motors are driven by the pulses transmitted from the host computer, that is, the actual control input to the joints, which are shown in Figure 6.10 through Figure 6.12. The pulses per second (PPS) are obtained by the following steps: (1) the joint velocity $\dot{\theta}_i$ is computed by using the presented algorithm (6.17) that solves (6.12)–(6.14), with $i = 1, 2, \cdots, 6$; (2) for the servomotor (i.e., Joint 1), $\mathrm{PPS}_1 = \kappa_1\dot{\theta}_1/(2\pi)$, and for the stepping

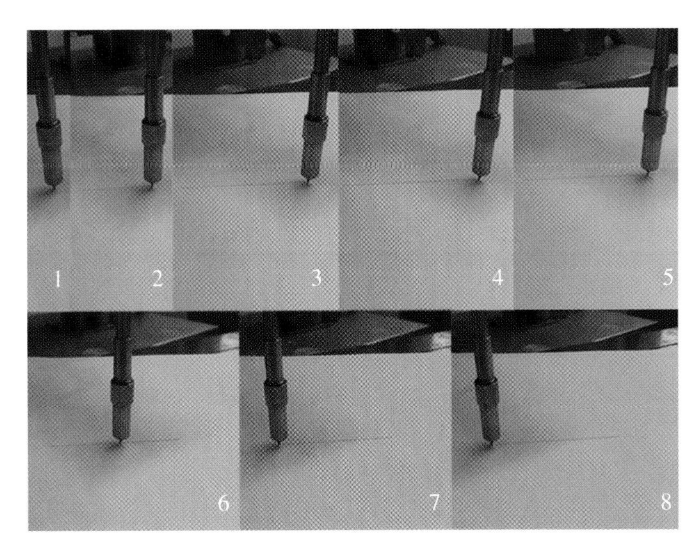

Figure 6.7 Snapshots for an actual task execution of a six-DOF planar robot manipulator synthesized by a VJVL-constrained MVN scheme when a robot end-effector tracks a line-segment path. Reproduced from Z. Zhang, and Y. Zhang, Variable Joint-Velocity Limits of Redundant Robot Manipulators Handled by Quadratic Programming, Figure 5, IEEE/ASME Trans. Mechatronics, Vol. 18, No. 2, pp. 674-686, 2013. *Source:* Zhang 2013. Reproduced with permission of IEEE.

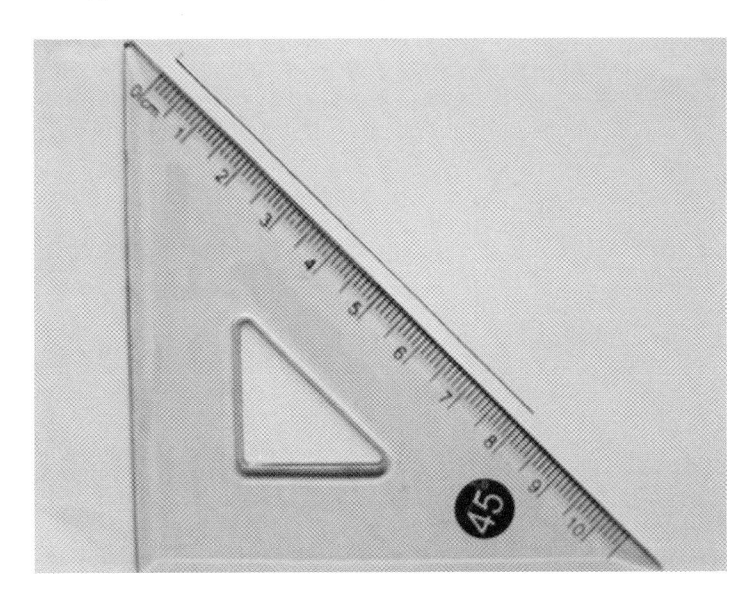

Figure 6.8 Actual end-effector trajectory generated by a six-DOF planar robot manipulator synthesized by a VJVL-constrained MVN scheme. *Source:* Zhang 2013. Reproduced with permission of IEEE.

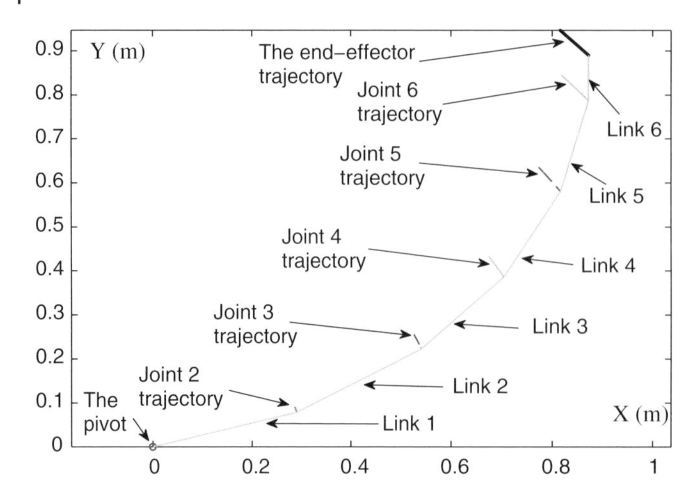

Figure 6.9 Simulated six-DOF planar robot manipulator and its joints' trajectories synthesized by a VJVL-constrained MVN scheme when the robot end-effector tracks a line-segment path.

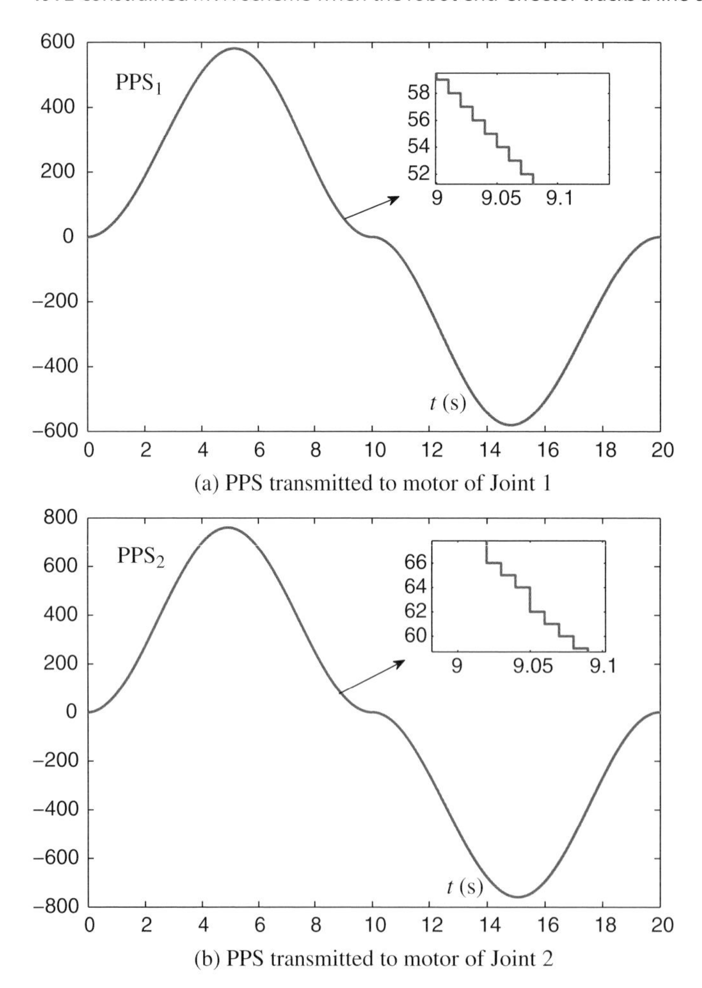

Figure 6.10 (a) PPS_1 and (b) PPS_2 transmitted to joint motors when the end-effector tracks a line-segment path.

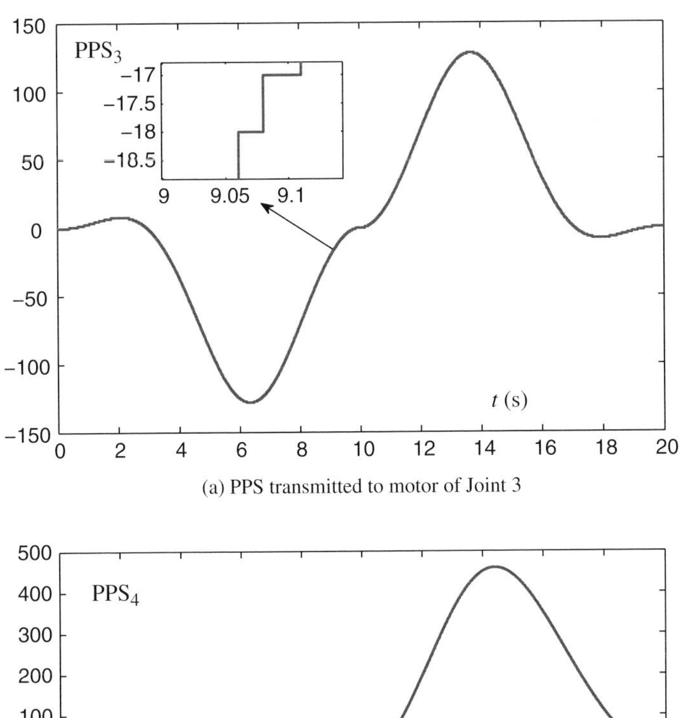

(a) PPS transmitted to motor of Joint 3

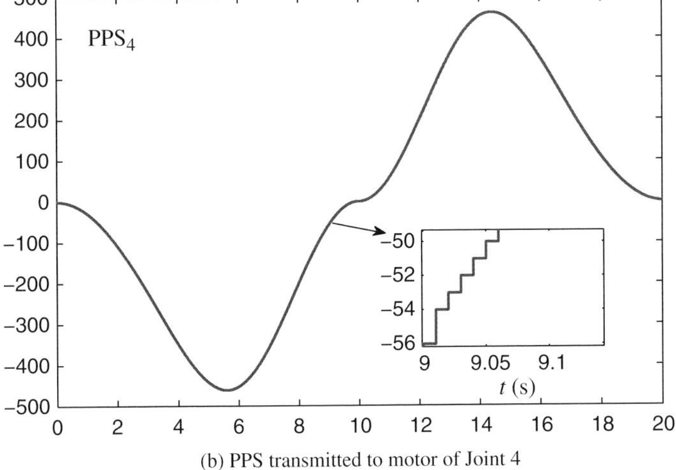

(b) PPS transmitted to motor of Joint 4

Figure 6.11 (a) PPS_3 and (b) PPS_4 transmitted to joint motors when the end-effector tracks a line-segment path.

motors (i.e., Joints 2 through 6), $\text{PPS}_i = \kappa_2 a_i b_i \dot{\theta}_i \cos\dot{\theta}_i/(s_i\sqrt{a_i^2 + b_i^2 + 2a_i b_i \sin\theta_i})$, with $i = 2, 3, \cdots, 6$; and (3) round the floating-point results of this calculation to integers that are the actual numbers of PPS. Here, κ_1 and κ_2 are the parameters related to the hardware system; specifically, $\kappa_1 = 3.2 \times 10^5$ and $\kappa_2 = 6.4 \times 10^3$.

Secondly, as shown in Figure 6.13, the joint-angle vector and the joint-velocity vector of the redundant robot manipulator are all smooth, which illustrates that the solution could be readily implemented in the robot hardware system. Moreover, the simulated position error between the desired path and the end-effector trajectory is shown in Figure 6.14, which is less than 2.5×10^{-6} m. In other words, the given line-segment path has been completed accurately.

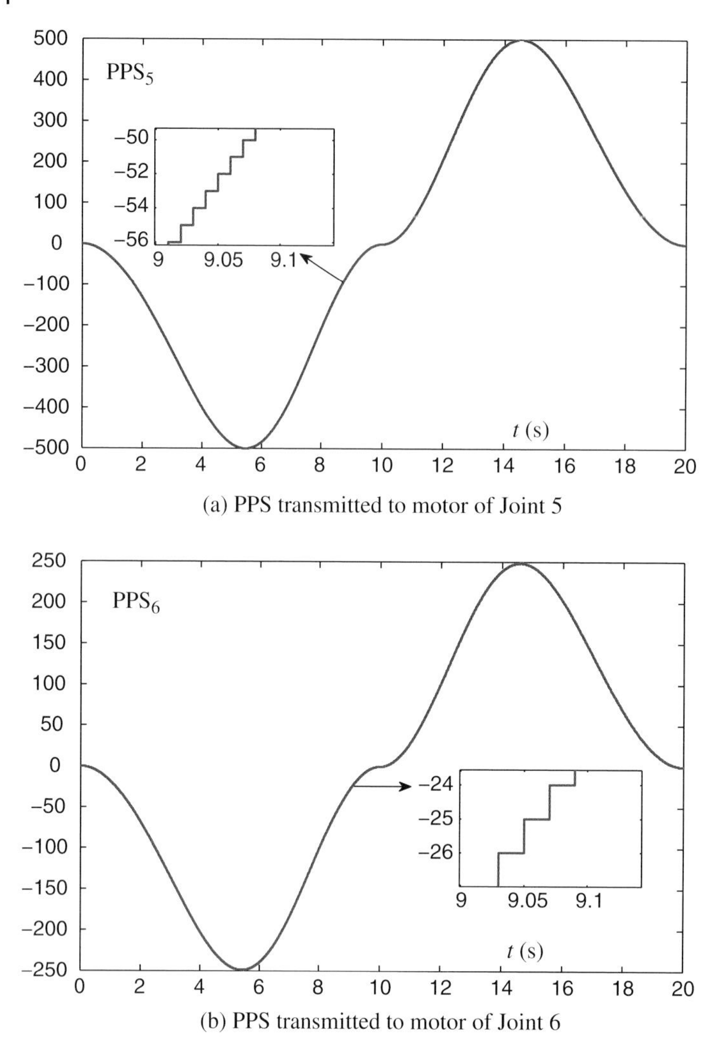

Figure 6.12 (a) PPS_5 and (b) PPS_6 transmitted to joint motors when the end-effector tracks a line-segment path.

Thirdly, Table 6.2 further illustrates the maximum and average position error of the end-effector, where $\boldsymbol{\varepsilon}(t) = \mathbf{r}_\mathrm{d}(t) - \mathbf{f}(\theta(t))$, and ε_X and ε_Y denote, respectively, the x- and y-axis components of the position error $\boldsymbol{\varepsilon}(t)$. From the line-segment path tracking results of the table, we can see that the maximum position error along the x-axis is 2.2968×10^{-6} m and the maximum position error along the y-axis is 2.3380×10^{-6} m, which are both very tiny. The error situation can also be monitored by the average error, which along the x-axis is 1.4107×10^{-6} m and along the y-axis is 1.4102×10^{-6} m. This error-analysis results illustrate the accuracy of the presented VJVL-constrained MVN scheme used in the six-DOF planar robot manipulator.

Fourthly, in order to further examine the computational efficiency and accuracy of the presented numerical algorithm (6.17) used in the experiment, the following data have

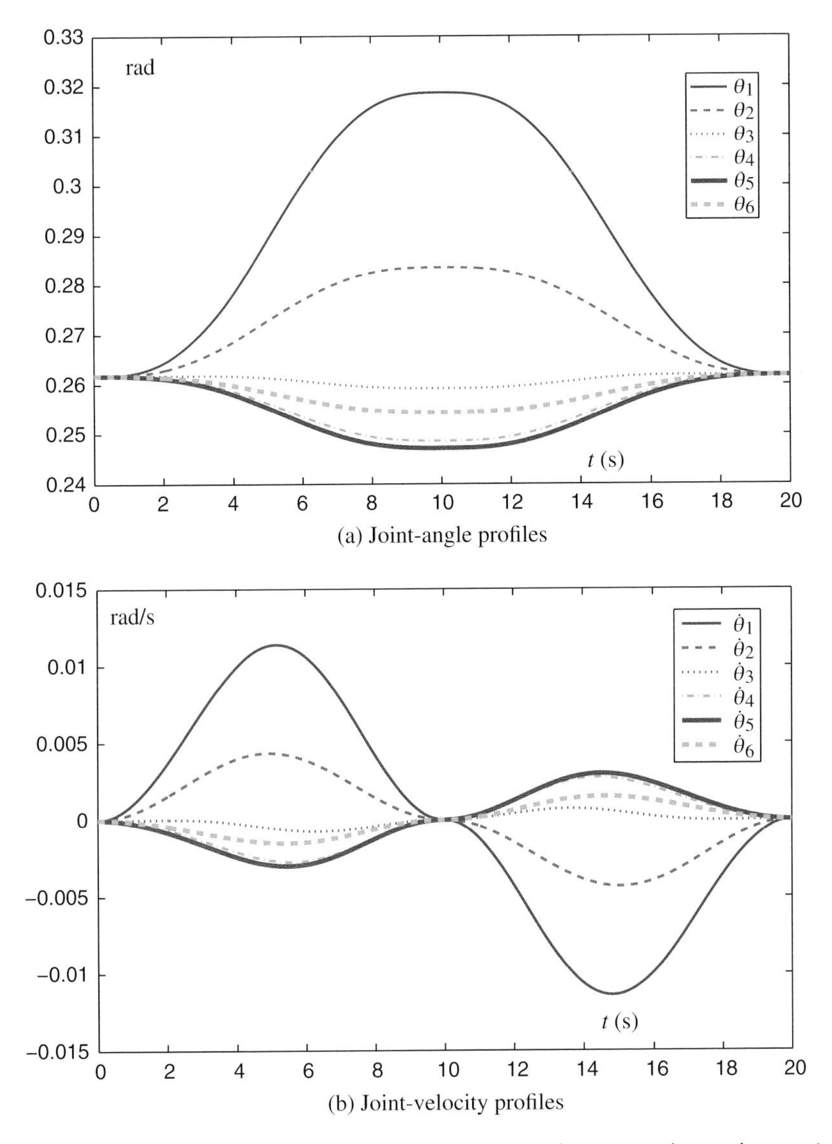

Figure 6.13 (a) Joint-angle and (b) joint-velocity profiles of a six-DOF planar robot manipulator synthesized by a VJVL-constrained MVN scheme when the robot end-effector tracks a line-segment path.

been measured, for example, the maximum and average numbers of iterations required within each sampling period (i.e., N_{max} and N_{ave}, respectively), the maximum and average computing time required within each sampling period (i.e., α_{max} and α_{ave}, respectively), and the maximum and average values of $\mathbf{e}(\mathbf{y})$ as the output of numerical algorithm (6.17) within each sampling period (i.e., $\mathbf{e}(\mathbf{y})_{\text{max}}$ and $\mathbf{e}(\mathbf{y})_{\text{ave}}$, respectively). The line-segment tracking experiment results are shown in the second row of Table 6.3, where $N_{\text{max}} = 74$, $N_{\text{ave}} = 56.75$, $\alpha_{\text{max}} = 0.003035828$ s, $\alpha_{\text{ave}} = 0.000164304$ s, and $\mathbf{e}(\mathbf{y})_{\text{ave}} \approx 9.77 \times 10^{-7}$, all being quite small. In addition, such a maximum computing time is less than one

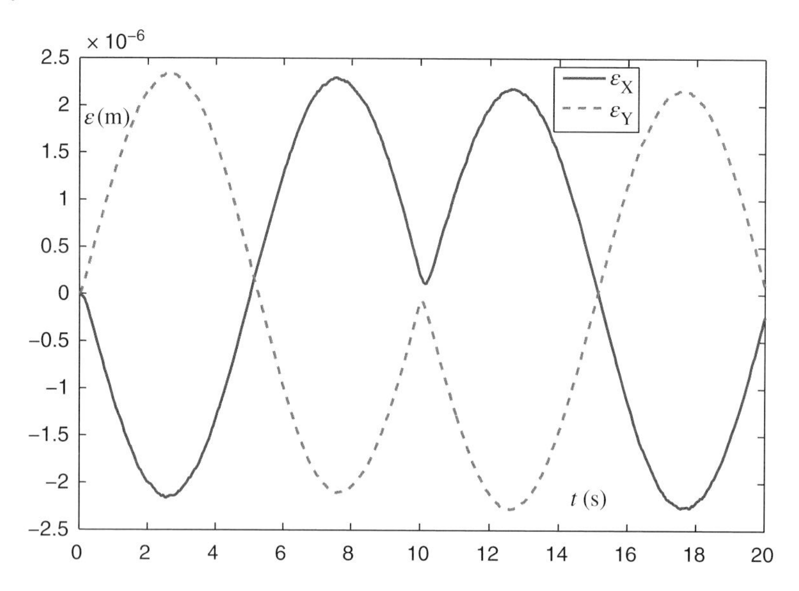

Figure 6.14 End-effector position error of a six-DOF planar robot manipulator synthesized by a VJVL-constrained MVN scheme when the robot end-effector tracks a line-segment path.

Table 6.2 Maximum and average position errors (m) of a six-DOF planar robot manipulator's end-effector for different paths tracked.

Path type	**Error**	$\lvert\varepsilon_X\rvert$ ($\times 10^{-6}$)	$\lvert\varepsilon_Y\rvert$ ($\times 10^{-6}$)
Line-segment	Maximum	2.2968	2.3380
	Average	1.4107	1.4102
Ellipse	Maximum	8.6191	3.2783
	Average	3.4339	1.1952

Table 6.3 Iteration number and computing time within each sampling period when a robot tracks a line-segment or an elliptical path.

Path type	Δt **(s)**	$N_{\max}$	N_{ave}	$\alpha_{\max}$ **(s)**	α_{ave} **(s)**	$\mathbf{e}(\eta)_{\max}$ **(rad/s)**	$\mathbf{e}(\eta)_{\text{ave}}$ **(rad/s)**
Line-segment	0.01	74	56.75	0.003035	0.0001643	1.0×10^{-6}	9.77×10^{-7}
Ellipse	0.01	115	89.38	0.002954	0.0002284	1.0×10^{-6}	9.80×10^{-7}

third of sampling period $\Delta t = 0.01$ s, which validates that the computing task can be fulfilled well within each sampling period of the line-segment path tracking experiment. These, including the small computing error $\mathbf{e}(\mathbf{y})_{\text{ave}}$, illustrate the efficiency, hardware realizability, and accuracy of the presented online redundancy-resolution scheme (i.e., the VJVL-constrained MVN scheme and the presented numerical algorithm).

In summary, the line-segment-path tracking experiment illustrates well the efficacy of the presented VJVL-constrained MVN scheme [depicted in (6.12)–(6.14)] and its numerical-solution method [depicted in (6.17)] for resolving the redundancy of this six-DOF planar robot manipulator. In addition, the position-error analysis further shows the accuracy of such a scheme for robots' path tracking. Furthermore, the measurement and analysis of the computing time required within each sampling period validate the real-time performance of the VJVL-constrained MVN scheme and numerical algorithm 94LVI.

6.4.2 Elliptical-Path Tracking Task

In this subsection, the tracking task of the robot arm is an elliptical path with the length of the major axis being $a = 0.03$ m and the length of the minor axis being $b = 0.01$ m. The x and y-axis velocity functions of the elliptical path are, respectively,

$$\dot{r}_X = -\gamma\pi^2 \sin(\varphi + 2\pi\sin^2(\pi t/(2T)))\sin(\pi t/T)/T$$
$$\dot{r}_Y = -\gamma\pi^2 \cos(\varphi + 2\pi\sin^2(\pi t/(2T)))\sin(\pi t/T)/(3T)$$

where $\gamma = 0.012\sqrt{X_0^2 + Y_0^2}$, and $\varphi = \arctan(Y_0/X_0)$ or $\pi/2$ (if $X_0 = 0$), with (X_0, Y_0) denoting the coordinate of the robot end-effector's initial position. In addition, the task duration here is $T = 10$ s, and the initial joint vector $\theta(0) = [\pi/12, \pi/12, \pi/12, \pi/12, \pi/12, \pi/12]^{\mathrm{T}}$ in radians. The sampling period Δt is also 0.01 s.

Firstly, we take snapshots every 2.5 s, similar to the manner as described in Subsection 6.4.1. The snapshots are shown in Figure 6.15, which illustrates that the end-effector task is completed well. The corresponding joint angles and joint velocities are further shown in Figure 6.16. Specifically, as seen from Figure 6.16(a), the curves of the six joint angles are continuous and smooth, which validates the feasibility of such a QP-based scheme, and Figure 6.16(b) shows that the joint velocities are smooth and of small values as well

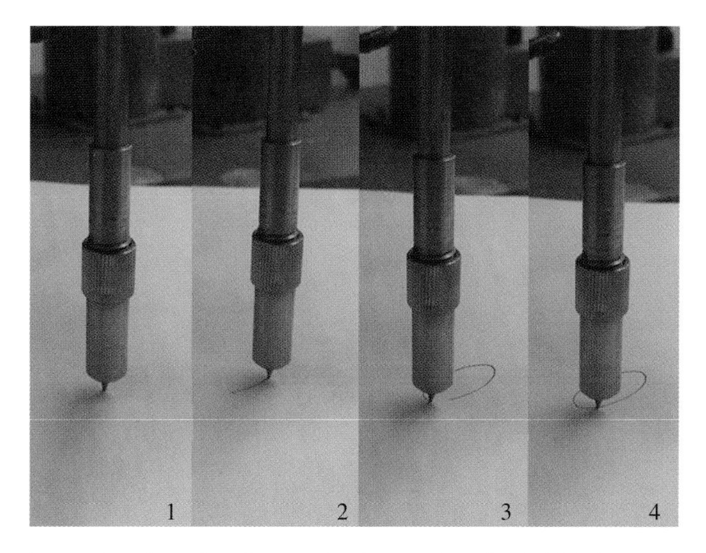

Figure 6.15 Snapshots for actual task execution of a six-DOF planar robot manipulator synthesized by a VJVL-constrained MVN scheme when the robot end-effector tracks an elliptical path. *Source:* Zhang 2013. Reproduced with permission of IEEE.

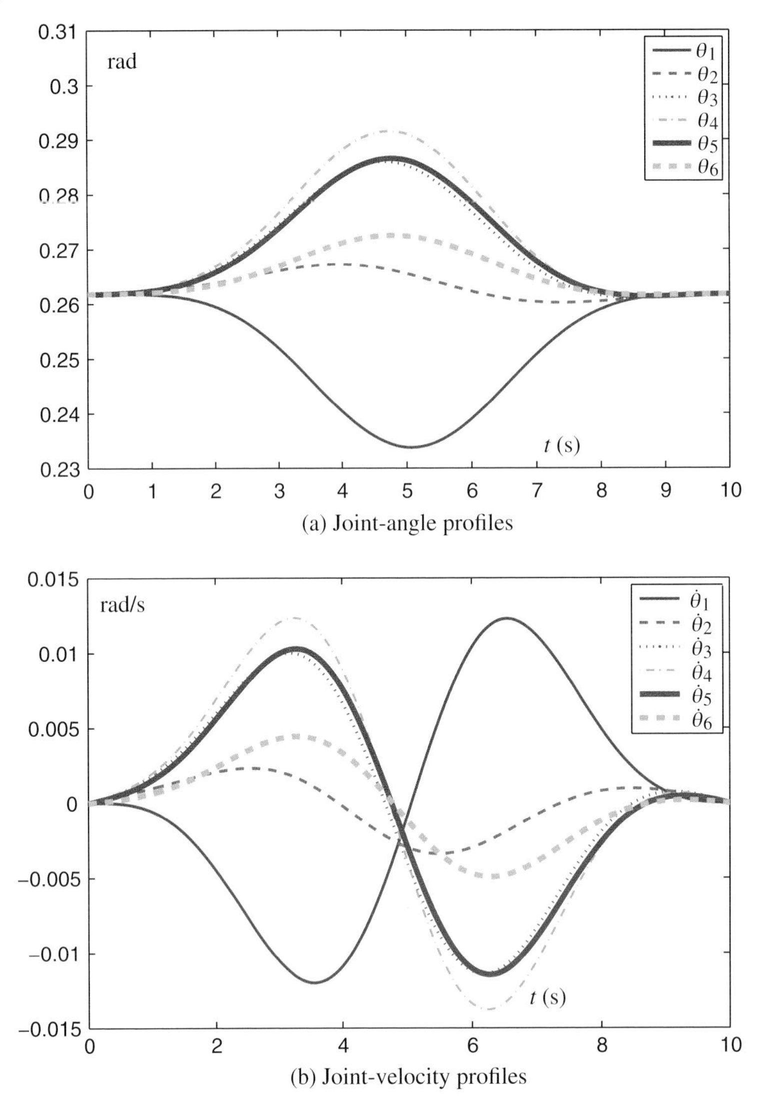

Figure 6.16 (a) Joint-angle and (b) joint-velocity profiles of a six-DOF planar robot manipulator synthesized by a VJVL-constrained MVN scheme when the robot end-effector tracks an elliptical path.

during the task execution. This illustrates that the solution could be readily implemented in the hardware system for the elliptical-path tracking task.

Secondly, regarding the error analysis, as seen in Figure 6.17, the position error between the desired path and the end-effector position trajectory are quite tiny (i.e., with a maximum value less than 1.0×10^{-5} m). This illustrates that the presented scheme is effective and accurate on the planar robot manipulator's redundancy-resolution. The error-analysis results are further shown in the lower part of Table 6.2 regarding the maximum and average errors of the end-effector position. From the table, we see that the maximum position errors along the x and the y-axis

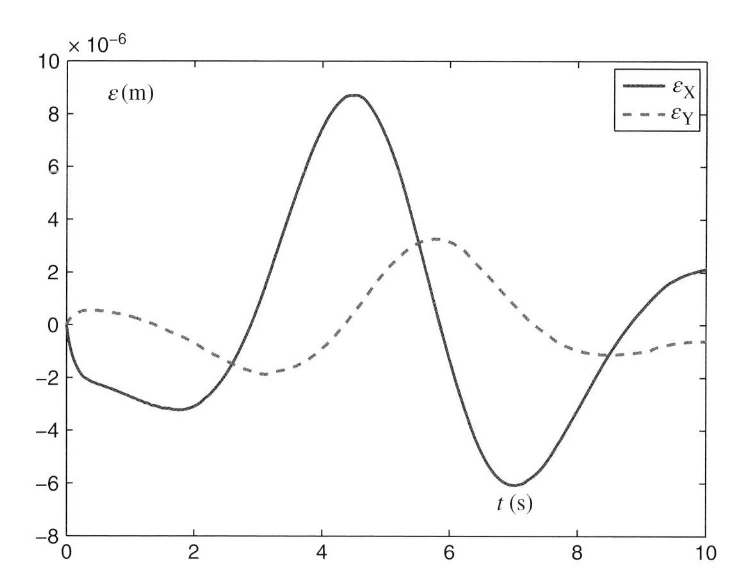

Figure 6.17 End-effector position error of a six-DOF planar robot manipulator synthesized by a VJVL-constrained MVN scheme when the robot end-effector tracks an elliptical path.

are 8.6191×10^{-6} m and 3.2783×10^{-6} m, respectively. These results, together with the average errors shown in Table 6.2, constitute a fine description of the well-performed ellipse-tracking task.

Thirdly, similar to that in the previous subsection, the third row of Table 6.3 shows the iteration numbers and computing time of numerical algorithm (6.17) required within each sampling period. From the table, we see that $N_{\max} = 115$ and $\alpha_{\max} = 0.002954386$ s, both being quite small. This substantiates that the joint velocities can be computed readily within each sampling period, and that the scheme can be applicable to real-time computing and online resolving redundancy of manipulators. In addition, such a computing method guarantees that the computed $\mathbf{e}(\mathbf{y})$ within each sampling period has quite a high level of precision (i.e., 1.0×10^{-6} m).

6.4.3 Simulations with Faster Tasks

In order to further investigate and illustrate the feasibility and efficiency of the presented method (i.e., the VJVL-constrained MVN scheme and its corresponding numerical solution method), two more simulations with faster end-effector tasks are performed. Note that the motion plane of the robot and the sampling period are the same as those in the previous two subsections.

6.4.3.1 Line-Segment-Path-Tracking Task

In this experiment, the task duration of the line-segment path tracking is 2 s, and the task length of the end-effector moving forward and backward along a line-segment path is 0.84 m. The slope of the line-segment is still −1. As the line-segment path tracking task is much faster than that in Subsection 6.4.1, the joints may exceed their limits if there is no protection or prevention measure taken. In this no-prevention situation, the manipulability of the robot may decrease, some end-effector tasks may not

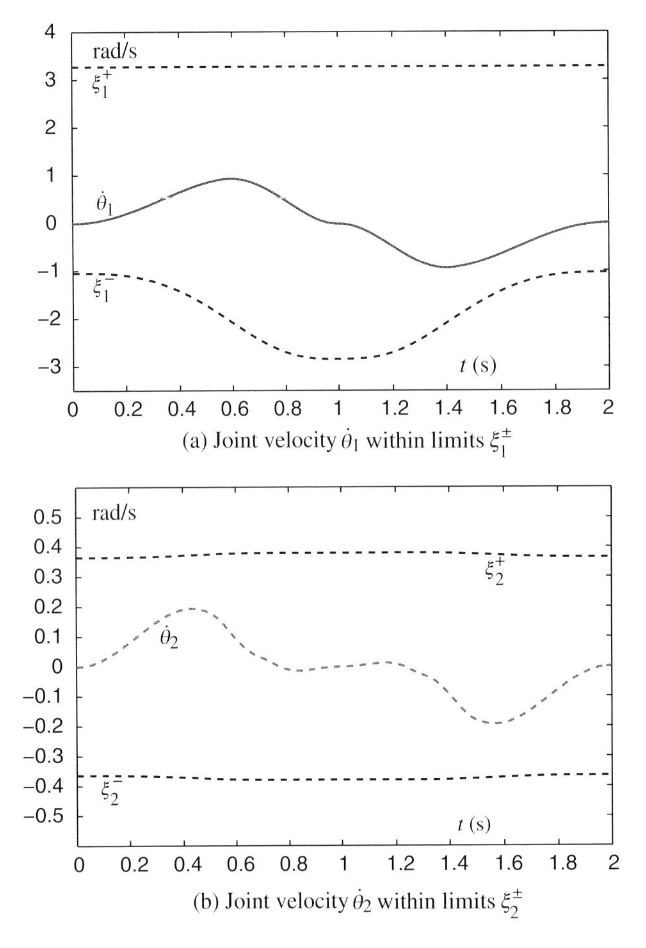

Figure 6.18 Joint-velocity profiles of (a) $\dot{\theta}_1$ and (b) $\dot{\theta}_2$ of a six-DOF planar robot manipulator synthesized by a VJVL-constrained MVN scheme when its end-effector tracks a line-segment path much faster.

be completed, and the robot could even get damaged. However, as synthesized by the presented VJVL-constrained MVN scheme, which incorporates the feature of avoiding joint physical limits, the dangerous situation can be prevented. The simulation results are presented firstly in Figure 6.18 through 6.20, where the velocities of Joints 3 through 5 reach their limits at some time periods (e.g., the ones around 0.6 s and 1.4 s), but never exceed the limits. That is to say, the bound constraint (6.14) becomes active at these time periods around 0.6 s and 1.4 s. Correspondingly, Figure 6.21 shows the small position errors during the faster line-segment path tracking task execution, of which the maximum value is 1.129×10^{-3} m. In addition, within each sampling period, the maximum computing time $\alpha_{\text{max}} = 0.003342167$ s and the average computing time $\alpha_{\text{ave}} = 0.001111253$ s, both of which are less than half of sampling period $\Delta t = 0.01$ s. These substantiate the feasibility, efficiency, and accuracy of the presented method, in addition to the feature of avoiding joint physical limits.

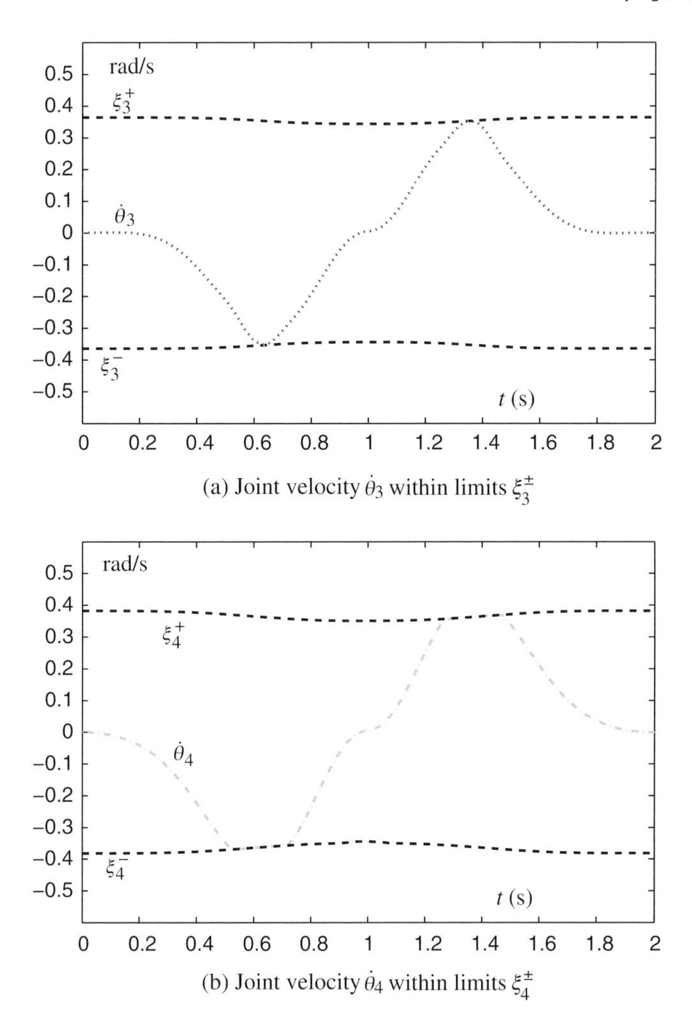

Figure 6.19 Joint-velocity profiles of (a) $\dot{\theta}_3$ and (b) $\dot{\theta}_4$ of a six-DOF planar robot manipulator synthesized by a VJVL-constrained MVN scheme when its end-effector tracks a line-segment path much faster.

6.4.3.2 Elliptical-Path-Tracking Task

In this section, the desired end-effector task is a faster elliptical path with task duration 1 s (compared to 10 s in Section 6.4.2). The length of the major axis of the desired ellipse is 0.0809 m, and the length of the minor major axis is 0.027 m. By applying the presented VJVL-constrained MVN scheme, the resultant joint-velocity profiles are shown in Figure 6.22, which are within the varying joint limits. It is further seen from the figure that the velocity-saturation phenomenon of Joint 4 happens at the time period around 0.62 s. The relatively small position errors are shown in Figure 6.23. In addition, the iteration number and the computing time per sampling period versus time are shown

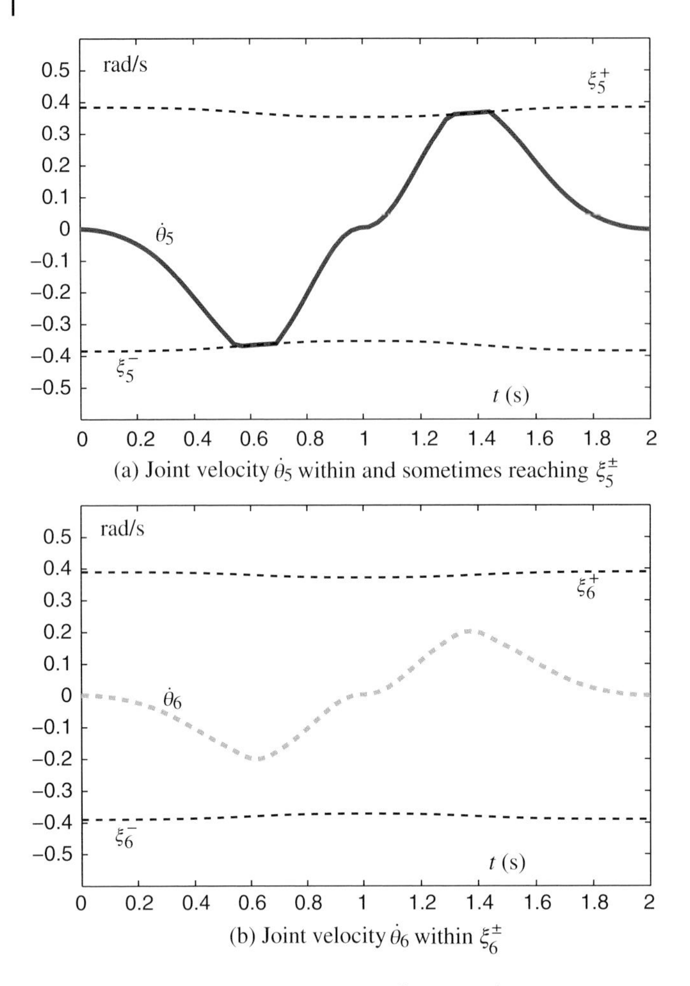

Figure 6.20 Joint-velocity profiles of (a) $\dot{\theta}_5$ and (b) $\dot{\theta}_6$ of a six-DOF planar robot manipulator synthesized by a the VJVL-constrained MVN scheme when its end-effector tracks a line-segment path much faster.

in Figure 6.24. From the upper plot of Figure 6.24, we see that the iteration number increases from the initial value 0 in the beginning to 228 and then fluctuate around it. On the other hand, as seen from the lower plot of Figure 6.24, the computing time of numerical algorithm (6.17) per sampling period in the beginning is 0.003232361 s (which may include an initial data-loading and memory-allocating time), then decreases to 0.000552997 s and fluctuates around it. Besides, these values are smaller than one third of the sampling period $\Delta t = 0.01$ s. These results validate again the feasibility and efficacy of the resolving redundancy method, in addition to the avoidance feature of joint physical limits. Before ending this chapter, it is worth mentioning that, in fact, the joint-saturation situation (i.e., the joints within and sometimes reaching their limits) is also undesirable and should be prevented in applications.

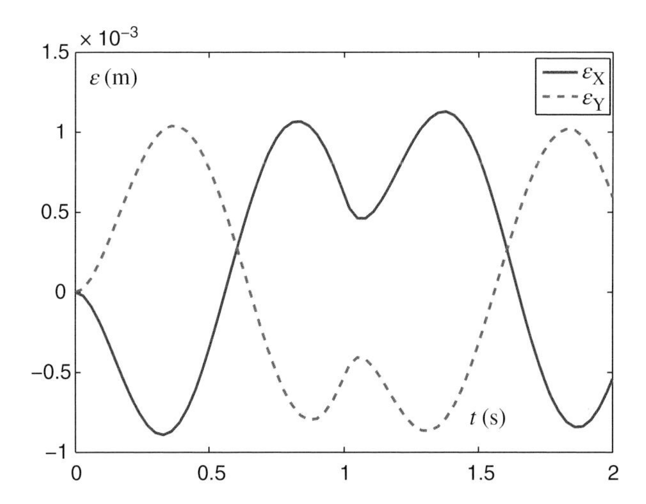

Figure 6.21 End-effector position error of a six-DOF planar robot manipulator synthesized by a VJVL-constrained MVN scheme when its end-effector tracks a line-segment path much faster.

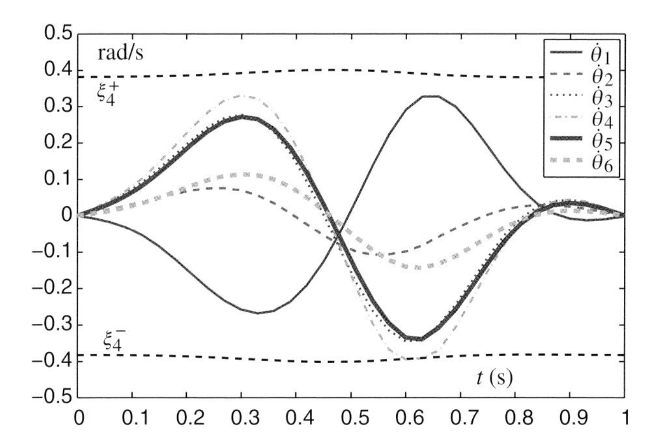

Figure 6.22 Joint-velocity profiles of a robot synthesized by a VJVL-constrained MVN scheme when its end-effector tracks an elliptical path much faster, which includes a profile of joint velocity $\dot{\theta}_4$ within and sometimes reaching limits of $\xi_4^{\pm}$.

In summary, the presented two examples, for example, the line-segment-path tracking and the elliptical-path tracking, have both illustrated the efficacy of the presented VJVL-constrained MVN scheme and its numerical-solution method for resolving the redundancy of manipulators. Furthermore, the error analyses have validated the accuracy of such a QP-based approach well. Besides, the VJVL-constrained MVN scheme should be applicable to not only planar robots (e.g., the six-DOF planar robot manipulator investigated in this chapter) but also the manipulators operating in three-dimensional space.

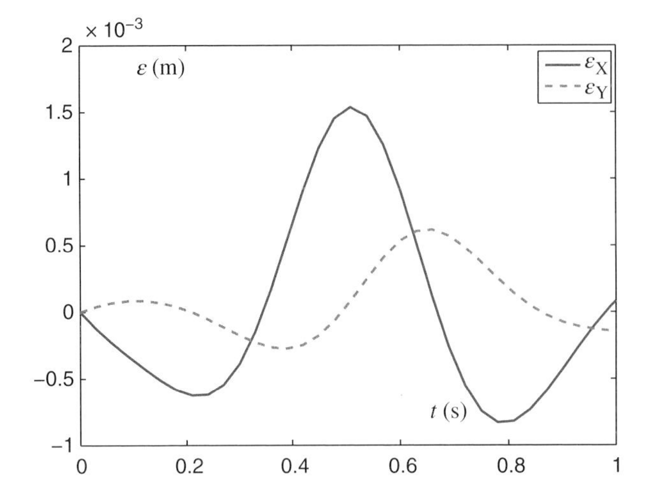

Figure 6.23 End-effector position error of a six-DOF planar robot manipulator synthesized by a VJVL-constrained MVN scheme when its end-effector tracks an elliptical path much faster.

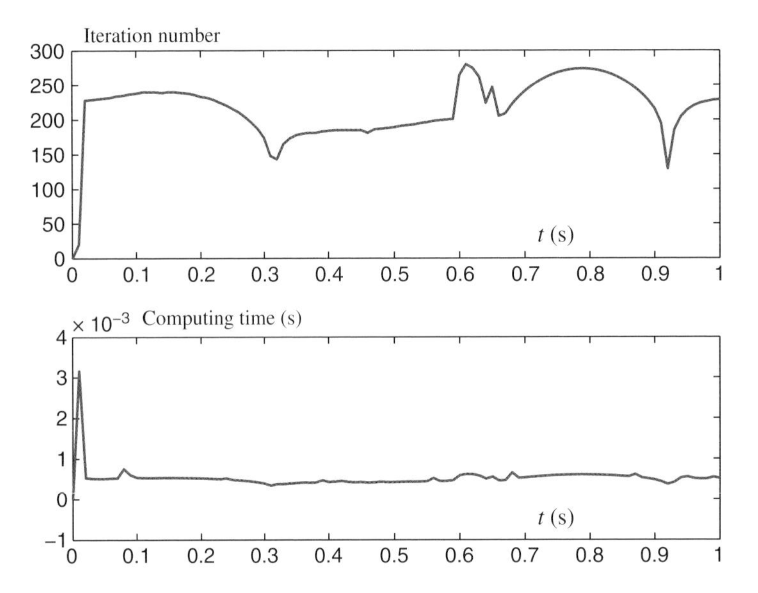

Figure 6.24 Iteration number and computing time of a numerical algorithm 94LVI (6.17) per sampling period in a much faster elliptical-path tracking task.

6.5 Chapter Summary

For operating a practical six-DOF planar robot manipulator, we formulate the redundancy-resolution problem as a VJVL-constrained MVN scheme. The push-rod joint is abstracted to a triangle, resulting in such a VJVL. The VJVL-constrained MVN scheme is then presented and reformulated as a QP, which is finally solved by an efficient numerical algorithm. The simulations and experiments regarding the line-segment and elliptical paths tracking for the six-DOF planar robot manipulator illustrate the efficacy of such a VJVL-constrained MVN scheme and the corresponding numerical QP solver well. Moreover, the end-effector position-error analysis results validate the accuracy of the presented resolving redundancy method well. Future research directions may lie in more complex examples and experiments, in addition to incorporating subtasks such as cyclic motion generation, obstacle avoidance, and singularity avoidance.

7

Feedback-Aided Minimum Joint Motion

7.1 Introduction

Kinematic redundancy can be obtained when a robot manipulator possesses more DOF than the minimum number required to execute a given end-effector task. If a robot manipulator is redundant, the inverse-kinematics problem may admit infinite solutions. How to obtain the required optimal solution, that is, the redundancy resolution problem, is an appealing and significant topic in robotics areas. In the past, many redundancy resolution schemes have been presented, and they can be divided into two main categories; that is, globally optimal control of redundancy and local/instantaneously optimal control of redundancy. The globally optimal concept and method were originally suggested by Whitney [140], and the early effective global schemes were presented by Uchiyama *et al.* [141] and Nakamura *et al.* [142]. The global optimal schemes can generate the global optimal solutions, but they require a large amount of computation and are not suitable for real-time kinematic resolution. The detailed discussion can be referred to [143]. Thus, the local/instantaneously optimal control schemes are widely investigated and used in robotics areas. The classic local optimal solution is pseudoinverse-type formulation, which is investigated in Chapter 1; that is, one minimum-norm particular solution plus a homogeneous solution. Based on such a pseudoinverse-type solution, many optimization performance criteria have been exploited in terms of manipulator configurations and interaction with the environment, such as joint-limits avoidance [144, 145], singularity avoidance [146], manipulability enhancement [147], and obstacle avoidance [87]. Recent research shows that the solutions to redundancy resolution problems can be enhanced by using optimization techniques based on quadratic program (QP) methods [90]. Compared with the classic pseudoinverse-based solutions, such QP-based methods do not need to compute the inverse of the Jacobian matrix, and are readily to deal with the inequality and/or bound constraints. Thus, new interests in QP-based methods have been generated. In [90], considering the physical limits, Cheng *et al.* presented a compact QP method to resolve the constrained kinematic redundancy problem. As a specific example in [90], a repetitive motion planning (RMP) scheme was presented to illustrate the effectiveness of the compact QP method. Note that such an RMP scheme is based on Euclidean norm (or two-norm), and used to remedy the non-repetitive problem of the closed end-effector path (which is different from the path tracked in this chapter).

Robot Manipulator Redundancy Resolution, First Edition. Yunong Zhang and Long Jin.

The two-norm was widely used because of its analytical tractability [148], but minimizing the two-norm solution was not an ideal choice [149]. Thus, some researchers began to investigate the optimization schemes based on the infinity norm and minimizing the infinity norm implied the determination of a minimum-effort solution as opposed to the minimum-energy criterion associated with the Euclidean norm [1, 149, 150]. The torque-level [151] and acceleration-level optimization schemes [84] have also been investigated. It is worth pointing out that various criteria are exploited to optimize different objectives according to the requirements in general.

In actual applications, especially in the environment where both the human being and the robot manipulator exist, it is necessary to keep the minimum movement for the robot manipulator to avoid colliding with the human being (such as the operators). For instance, in the medical treatment assisted by the robot (i.e., the medical robot), the robot needs to fulfill the end-effector task while keeping movement to a minimum (of the manipulator) as far as possible. Therefore, given the end-effector's trajectories in the task space, we expect to generate the minimum joint motion in a certain criterion.

It is worth mentioning that, among the published papers, the QP solvers adopt mainly the packaged routines and continuous-time models (e.g., the neural networks). Cheng *et al.* employed the "QPROG" in the IMSL library to solve the compact QP problem [71]. In [1], Zhang *et al.* used the "QUADPROG" in MATLAB optimization toolbox to solve the QP problem. The "QPROG" and "QUADPROG" are already exiting routines, but they are complex, time consuming, and packaged, which may be difficult to perform on a physical robot. For constrained kinematic redundancy resolution, various neural networks, that is, feedforward neural networks [152] and recurrent neural networks [15, 76], are generated. In [152], Dermatas *et al.* presented the error back-propagation neural network to solve the inverse-kinematics problem of redundant manipulators. Different from feedforward neural networks, recurrent neural networks do not need offline supervised learning and thus are more suitable for online robot control. In [76], Ding and Tso applied the Tank–Hopfield neural network to the kinematic resolution problem for online resolution. To overcome the disadvantage of the penalty parameter that arose in the Tank–Hopfield network, Wang *et al.* presented a Lagrangian network [15], but this may have a premature defect when solving inequality-constrained QP problems [15]. To always obtain optimal/exact solutions, traditional primal-dual neural networks have been presented based on the Karush–Kuhn–Tacker condition and projection operator [153]. However, due to minimizing the duality gap by gradient descent methods, the dynamic equations of such primal-dual networks are usually complicated, even containing high-order nonlinear terms [153]. In [148], Xia and Wang presented a dual neural network to effectively handle the physical limit constraints. However, the disadvantages of the dual neural network are that an inverse matrix operation is required, and the exponential convergence cannot always be guaranteed. With simple piecewise linear dynamics, a linear variational inequality (LVI) aided primal-dual neural network (PDNN) and its simplified model (called simplified LVIAPDNN) were presented [97], which have the global convergence. In [154], a primal-dual neural network with a one-layer structure for online resolution was exploited, which has an exponential convergence rate without any additional assumption. Moreover, many existing QP-based redundancy resolution schemes [107, 149, 150] do not consider feedback control, which may be undesirable in engineering applications.

This chapter presents a QP-based feedback-aided minimum joint motion (FAMJM) scheme and performs the scheme on a physical redundant robot manipulator. To avoid some joint(s) being locked by safety devices in practical applications, a new bound constraint with sufficient joint-limit margins has been designed and exploited. In addition, numerical algorithm 94LVI presented in Chapter 6 is employed as the QP solver. Moreover, the position-error feedback is considered in the presented QP-based FAMJM scheme, which can decrease effectively the model disturbance and computational round-off errors.

7.2 Preliminaries and Problem Formulation

In this section, to keep the minimum movement of the joints of redundant manipulators in the task duration, an FAMJM index is designed via the ZD method presented in Chapter 1. Based on Lyapunov method, the convergence performance of the feed-aided kinematics equation is proved. Then, considering the physical limits, an FAMJM scheme subject to varying joint-velocity limits (VJVL) with joint-limit margins is presented, which is solved by numerical algorithm 94LVI.

7.2.1 Minimum Joint Motion Performance Index

To lay a basis for further discussion, the following essential equations for redundant robot manipulators are given here [153]:

$$\mathbf{f}(\theta) = \mathbf{r}_{\mathrm{e}}, \tag{7.1}$$

$$J(\theta)\dot{\theta} = \dot{\mathbf{r}}_{\mathrm{e}}. \tag{7.2}$$

The detailed descriptions on Equation (7.1) and Equation (7.2) have been presented in previous chapters and thus are omitted here. Since the manipulator system is redundant (i.e., $m < n$), Equation (7.1) and Equation (7.2) are all under-determined and generally admit an infinite number of solutions in terms of inverse kinematics. For instance, in this chapter, the redundant manipulator is a six-link planar manipulator, and we consider only its end-effector positioning; thus, with $n = 6$ and $m = 2$, the redundant DOF is 4.

Remark 7.1 The general redundancy resolution process of the QP-based scheme at the joint-velocity level is divided into the following five steps. Firstly, we need to determine the optimization objective according to the requirements. Generally speaking, different requirements determine the different optimization objectives, which would lead to different redundancy resolution results. Secondly, the desired path $\mathbf{r}_{\mathrm{d}}$ is given and differentiated with respect to t. Then, the resultant $\dot{\mathbf{r}}_{\mathrm{d}}$ and the necessary feedback term (e.g., the position-error feedback $\mathbf{r}_{\mathrm{d}} - \mathbf{f}(\theta)$), as the right-hand of the kinematics equation at the joint-velocity level, are incorporated into the designed scheme. Thirdly, the scheme is reformulated as a unified QP problem. Fourthly, a QP-solver (e.g., numerical algorithm 94LVI) is used to solve the resultant QP problem. Finally, the computed data (i.e., joint angle θ and joint velocity $\dot{\theta}$) are used to control the manipulator to fulfill the desired path tracking task.

Theorem 7.1 For keeping the minimum movement of the manipulator in the task duration, and considering the position-error feedback, the following QP-based scheme can be used:

$$\text{minimize} \quad \|\dot{\theta} + \mathbf{q}\|_2^2/2, \tag{7.3}$$

$$\text{subject to} \quad J(\theta)\dot{\theta}(t) = \dot{\mathbf{r}}_\text{d} + \kappa_\text{p}(\mathbf{r}_\text{d} - \mathbf{f}(\theta)), \tag{7.4}$$

$$\text{with} \quad \mathbf{q} = \gamma(\theta(t) - \theta(0)), \tag{7.5}$$

where $\kappa_\text{p} > 0 \in \mathbb{R}$ is the position-error feedback gain (in the ensuing simulations and experiments, $\kappa_\text{p} = 4$); $\theta(0)$ is the initial state of the joint variable vector; and the design parameter, $\gamma > 0$, is used to scale the magnitude of the manipulator response to the joint displacement.

Proof: To minimize the joint displacement $(\theta(t) - \theta(0))$ between the current state $\theta(t)$ and the initial state $\theta(0)$ of the manipulator at each time instant t, we firstly define a vector-valued displacement-function, which is to be kept minimum over the whole range:

$$\mathcal{E}(t) = \theta(t) - \theta(0) \in \mathbb{R}^n. \tag{7.6}$$

Based on the ZD method presented in Chapter 1, in order to minimize $\mathcal{E}(t)$ (even converge exponentially to zero) at each time instant, the following equation can be obtained:

$$\dot{\mathcal{E}}(t) = -\gamma\mathcal{E}(t), \tag{7.7}$$

where $\gamma > 0 \in \mathbb{R}$ denotes the exponential convergence rate of $\mathcal{E}(t)$. According to the theory of first-order differential equation, the solution to Equation (7.7) is $\mathcal{E}(t) = \exp(-\gamma t)\mathcal{E}(0)$.

In view of Equation (7.7), the derivative of $\mathcal{E}(t)$ with respect to t is $\dot{\theta}(t)$. Substituting $\dot{\mathcal{E}}(t) = \dot{\theta}(t)$ and Equation (7.7) to Equation (7.6), we can get

$$\dot{\theta}(t) = -\gamma(\theta(t) - \theta(0)),$$

which can be rewritten as

$$\dot{\theta}(t) + \gamma(\theta(t) - \theta(0)) = 0. \tag{7.8}$$

In practice, in view of potential requirements on robot motion planning and control (such as joint-angle limits and joint-velocity limits), minimizing $\|\dot{\theta}(t) + \gamma\left(\theta(t) - \theta(0)\right)\|_2^2/2$ is better than forcing $\dot{\theta}(t) + \gamma(\theta(t) - \theta(0)) = 0$ directly. Considering Equation (7.2), with $\mathbf{q} = \gamma(\theta(t) - \theta(0))$, we can obtain the minimum joint motion scheme for redundancy resolution of robot manipulators:

$$\text{minimize} \quad \|\dot{\theta} + \mathbf{q}\|_2^2/2, \tag{7.9}$$

$$\text{subject to} \quad J(\theta)\dot{\theta}(t) = \dot{\mathbf{r}}_\text{d}. \tag{7.10}$$

Evidently, schemes (7.9) through (7.10) do not consider the position-error feedback.

In order to further improve the accuracy of the redundancy resolution, the position-error feedback should be considered in the scheme. According to the ZD method [155], similar to (7.6), we define a vector-valued error-function:

$$\epsilon(t) = \mathbf{r}_\text{d} - \mathbf{f}(\theta) \in \mathbb{R}^n. \tag{7.11}$$

Based on the ZD method, we can simply set:

$$\dot{\epsilon}(t) = -\kappa_\text{p}\epsilon(t), \tag{7.12}$$

where κ_p is a positive position-error feedback gain. In view of Equation (7.2), Equation (7.12) can be written as

$$\dot{\mathbf{r}}_\mathrm{d} - J(\theta)\dot{\theta}(t) = -\kappa_\mathrm{p}(\mathbf{r}_\mathrm{d} - \mathbf{f}(\theta)). \tag{7.13}$$

That is

$$J(\theta)\dot{\theta}(t) = \dot{\mathbf{r}}_\mathrm{d} + \kappa_\mathrm{p}(\mathbf{r}_\mathrm{d} - \mathbf{f}(\theta)). \tag{7.14}$$

Replacing Equation (7.10) in the scheme with Equation (7.14), we can get the minimum joint motion scheme with position-error feedback considered; that is, Equation (7.3) through Equation (7.5). The proof is thus completed. □

Theorem 7.2 Starting from any initial joint state $\theta(0)$ at the joint feasible region, the computed joint-velocity $\dot{\theta}(t)$ of the kinematics equation with position-error feedback considered, Equation (7.14) guarantees that the resolved end-effector state $\mathbf{f}(\theta)$ of the manipulator can be convergent to the desired state $\mathbf{r}_\mathrm{d}$.

Proof: We firstly define error vector $\epsilon(t) = \mathbf{r}_\mathrm{d} - \mathbf{f}(\theta)$, and the Lyapunov candidate

$$V = \frac{1}{2}(\epsilon(t))^\mathrm{T}\kappa_\mathrm{p}(\epsilon(t)) \geqslant 0. \tag{7.15}$$

Its time derivative along the system trajectories is

$$\begin{aligned} \dot{V} &= \frac{1}{2}\left((\dot{\epsilon}(t))^\mathrm{T}\kappa_\mathrm{p}(\dot{\epsilon}(t)) + (\epsilon(t))^\mathrm{T}\kappa_\mathrm{p}\dot{\epsilon}(t)\right) \\ &= \frac{1}{2}\left(\kappa_\mathrm{p}\sum_{i=1}^{n}\dot{\epsilon}_i(t)\epsilon_i(t) + \kappa_\mathrm{p}\sum_{i=1}^{n}\epsilon_i(t)\dot{\epsilon}_i(t)\right). \end{aligned} \tag{7.16}$$

According to the design principle of position-error feedback; that is, $\dot{\epsilon}(t) = -\kappa_\mathrm{p}\epsilon(t)$, Equation (7.16) can be reformulated as

$$\begin{aligned} \dot{V} &= \frac{1}{2}\left(\kappa_\mathrm{p}\sum_{i=1}^{n}(-\kappa_\mathrm{p})\epsilon_i(t)\epsilon_i(t) + \kappa_\mathrm{p}\sum_{i=1}^{n}\epsilon_i(t)(-\kappa_\mathrm{p})\epsilon_i(t)\right) \\ &= -\frac{1}{2}(\kappa_\mathrm{p}^2)\left(\sum_{i=1}^{n}\epsilon_i(t)\epsilon_i(t) + \sum_{i=1}^{n}\epsilon_i(t)\epsilon_i(t)\right) \\ &= -\kappa_\mathrm{p}^2(\epsilon(t))^\mathrm{T}(\epsilon(t)) \leqslant 0. \end{aligned} \tag{7.17}$$

V is zero if and only if $\epsilon(t) = 0$. The proof is thus completed. □

The self regulating process of the position-error feedback is discussed as in the following three cases.

1) When $\mathbf{f}(\theta) < \mathbf{r}_\mathrm{d}$, that is, the actual end-effector's movement is less than the desired movement, $\mathbf{r}_\mathrm{d} - \mathbf{f}(\theta) > 0$. Since the redundancy resolution is resolved at the velocity level, the positive position-error $\kappa_\mathrm{p}(\mathbf{r}_\mathrm{d} - \mathbf{f}(\theta))$ should be added to the right-hand side of the differential relation $J(\theta)\dot{\theta}(t) = \dot{\mathbf{r}}_\mathrm{d}$. Consequently, the resultant desired end-effector velocity would be enhanced, which can generate the larger joint velocity via the scheme and the responding QP solver. Such a larger joint velocity can drive the controller more strongly to decrease or eliminate the positive position-error.

2) When $\mathbf{f}(\theta) > \mathbf{r}_d$, that is, the actual end-effector's movement is larger than the desired one, $\mathbf{r}_d(t) - \mathbf{f}(\theta) < 0$. The negative position-error feedback should be added to the desired end-effector velocity $\dot{\mathbf{r}}_d$, which can generate smaller joint velocities via the scheme and the QP solver. Such a smaller joint velocity would drive the controller to slow down the end-effector movement so as to decrease the difference between the actual end-effector position and the desired position.
3) When $\mathbf{f}(\theta) = \mathbf{r}_d(t)$; that is, the actual end-effector's motion is exactly the desired one. The feedback system does not react on the original open control system.

7.2.2 Varying Joint-Velocity Limits

In practical applications, the physical limits always exist. For instance, in the presented six-DOF planar robot manipulator, the joint-velocity limits are functions of the joint angle θ. That is to say, the joint-angle and joint-velocity limits should be considered in the presented FAMJM scheme, that is,

$$\text{minimize} \quad \|\dot{\theta} + \mathbf{q}\|_2^2/2, \tag{7.18}$$

$$\text{subject to} \quad J(\theta)\dot{\theta} = \dot{\mathbf{r}}_d + \kappa_p(\mathbf{r}_d - \mathbf{f}(\theta)), \tag{7.19}$$

$$\theta^- \leqslant \theta \leqslant \theta^+, \tag{7.20}$$

$$\dot{\theta}^-(\theta) \leqslant \dot{\theta} \leqslant \dot{\theta}^+(\theta), \tag{7.21}$$

$$\text{with} \quad \mathbf{q} = \gamma(\theta(t) - \theta(0)).$$

Through the results presented in previous chapters, the bounds of (7.21) are obtained. In addition, as the redundancy is resolved at the joint-velocity level, the joint physical limits (7.20) and (7.21) have to be transformed and incorporated into a bound constraint in terms of $\dot{\theta}$, i.e., $\zeta^-(\theta) \leqslant \dot{\theta} \leqslant \zeta^+(\theta)$.

As presented in Chapter 6, a joint-limit conversion method is thus exploited:

$$\begin{aligned} \zeta_i^-(\theta_i) &= \max\{\dot{\theta}_i^-(\theta_i), \mu(\vartheta_i^- - \theta_i)\}, \\ \zeta_i^+(\theta_i) &= \min\{\dot{\theta}_i^+(\theta_i), \mu(\vartheta_i^+ - \theta_i)\}, \end{aligned} \tag{7.22}$$

where $\vartheta_i^\pm$ denotes margins-considered joint limits; and we can set $\vartheta_i^- = \theta_i^- + \kappa\pi/180$, and $\vartheta_i^+ = \theta_i^+ - \kappa\pi/180$ in applications. In addition, $\mu = 4$ is used to scale the feasible region of joint-velocity $\dot{\theta}$; κ is a constant vector used to scale the safety region, which is set to 2 degrees in simulations and experiments in this chapter. It is worth mentioning that, since the final bounds $\zeta_i^\pm$ [i.e., (7.22)] include $\dot{\theta}_i^\pm(\theta_i)$ and θ_i, the bounds $\zeta_i^\pm$ are time-varying, being functions of joint angles and being in accordance with the push-rod-joint structure. Therefore, the FAMJM redundancy resolution scheme for the six-DOF planar robot manipulator can be reformulated finally as the following QP:

$$\text{minimize} \quad \mathbf{x}^{\mathrm{T}} W \mathbf{x}/2 + \mathbf{q}^{\mathrm{T}}\mathbf{x}, \tag{7.23}$$

$$\text{subject to} \quad J\mathbf{x} = \mathbf{b}, \tag{7.24}$$

$$\zeta^-(\theta) \leqslant \mathbf{x} \leqslant \zeta^+(\theta), \tag{7.25}$$

$$\text{with} \quad W = I,\ \mathbf{b} = \dot{\mathbf{r}}_d + \kappa_p(\mathbf{r}_d - \mathbf{f}(\theta)),$$

$$\mathbf{q} = \gamma(\theta(t) - \theta(0)),$$

where decision variable vector $\mathbf{x} \in \mathbb{R}^n$ denotes the joint-velocity vector $\dot{\theta}$. Performance Index (7.23) is obtained from the derivation of (7.18). The equality constraint (7.24) expresses a linear relation between the desired Cartesian velocity $\dot{\mathbf{r}}_{\mathrm{d}} \in \mathbb{R}^m$ and the joint velocity $\dot{\theta} \in \mathbb{R}^n$ with position-error feedback $\kappa_{\mathrm{p}}(\mathbf{r}_{\mathrm{d}} - \mathbf{f}(\theta))$ included. To keep all joint variables within their mechanical ranges, it is straightforward and concise to use the bound constraint (7.25).

7.3 Computer Simulations and Physical Experiments

For illustrating the efficacy and accuracy of the FAMJM method and the corresponding numerical algorithm 94LVI, computer simulations and experiments are performed by using the six-DOF planar robot manipulator to track letters "M"- and "P"-shaped paths in the two-dimensional horizontal plane. Besides, the actual joint-angle limits used in this chapter of the six-DOF planar robot manipulator are listed in Table 7.1.

7.3.1 "M"-Shaped Path-Tracking Task

In this subsection, the end-effector of the six-DOF robot manipulator is expected to track a letter "M"-shaped path as shown in Figure 7.1(a), which is the initial letter of English word "MOVEMENT". For the convenience of observation and analysis, we also present a simulated manipulator trajectories of the six-DOF robot manipulator tracking a letter "M"-shaped path as shown in Figure 7.1(b). The x and the y-axis velocity functions of the "M"-shaped path are, respectively,

$$r_{\mathrm{dX}}(t) = \begin{cases} (s/T)\big(t - (T/(2\pi))\sin(2\pi t/T)\big) + f_{\mathrm{X}}(\theta(0)), & \forall t \in [0, T] \\ -(\sqrt{3}s/(2T))\big((t-T) - (T/(2\pi))\sin(2\pi(t-T)/T)\big) & \\ +f_{\mathrm{X}}(\theta(0)) + s, & \forall t \in [T, 2T] \\ (\sqrt{3}s/(2T))\big((t-2T) - (T/(2\pi))\sin(2\pi(t-2T)/T)\big) & \\ -\sqrt{3}s/2 + s + f_{\mathrm{X}}(\theta(0)), & \forall t \in [2T, 3T] \\ s + f_{\mathrm{X}}(\theta(0)) - (s/T)\big((t-3T) & \\ -(T/(2\pi))\sin(2\pi(t-3T)/T)\big), & \forall t \in [3T, 4T] \end{cases}$$

$$r_{\mathrm{dY}}(t) = \begin{cases} f_{\mathrm{Y}}(\theta(0)), & \forall t \in [0, T] \\ -(s/(2T))\big((t-T) - (T/(2\pi))\sin(2\pi(t-T)/T)\big) & \\ +f_{\mathrm{Y}}(\theta(0)), & \forall t \in [T, 2T] \\ (s/(2T))\big((t-2T) - (T/(2\pi))\sin(2\pi(t-2T)/T)\big) & \\ +s/2 + f_{\mathrm{Y}}(\theta(0)), & \forall t \in [2T, 3T] \\ s + f_{\mathrm{Y}}(\theta(0)), & \forall t \in [3T, 4T] \end{cases}$$

where s is the design parameter that depends on the given task; $f_{\mathrm{X}}(\theta(0))$ and $f_{\mathrm{Y}}(\theta(0))$ denote, respectively, the x- and y-axis components of the initial position vector $\mathbf{f}(\theta(0))$ of the end-effector; T is a time parameter related to the task duration. In the simulations and experiment of this subsection, $s = 6$ cm, and the task duration $4T$ is 20 s (i.e., $T = 5$ s). The initial joint state $\theta(0) = [\pi/6, \pi/12, \pi/12, \pi/12, \pi/12, \pi/12]^{\mathrm{T}}$ rad (i.e., $\theta(0) = [30, 15, 15, 15, 15, 15]^{\mathrm{T}}$ in degrees).

7.3.1.1 Simulation Comparisons with Different κ_{p}

In order to test the effectiveness of the position-error feedback parameter κ_{p}, some computer simulations are conducted when κ_{p} increases from 0 to 100. Without loss of

Table 7.1 Actual joint-angle limits used in this chapter of six-DOF planar robot manipulator.

#	$i = 1$	$i = 2$	$i = 3$	$i = 4$	$i = 5$	$i = 6$
θ_i^-(rad)	−1.536	0.052	0.026	0.066	0.017	0.009
θ_i^+(rad)	1.431	0.785	0.611	0.576	0.559	0.445

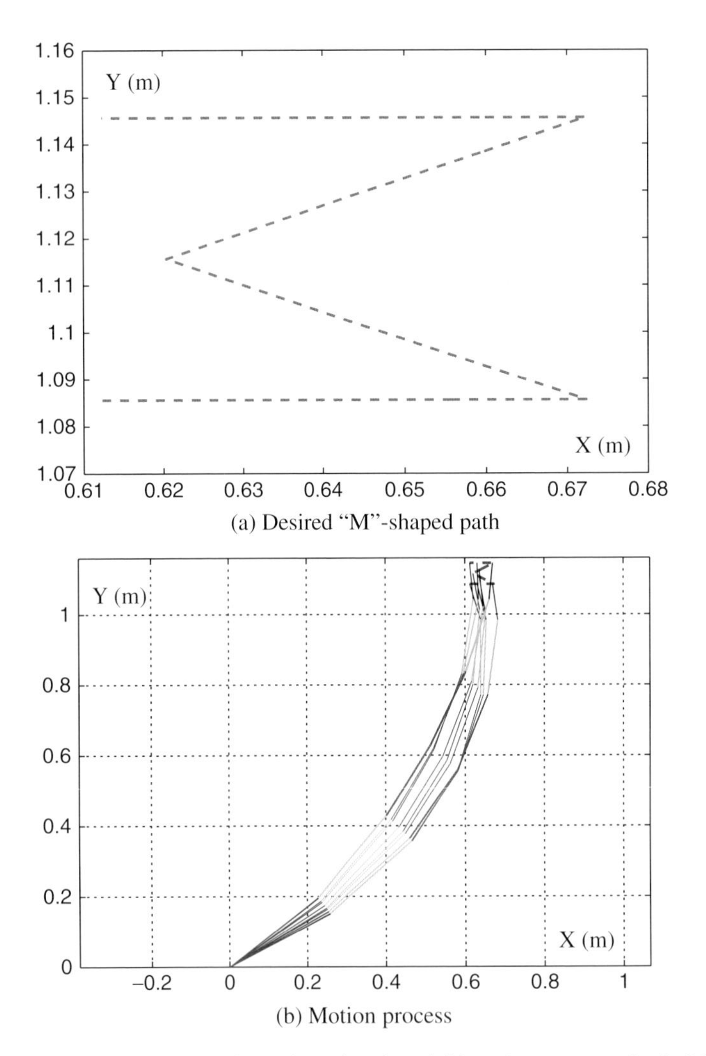

Figure 7.1 (a) Desired "M"-shaped path and (b) motion process of a six-DOF planar robot manipulator synthesized by a FAMJM scheme (7.23)–(7.25).

Table 7.2 Maximum errors of a six-DOF planar robot manipulator synthesized by FAMJM scheme (7.23)–(7.25) when κ_p increases from 0 to 100 during task execution of "M"-shaped-path tracking.

Parameters	Maximum Error (m)
$(\gamma, \kappa_p) = (4, 0)$	$\lvert\epsilon_X\rvert_{max} = 1.2114 \times 10^{-4}$, $\lvert\epsilon_Y\rvert_{max} = 1.0128 \times 10^{-4}$
$(\gamma, \kappa_p) = (4, 2)$	$\lvert\epsilon_X\rvert_{max} = 3.2780 \times 10^{-5}$, $\lvert\epsilon_Y\rvert_{max} = 1.8947 \times 10^{-5}$
$(\gamma, \kappa_p) = (4, 4)$	$\lvert\epsilon_X\rvert_{max} = 1.8115 \times 10^{-5}$, $\lvert\epsilon_Y\rvert_{max} = 1.0353 \times 10^{-5}$
$(\gamma, \kappa_p) = (4, 6)$	$\lvert\epsilon_X\rvert_{max} = 1.2368 \times 10^{-5}$, $\lvert\epsilon_Y\rvert_{max} = 7.0297 \times 10^{-6}$
$(\gamma, \kappa_p) = (4, 8)$	$\lvert\epsilon_X\rvert_{max} = 9.3626 \times 10^{-6}$, $\lvert\epsilon_Y\rvert_{max} = 5.3050 \times 10^{-6}$
$(\gamma, \kappa_p) = (4, 10)$	$\lvert\epsilon_X\rvert_{max} = 7.5230 \times 10^{-6}$, $\lvert\epsilon_Y\rvert_{max} = 4.2586 \times 10^{-6}$
$(\gamma, \kappa_p) = (4, 15)$	$\lvert\epsilon_X\rvert_{max} = 5.0388 \times 10^{-6}$, $\lvert\epsilon_Y\rvert_{max} = 2.8476 \times 10^{-6}$
$(\gamma, \kappa_p) = (4, 20)$	$\lvert\epsilon_X\rvert_{max} = 3.7843 \times 10^{-6}$, $\lvert\epsilon_Y\rvert_{max} = 2.1381 \times 10^{-6}$
$(\gamma, \kappa_p) = (4, 25)$	$\lvert\epsilon_X\rvert_{max} = 3.0294 \times 10^{-6}$, $\lvert\epsilon_Y\rvert_{max} = 1.7114 \times 10^{-6}$
$(\gamma, \kappa_p) = (4, 30)$	$\lvert\epsilon_X\rvert_{max} = 2.5266 \times 10^{-6}$, $\lvert\epsilon_Y\rvert_{max} = 1.4266 \times 10^{-6}$
$(\gamma, \kappa_p) = (4, 40)$	$\lvert\epsilon_X\rvert_{max} = 1.8975 \times 10^{-6}$, $\lvert\epsilon_Y\rvert_{max} = 1.0704 \times 10^{-6}$
$(\gamma, \kappa_p) = (4, 50)$	$\lvert\epsilon_X\rvert_{max} = 1.5196 \times 10^{-6}$, $\lvert\epsilon_Y\rvert_{max} = 5.5669 \times 10^{-7}$
$(\gamma, \kappa_p) = (4, 60)$	$\lvert\epsilon_X\rvert_{max} = 1.2675 \times 10^{-6}$, $\lvert\epsilon_Y\rvert_{max} = 7.1416 \times 10^{-7}$
$(\gamma, \kappa_p) = (4, 80)$	$\lvert\epsilon_X\rvert_{max} = 9.5190 \times 10^{-7}$, $\lvert\epsilon_Y\rvert_{max} = 5.3615 \times 10^{-7}$
$(\gamma, \kappa_p) = (4,100)$	$\lvert\epsilon_X\rvert_{max} = 7.6187 \times 10^{-7}$, $\lvert\epsilon_Y\rvert_{max} = 4.2958 \times 10^{-7}$

generality, we set $\gamma = 4$ in the tests. The simulative results are shown in Table 7.2 and Figure 7.2. From Table 7.2, we can see that the maximum position errors (i.e., $\lvert\epsilon_Y\rvert_{max}$ and $\lvert\epsilon_Y\rvert_{max}$) of the end-effector are small (i.e., less than 1.0128×10^{-4} m) even the feedback gain κ_p is not considered. This illustrates that the FAMJM scheme and the corresponding numerical algorithm 94LVI are effectiveness and accuracy. In order to improve the accuracy further, the position-error feedback can be incorporated into the FAMJM scheme. From Table 7.2 and Figure 7.2, we can see that, with feedback gain κ_p increasing from 0 to 100, the maximum position error decreases from 1.0128×10^{-4} m to 7.6187×10^{-7} m. The simulative results substantiate the effectiveness of the position-error feedback in the FAMJM scheme, as well as the accuracy of the presented FAMJM scheme (7.23) through (7.25).

7.3.1.2 Simulation Comparisons with Different γ

In this subsection, to test the effectiveness of the minimum joint motion parameter γ in the FAMJM scheme, we conduct simulations with different γ; that is, from 0 to 100.

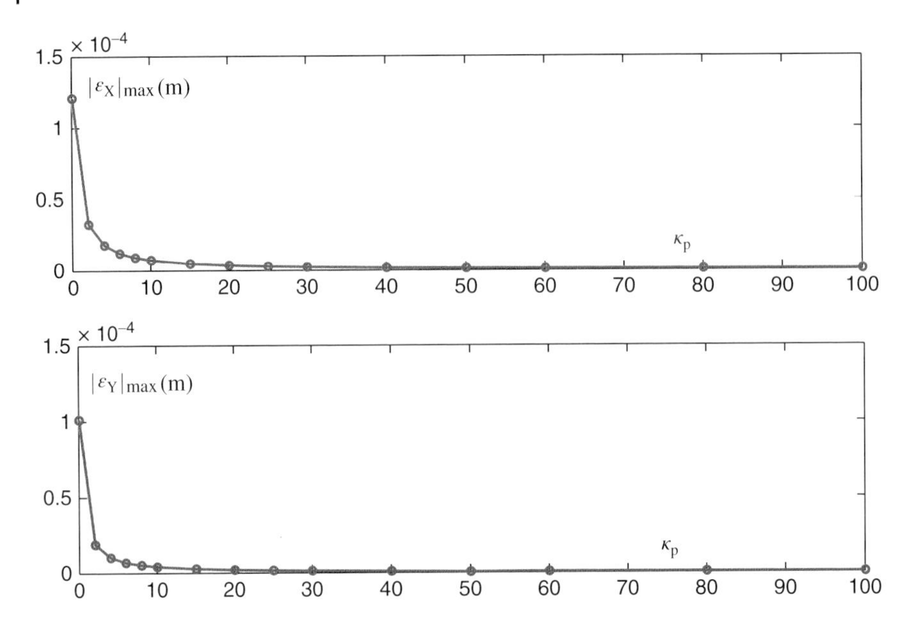

Figure 7.2 Maximum error variation tendency of a six-DOF planar robot manipulator synthesized by FAMJM scheme (7.23)–(7.25) when κ_p increases from 0 to 100 during "M"-shaped-path-tracking execution.

For ruling out the influence of the feedback control, we first set $\kappa_p = 0$. The simulative results are presented in Figure 7.3. Figure 7.3(a) shows the variation tendency of the global displacement during the task execution when γ increases from 0 to 100. It is worth mentioning that $N = T/\delta + 1$ with $\delta = 0.01$ denoting the sampling interval. Figure 7.3(b) shows the final state displacement of the joints after fulfilling the task.

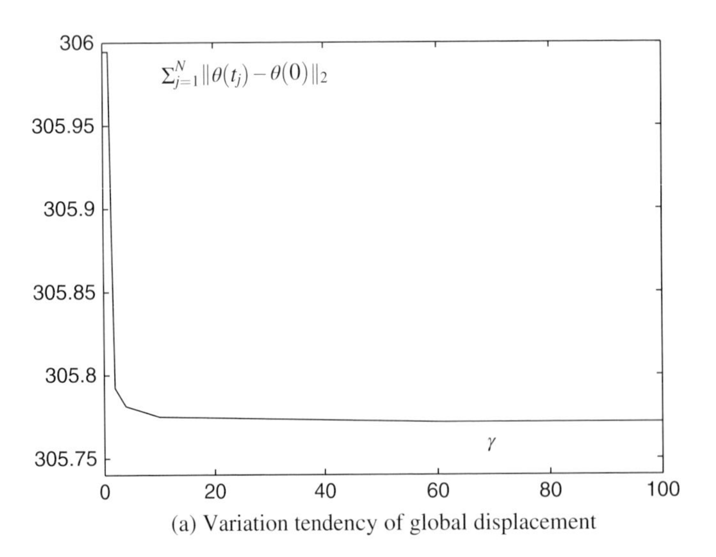

Figure 7.3 Joint displacement variation tendency of a six-DOF planar robot manipulator synthesized by FAMJM scheme (7.23)–(7.25) when a robot end-effector tracks an "M"-shaped path.

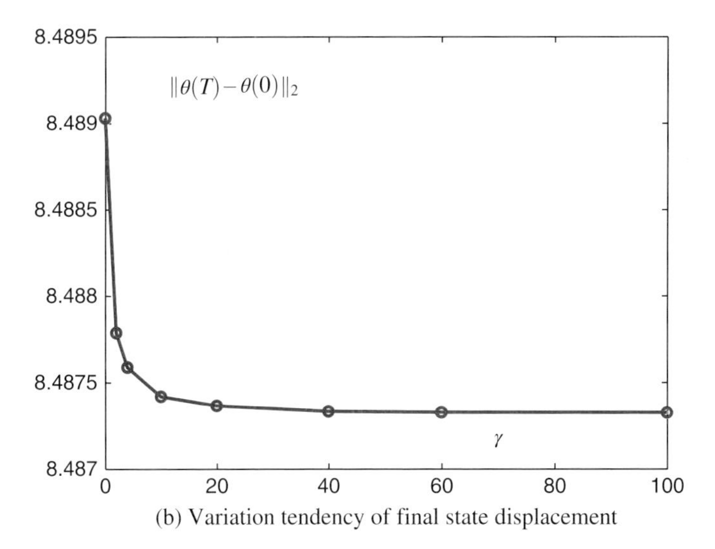

(b) Variation tendency of final state displacement

Figure 7.3 *(Continued)*

From Figure 7.3, we can see that both the global displacement of the joints and the final state displacement of the joints decrease when the minimum joint motion parameter γ increases from 0 to 100. Specifically, when $\gamma = 0$, that is, the minimum joint motion parameter is not considered, the obtained joint state during the task execution has the maximum drift from the initial state. When $\gamma \neq 0$, that is, the minimum joint motion parameter γ is active, the joint displacement will decrease. It is worth mentioning that, when $\gamma = 0$ and $\kappa_{\mathrm{p}} = 0$, the scheme is reduced to the minimum velocity norm (MVN) scheme. These simulative results illustrate that the minimum joint motion norm scheme is effective.

7.3.1.3 Simulative and Experimental Verifications of FAMJM Scheme

The simulative comparisons presented here illustrate well the efficacy of the minimum joint motion parameter γ and the position-error feedback gain κ_{p} in the FAMJM scheme. In this subsection, for validating the physical realizability of the FAMJM scheme, we perform the scheme on a six-DOF planar robot manipulator. The tracking path, duration, and the initial joint state are the same as the earlier subsections, and $\gamma = 4$ as well as $\kappa_{\mathrm{p}} = 6$ without loss of generality.

The actual task execution and experimental result of the six-DOF robot manipulator synthesized by the FAMJM scheme is shown in Figure 7.4. Snapshots 1 through 11 of Figure 7.4 illustrate the task execution process and Snapshot 12 of Figure 7.4 shows the experimental result. From Figure 7.4, we can see that the end-effector is fulfilled very well, which illustrates that such an FAMJM scheme (7.23)–(7.25) is effective and accurate on the redundancy resolution of the redundant robot manipulator. Correspondingly, the joint-angle profiles and the joint-velocity profiles are shown in Figure 7.5(a) and Figure 7.5(b). The desired tracking path and the end-effector trajectory are presented in Figure 7.6(a), which shows that they show good coincidence with each other. The tiny position error is less than 2×10^{-5} m as shown in Figure 7.6(b).

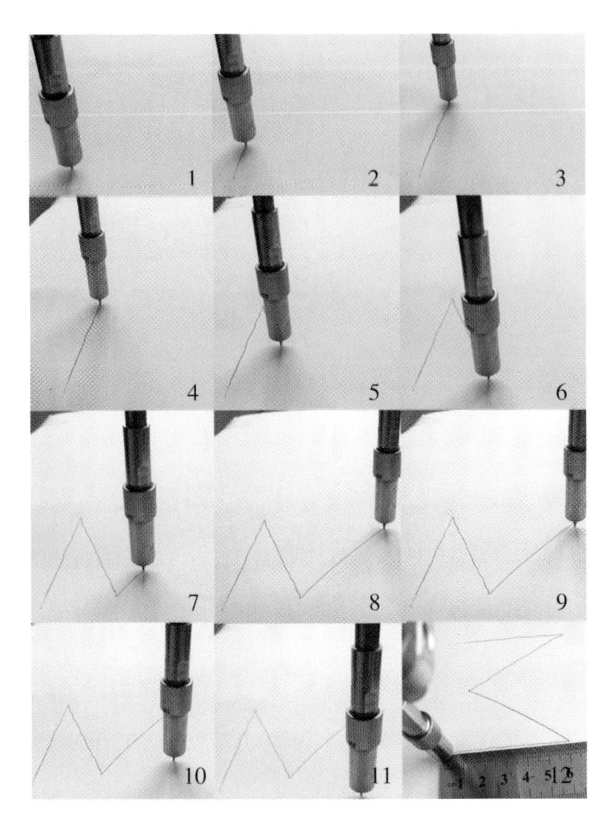

Figure 7.4 Snapshots for actual task execution of a six-DOF planar robot manipulator synthesized by FAMJM scheme (7.23)–(7.25) when a robot end-effector tracks an "M"-shaped path. Source: Zhang 2013. Reproduced with permission of permission of TCCT (License Number 201611080923).

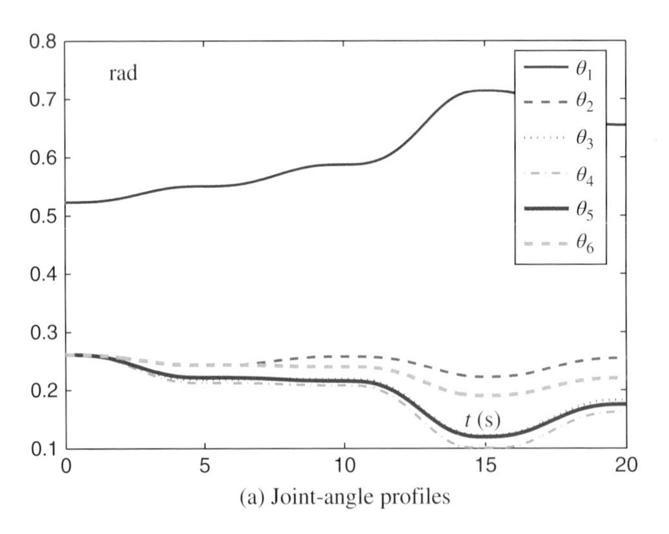

Figure 7.5 (a) Joint-angle and (b) joint-velocity profiles of a six-DOF planar robot manipulator synthesized by FAMJM scheme (7.23)–(7.25) with $\gamma = 4$ and $\kappa_p = 6$ when a robot end-effector tracks an "M"-shaped path.

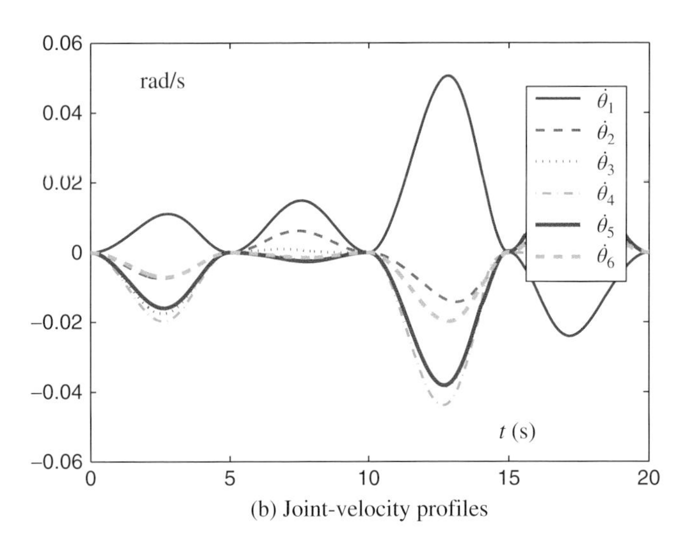

(b) Joint-velocity profiles

Figure 7.5 (*Continued*)

In summary, the letter "M"-shaped path-tracking experiment illustrates the efficacy of the presented FAMJM scheme [depicted in (7.23) through (7.25)] and numerical algorithm 94LVI depicted in (6.17) on the redundancy resolution of the six-DOF planar robot manipulator well. In addition, the position-error analysis further shows the accuracy of such a QP-based FAMJM scheme for robots' path tracking.

7.3.2 "P"-Shaped Path Tracking Task

To further verify the validity of the presented FAMJM scheme, a letter "P"-shaped path with the length being 0.12 cm is given for the six-DOF planar robot manipulator to track.

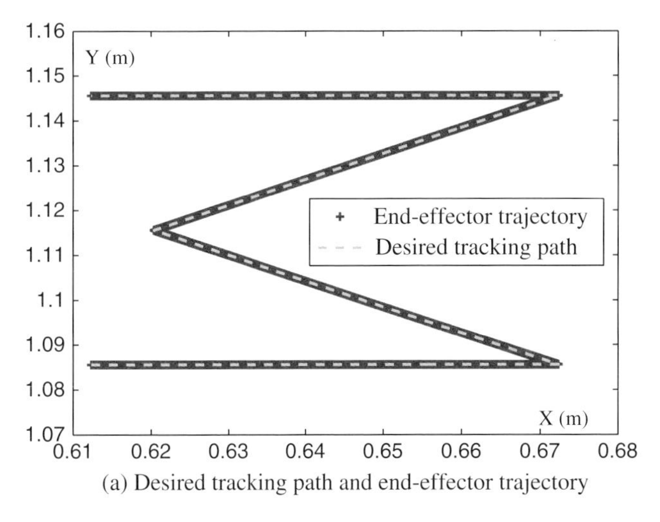

(a) Desired tracking path and end-effector trajectory

Figure 7.6 (a) Desired path, end-effector trajectory, and (b) position error for actual task execution of a six-DOF planar robot manipulator synthesized by FAMJM scheme (7.23)–(7.25) with $\gamma = 4$ and $\kappa_{\text{p}} = 6$ when a robot end-effector tracks an "M"-shaped path.

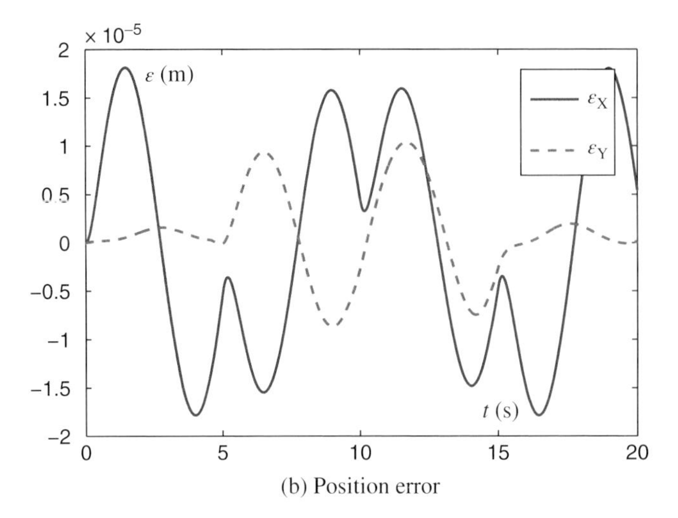

(b) Position error

Figure 7.6 (*Continued*)

The task duration, initial joint state $\theta(0)$ and other design parameters (i.e., γ and $\kappa_{\rm p}$) are set the same as the letter "M"-shaped path-tracking experiment before. The snapshots of the task execution are illustrated in Figure 7.7. From the figure, we see that the letter "P"-shaped task is completed successfully. This experimental result further substantiates the physical realizability and effectiveness of the FAMJM scheme.

7.3.3 Comparisons with Pseudoinverse-Based Approach

In order to further illustrate the superiority of the presented FAMJM scheme, comparisons between the FAMJM scheme and the classic pseudoinverse-based approach are illustrated in this subsection.

In the existing approaches, the classic solution to the redundancy resolution problem is the pseudoinverse-based formulation, that is, one minimum-norm particular solution plus a homogeneous solution. Specifically, at the joint-velocity level, the pseudoinverse-based solution can be formulated as

$$\dot{\theta} = J^{\dagger}(\theta)\dot{\mathbf{r}}_{\rm d} + (I - J^{\dagger}(\theta)J(\theta))\mathbf{w}. \tag{7.26}$$

where $J^{\dagger}(\theta) \in \mathbb{R}^{n\times m}$ denotes the pseudoinverse of the Jacobian matrix $J(\theta)$, $I \in \mathbb{R}^{n\times n}$ is the identity matrix, and $\mathbf{w} \in \mathbb{R}^{n}$ is an arbitrary vector usually selected by using some optimization criteria. The first term of the right-hand of (7.26) is the particular solution (i.e., the minimum norm solution), and the second term is the homogeneous solution corresponding to the manipulator's self-motion (without having an affect on the end-effector motion). Generally speaking, different ν values can be chosen according to different optimization objectives (i.e., the secondary tasks). With $\mathbf{w} = -\nu(\theta(t) - \theta(0))$, a classic pseudoinverse-based minimum joint motion (PBMJM) scheme can be formulated as [156]:

$$\dot{\theta} = J^{\dagger}(\theta)\dot{\mathbf{r}}_{\rm d} - \nu(I - J^{\dagger}(\theta)J(\theta))(\theta(t) - \theta(0)). \tag{7.27}$$

where $\nu \in \mathbb{R}$ is a nonnegative constant, and is set at 4 in the simulations in this subsection.

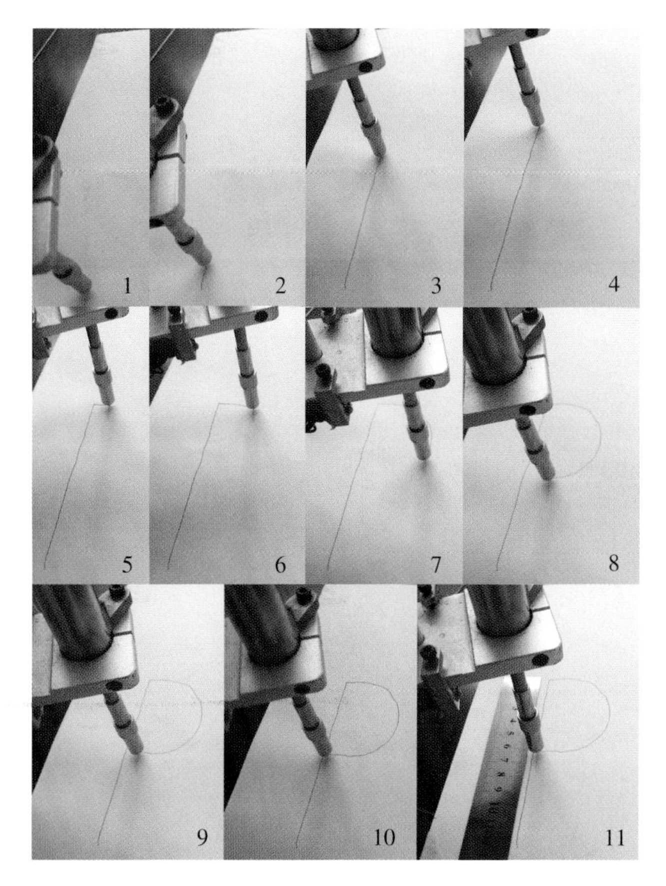

Figure 7.7 Snapshots of the actual task execution of a six-DOF planar robot manipulator synthesized by FAMJM scheme (7.23)–(7.25) when a robot end-effector tracks a "P"-shaped path. *Source*: Zhang 2013. Reproduced with permission of permission of TCCT (License Number 201611080923).

Compared the FAMJM scheme (7.23)–(7.25) with the classic PBMJM approach (7.27), such an FAMJM scheme possesses obvious superiorities. For instance, there is no explicit need to invert $J(\theta)$ in the FAMJM scheme. Another advantage is that the additional constraints such as joint physical limits are considered in the FAMJM scheme. On the contrary, the classic PBMJM scheme can not consider the joint physical limits of the robot in its formulation. That is to say, the robot synthesized by such a classic PBMJM scheme may hit/exceed the joint physical limits. Generally speaking, this is not applicable in the actual engineering applications. For comparison and illustration, tracking two larger paths examples, that is, tracking larger "M"- and "P"-shaped paths (as compared with the ones shown in the previous subsections), are investigated, illustrated, and compared in the ensuing parts of this subsection. Both the classic PBMJM scheme (7.27) and the presented FAMJM scheme (7.23) through (7.25) are applied to the planar six-DOF redundant robot manipulator. It is worth pointing out that the parameters of the FAMJM scheme and the corresponding numerical algorithm 94LVI are set the same as those in Subsections 7.3.1 and 7.3.2, and the simulative results are shown in Figures 7.8 through 7.19.

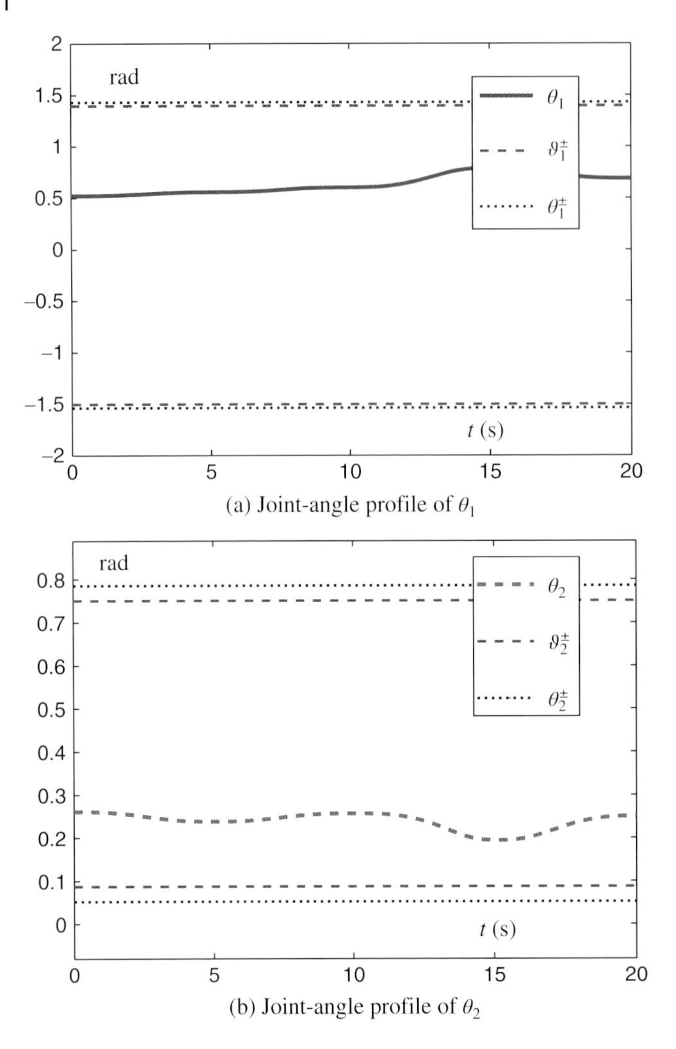

Figure 7.8 Joint-angle profiles of (a) θ_1 and (b) θ_2 synthesized by PBMJM scheme (7.27) when a robot end-effector tracks a larger "M"-shaped path.

7.3.3.1 Comparison with Tracking Task of Larger "M"-Shaped Path

In this example, the task duration of tracking a larger "M"-shaped path is 20 s (i.e., $T = 5$ s), and $s = 7.2$ cm that reflects the task size. Figures 7.8 through 7.10 show the joint-angle profiles of the planar six-DOF planar robot manipulator synthesized by PBMJM scheme (7.27). Note that, in Figures 7.8 through 7.10, $\theta_i^{\pm}$ (with $i = 1, 2, \cdots, 6$) denotes the original-joint-angle limit (as listed in Table 7.1), and $\vartheta_i^{\pm}$ (with $i = 1, 2, \cdots, 6$) denotes the margins-considered joint-angle limit, which leaves some margins as the "safety region." As shown in Figure 7.9 (b), joint angle θ_4 exceeds its lower limit θ_4^-, which is not expected. That is because the joint would be locked by the hardware safety device of the six-DOF planar robot manipulator if the corresponding resolved

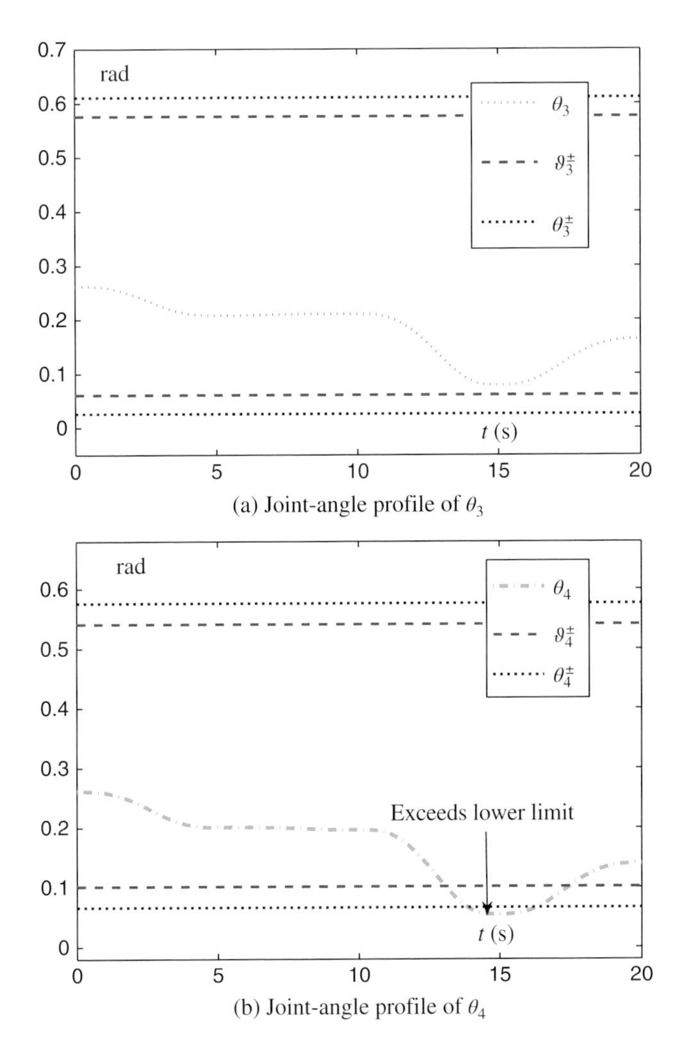

Figure 7.9 Joint-angle profiles of (a) θ_3 and (b) θ_4 synthesized by PBMJM scheme (7.27) when a robot end-effector tracks a larger "M"-shaped path.

joint angle exceeds its joint physical limits. Such unexpected situation would result in the failure of the end-effector task execution. What is more, in some cases, the robot manipulator would be damaged if there are no hardware safety devices in the robot. For comparison and illustration, Figures 7.11 through 7.13 show the joint-angle profiles of the planar six-DOF planar robot manipulator synthesized by FAMJM schemes (7.23) through (7.25). From Figures 7.11 through 7.13, we can see that all the joint angles are within their margins-considered joint-angle limits $\vartheta_i^{\pm}$ (with $i = 1, 2, \cdots, 6$), and never cross the "safety regions". Besides, joint angles θ_3, θ_4, and θ_5 are constrained by their margins-considered lower limits ϑ_i^{-} with $i = 3, 4, 5$, but never exceed them. These compared results illustrate well the superiority of the presented FAMJM scheme for handling the redundancy resolution problem.

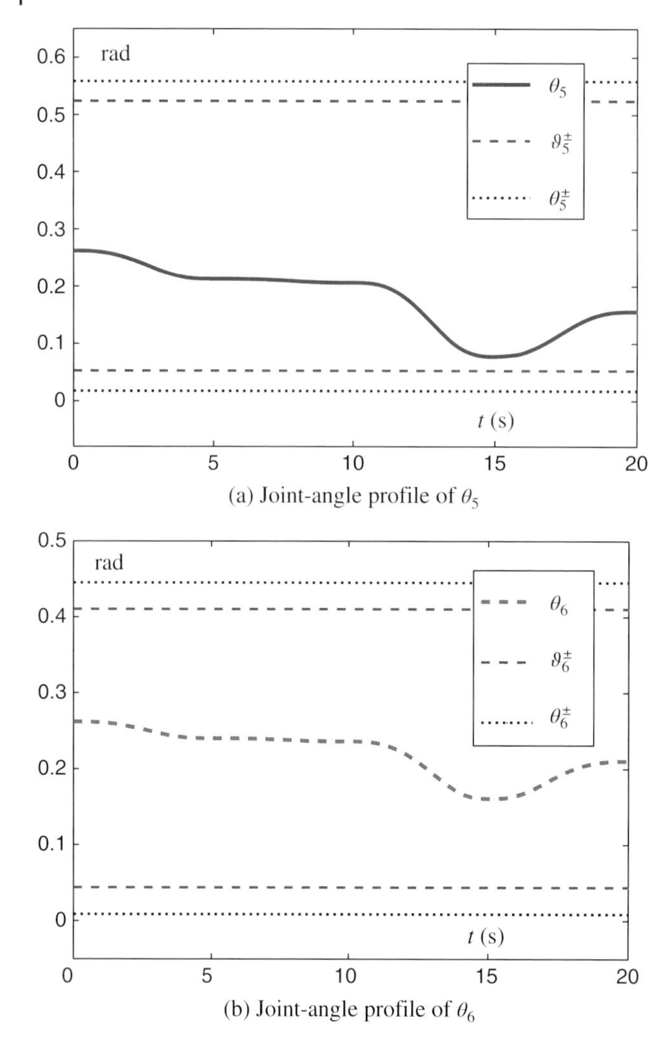

Figure 7.10 Joint-angle profiles of (a) θ_5 and (b) θ_6 synthesized by PBMJM scheme (7.27) when a robot end-effector tracks a larger "M"-shaped path.

7.3.3.2 Comparison with Tracking Task of Larger "P"-Shaped Path

In this example, the task duration of tracking a larger "M"-shaped path is 15 s (i.e., $T = 5$ s), and the length of "M" is 18.5 cm. Similar to the previous example, the joint-angle profiles synthesized by PBMJM scheme (7.27) and FAMJM scheme (7.23) through (7.25) are shown in Figures 7.14 through 7.19, respectively. Since the PBMJM scheme does not consider the joint physical constraints, joint angle θ_4 exceeds its lower limit θ_4^- at around $t = 5$ s. In contrast, joint angles θ_3, θ_4, and θ_5 are constrained, respectively, by the margin-considered joint-angle limits ϑ_3^-, ϑ_4^- and ϑ_5^-, because the

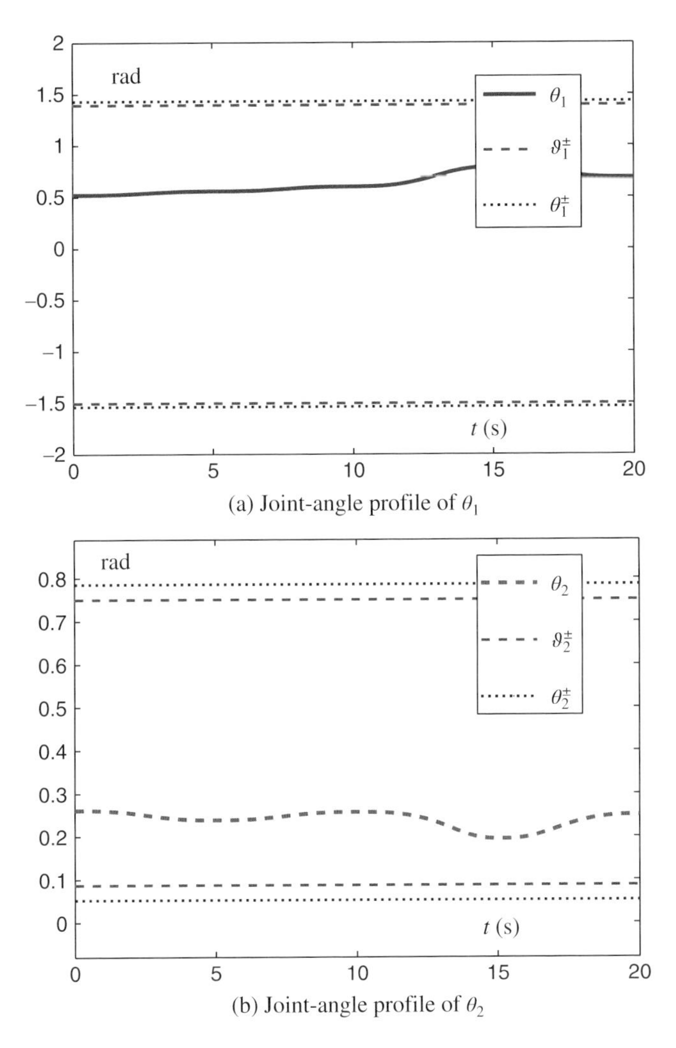

Figure 7.11 Joint-angle profiles of (a) θ_1 and (b) θ_2 synthesized by FAMJM scheme (7.23)–(7.25) when a robot end-effector tracks a larger "M"-shaped path.

joint physical limits and their margins are considered in the FAMJM scheme. Evidently, the presented FAMJM scheme with the joint physical limits [i.e., (7.23) through (7.25)] has a powerful advantage over the PBMJM scheme (7.27), and it is preferred and more applicable in engineering practice.

In summary, the presented two examples, that is, following the "M"- and "P"-shaped paths, have both illustrated the efficacy of the presented FAMJM scheme and the corresponding numerical algorithm 94LVI on the redundancy resolution of robot manipulators. Furthermore, the error analysis has validated well the accuracy of such a QP-based inverse kinematic solving method.

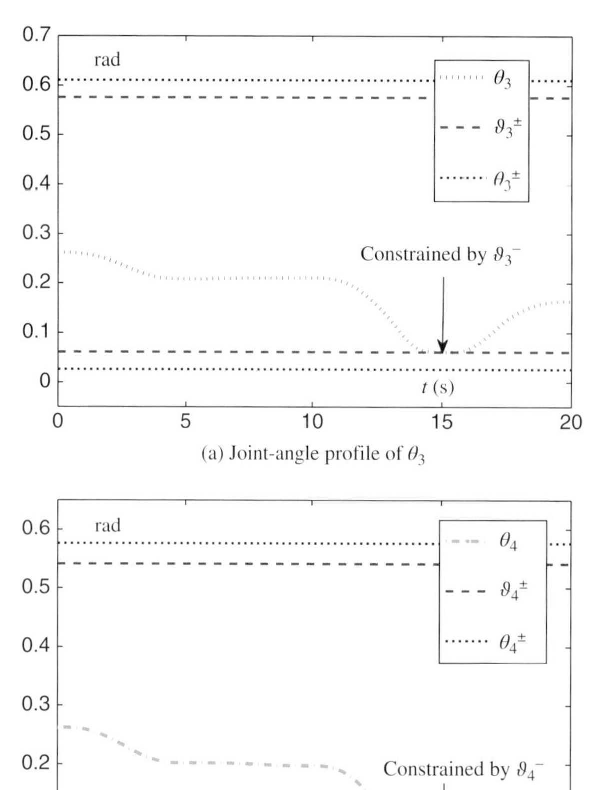

Figure 7.12 Joint-angle profiles of (a) θ_3 and (b) θ_4 synthesized by FAMJM scheme (7.23)–(7.25) when a robot end-effector tracks a larger "M"-shaped path.

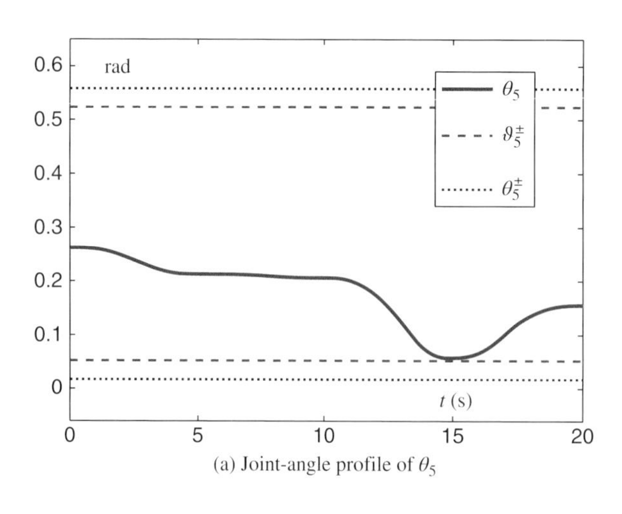

Figure 7.13 Joint-angle profiles of (a) θ_5 and (b) θ_6 synthesized by FAMJM scheme (7.23)–(7.25) when a robot end-effector tracks a larger "M"-shaped path.

Figure 7.13 (*Continued*)

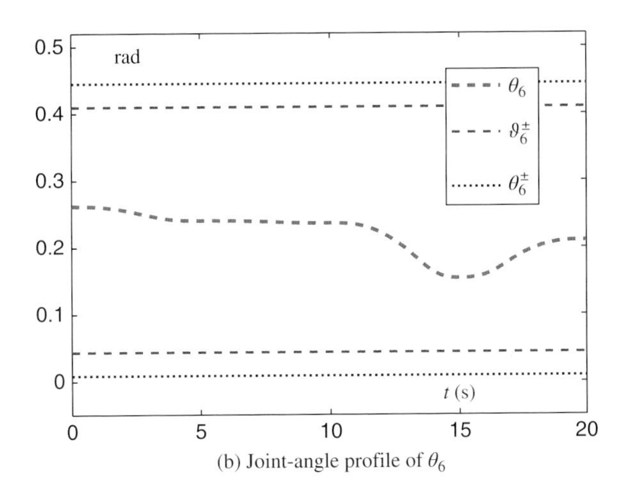

(b) Joint-angle profile of θ_6

Figure 7.14 Joint-angle profiles of (a) θ_1 and (b) θ_2 synthesized by PBMJM scheme (7.27) when robot end-effector tracks a larger "P"-shaped path.

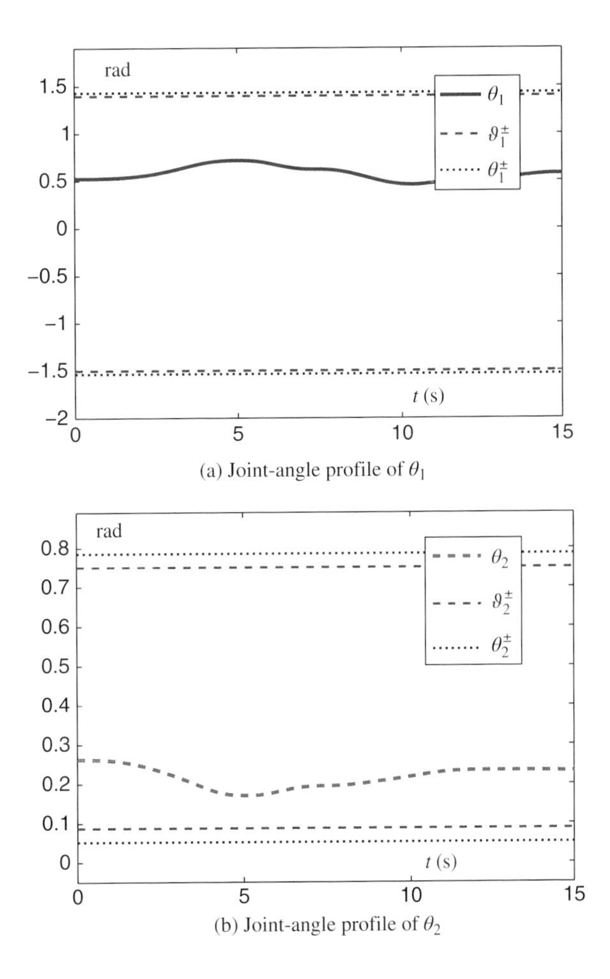

(a) Joint-angle profile of θ_1

(b) Joint-angle profile of θ_2

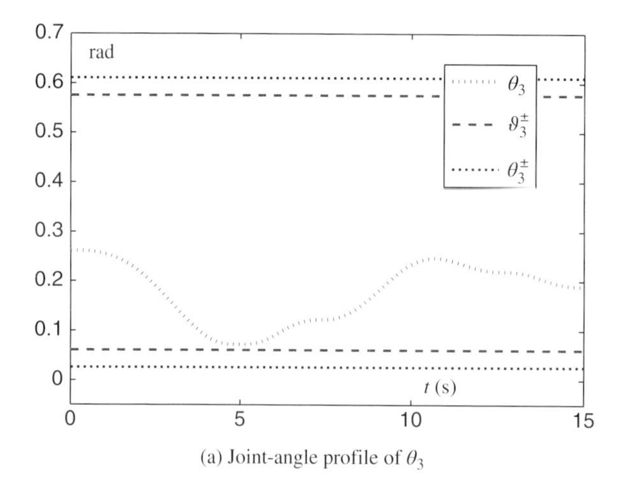

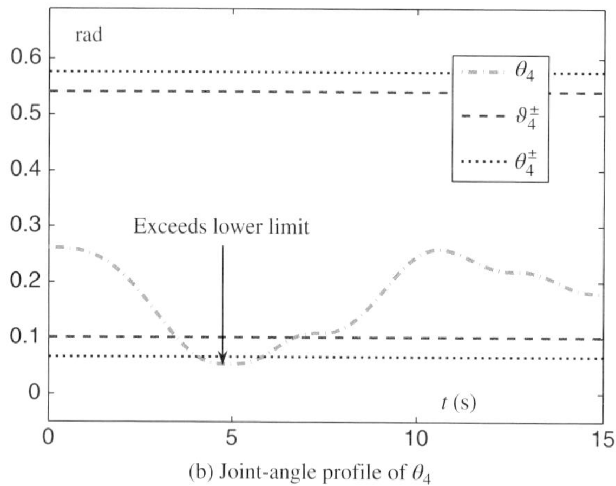

Figure 7.15 Joint-angle profiles of (a) θ_3 and (b) θ_4 synthesized by PBMJM scheme (7.27) when a robot end-effector tracks a larger "P"-shaped path.

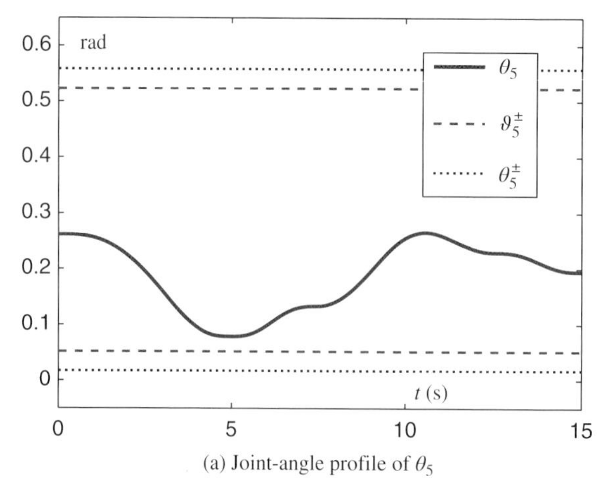

Figure 7.16 Joint-angle profiles of (a) θ_5 and (b) θ_6 synthesized by PBMJM scheme (7.27) when a robot end-effector tracks a larger "P"-shaped path.

Figure 7.16 (*Continued*)

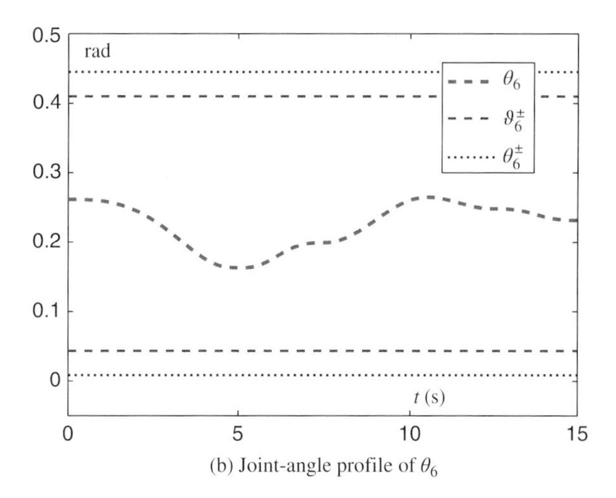

(b) Joint-angle profile of θ_6

Figure 7.17 Joint-angle profiles of (a) θ_1 and (b) θ_2 synthesized by FAMJM scheme (7.23)–(7.25) when a robot end-effector tracks a larger "P"-shaped path.

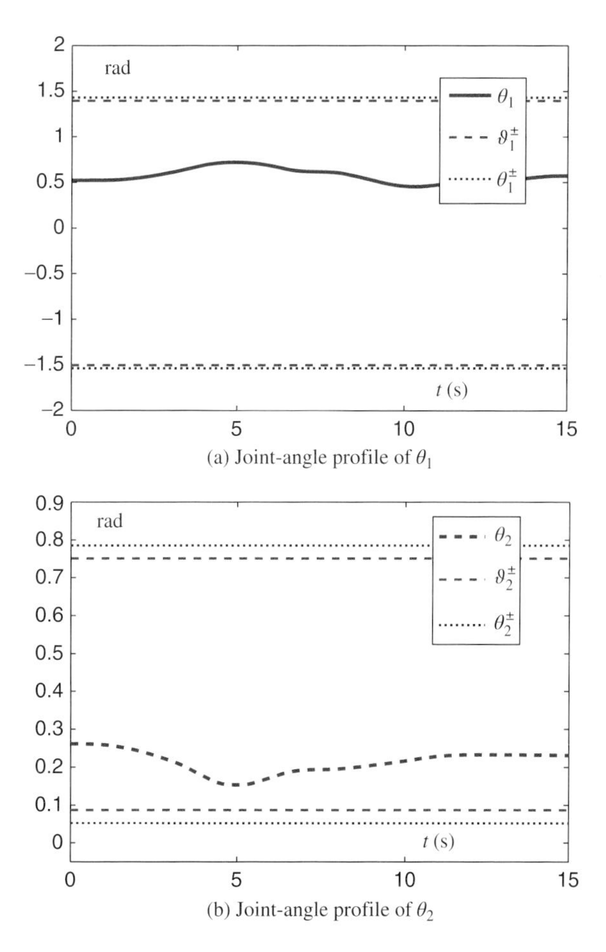

(b) Joint-angle profile of θ_2

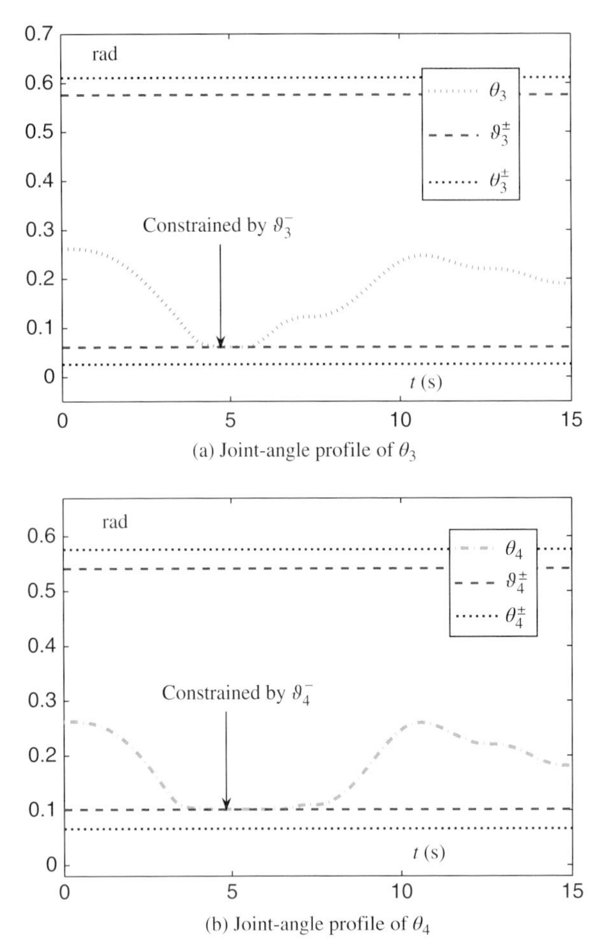

Figure 7.18 Joint-angle profiles of (a) θ_3 and (b) θ_4 synthesized by FAMJM scheme (7.23)–(7.25) when a robot end-effector tracks a larger "P"-shaped path.

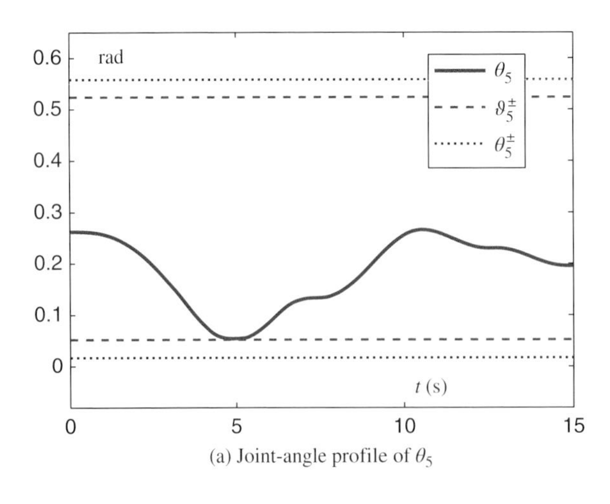

Figure 7.19 Joint-angle profiles of (a) θ_5 and (b) θ_6 synthesized by FAMJM scheme (7.23)–(7.25) when a robot end-effector tracks a larger "P"-shaped path.

Figure 7.19 (*Continued*)

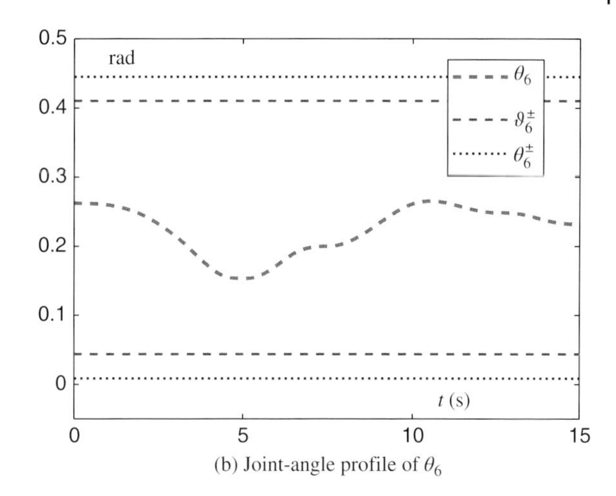

(b) Joint-angle profile of θ_6

7.4 Chapter Summary

For operating a practical planar six-DOF redundant robot manipulator, the redundancy resolution problem has been formulated as an FAMJM scheme in this chapter. Then, the presented FAMJM scheme has been presented and reformulated as a QP, solved by the efficient numerical algorithm 94LVI. The simulations and experiments of tracking the letter "M"- and "P"-shaped paths on the planar six-DOF robot manipulator have illustrated well the efficacy of such an FAMJM scheme and numerical algorithm 94LVI. Furthermore, the end-effector position-error analysis results have validated well the accuracy of the presented redundancy resolution method.

8

QP Based Manipulator State Adjustment

8.1 Introduction

A manipulator is said to be redundant when more DOF are available than necessary for a given end-effector path-tracking task. This implies that redundancy can be established simply with respect to some particular tasks [62]. For non-redundant manipulators, the joint motion is uniquely determined by a prescribed end-effector primary task, and thus there are no redundant freedoms left for executing some secondary tasks, such as handling joint physical limits, environmental constraints, and configuration singularities. In contrast, redundant manipulators have wider operational space and meet more functional constraints, because an infinite number of feasible joint configurations can be available. They have thus been widely applied, for example, in cleanup and remediation of nuclear and hazardous materials, and in space or sea exploration [62, 77, 78, 94, 100, 157–165]. Therefore, a lot of research into redundant manipulators has been carried out, and much attention has been paid by the research community to motion planning and control of robot manipulators [15, 62, 76–78, 94, 100, 150, 157–167].

A redundancy-resolution scheme is a method or an algorithm that selects one joint-space solution from an infinite number of possible solutions, given the end-effector primary task of following a desired workspace trajectory. This kind of selection is usually used to accomplish some secondary tasks for robot manipulators or generate an optimal solution (to achieve some performance criteria). The redundancy-resolution scheme often makes use of optimization techniques, especially QP. More specifically, such QP are equivalent to systems of linear variational inequality (LVI), and the LVI can then be solved by many algorithms, methods, and techniques efficiently, such as numerical methods [136, 168] and recurrent neural networks [15, 76, 77, 150, 161, 169].

One of the important issues in controlling actual robot manipulators is the state-adjustment problem, that is, from one state to another. State adjustment appears in many situations. For example, to initialize the execution of a new end-effector task, a robot manipulator has to move from an arbitrary state to the desired starting state of a new task. In addition, in robot motion planning, different end-effector tasks possibly need different starting joint states [15, 76–78, 100, 150, 158–162, 167], thus it is necessary to adjust the state of the robot manipulator from the final state of the previous task to the desired starting state of the next task. Furthermore, in repetitive motion planning [163, 170], the task of multiple cycles needs to start from the same initial states;

Robot Manipulator Redundancy Resolution, First Edition. Yunong Zhang and Long Jin.

however, due to the existence of non-repetitive phenomena, the consistency of the initial and final states usually can not be guaranteed. In order to complete the repetitive motion task, the states should be adjusted at the end of every cycle. To the best of the authors' knowledge, the QP-based method of state adjustment has not been investigated in present literature, which theoretically guarantees that the resultant movement is within its physical limits. Besides, the state adjustment of the robot manipulator can be done manually (i.e., adjusting the joints one by one), but it is evidently much less effective and more time-consuming than an automatic state-adjustment scheme. Simply put, the manual scheme is not suitable for highly automated operations and manufacturing (unless it is necessary).

To make the robot manipulator execute tasks more readily and efficiently, a scheme for state adjustment of the redundant robot manipulator (termed a state-adjustment scheme) is presented with no end-effector task explicitly assigned, and then unified into a QP in this chapter. Based on the conversion technique of QP to LVI, a numerical algorithm is finally developed and employed for solving such a QP as well as the original robot scheme.

8.2 Preliminaries and Scheme Formulation

It is shown in [15, 76–78, 100, 150, 158–162, 167, 171] that, in the robot motion planning, different tasks possibly need to start from different initial joint angles, and thus it is necessary to adjust the state of the manipulator from one configuration to another before performing the next task. In addition to this case, the repetitive motion planning [163] often needs state adjustment due to the existence of the non-repetitive motion problem. According to previous research on QP-based redundant-resolution and planning [15, 169], a state-adjustment scheme is presented, which minimizes the joint displacement between the current and the desired state. The state-adjustment scheme is thus formulated as follows:

$$\text{minimize} \quad \frac{1}{2}\left((\dot{\theta} + \mathbf{z})^{\mathrm{T}}(\dot{\theta} + \mathbf{z})\right) \quad \text{with } \mathbf{z} = \gamma(\theta - \theta_{\mathrm{d}}) \tag{8.1}$$

$$\text{subject to} \quad \theta^- \leqslant \theta \leqslant \theta^+, \tag{8.2}$$

$$\dot{\theta}^- \leqslant \dot{\theta} \leqslant \dot{\theta}^+, \tag{8.3}$$

where $\theta \in \mathbb{R}^n$ denotes the joint-space vector of the manipulator, $\dot{\theta} \in \mathbb{R}^n$ denotes the joint velocity vector defined as the time-derivative of θ, and $\theta_{\mathrm{d}} \in \mathbb{R}^n$ denotes the desired state (or target state). Design parameter $\gamma > 0$ is used to scale the convergence rate of the state-adjustment scheme, and superscript ^{T} denotes the transpose of a matrix/vector. In addition, $\theta^{\pm}$ and $\dot{\theta}^{\pm}$ denote the upper and lower limits of the joint-angle vector and joint-velocity vector, respectively.

The minimization of the performance index $(\dot{\theta} + \mathbf{z})^{\mathrm{T}}(\dot{\theta} + \mathbf{z})/2$ (i.e., (8.1)) of the state-adjustment scheme with a feature of global exponential convergence can be described in the following theorem.

Theorem 8.1 Without considering the physical joint limits, starting from any initial state $\theta(0)$, the state vector $\theta(t)$ of the minimization (8.1) converges to the desired state θ_{d}. Moreover, the exponential convergence can be achieved for the minimization

(8.1) with convergence rate γ. In addition, the joint velocity vector $\dot{\theta}(t)$ globally exponentially converges to 0 with convergence rate γ, and the initial joint velocity $\dot{\theta}(0)$ is $-\gamma(\theta(0)-\theta_{\mathrm{d}})$.

Proof: By following the ZD method [170, 172] presented in Chapter 1, the proof is given as the following steps.

Step 1. We define the following vector-valued error function $\mathbf{e}(t) \in \mathbb{R}^n$; in mathematics,

$$\mathbf{e}(t) = \theta(t) - \theta_{\mathrm{d}}. \tag{8.4}$$

Step 2. To make the error-function (8.4) converge to zero, the time derivative $\dot{\mathbf{e}}(t)$ of $\mathbf{e}(t)$ can be

$$\dot{\mathbf{e}}(t) = \frac{\mathrm{d}\mathbf{e}(t)}{\mathrm{d}t} = -\gamma \mathbf{e}(t). \tag{8.5}$$

Step 3. Since the time derivative of (8.4) is $\dot{\mathbf{e}}(t) = \dot{\theta}(t)$, substituting it into (8.5) yields

$$\dot{\theta}(t) = -\gamma \mathbf{e}(t) = -\gamma(\theta(t) - \theta_{\mathrm{d}}), \tag{8.6}$$

that is, $\dot{\theta}(t) + \gamma(\theta(t) - \theta_{\mathrm{d}}) = 0$, which is exactly the theoretical solution of the minimization (8.1). Besides, it follows from Equation (8.5) that $\mathbf{e}(t) = \mathbf{e}(0)\exp(-\gamma t)$, and

$$\theta(t) - \theta_{\mathrm{d}} = (\theta(0) - \theta_{\mathrm{d}})\exp(-\gamma t). \tag{8.7}$$

Then, we have the following θ_{d}-difference ratio $\delta(t)$ defined as

$$\delta(t) = \frac{\|\theta(t) - \theta_{\mathrm{d}}\|_2}{\|\theta(0) - \theta_{\mathrm{d}}\|_2} = \exp(-\gamma t), \tag{8.8}$$

where $\|\cdot\|_2$ denotes the two norm of a vector. That is, $\theta(t)$ converges globally exponentially to θ_{d} with convergence rate γ.

Moreover, differentiating Equation (8.7) with respect to time t, we have the following explicit expression about joint-velocity $\dot{\theta}$:

$$\dot{\theta}(t) = -\gamma(\theta(0) - \theta_{\mathrm{d}})\exp(-\gamma t). \tag{8.9}$$

It follows from Equation (8.9) that joint velocity $\dot{\theta}(t)$ globally exponentially converges to 0 with the convergence rate being γ as well, and the initial joint velocity $\dot{\theta}(0)$ is $-\gamma(\theta(0) - \theta_{\mathrm{d}})$. Note that, in practice, in view of potential requirements on robot motion planning and control (such as joint limits and joint-velocity limits), it is better to minimize $\|\dot{\theta}(t) + \gamma(\theta(t) - \theta_{\mathrm{d}})\|_2^2/2$ than forcing $\dot{\theta}(t) + \gamma(\theta(t) - \theta_{\mathrm{d}}) = 0$ directly, and minimizing $\|\dot{\theta}(t) + \gamma(\theta(t) - \theta_{\mathrm{d}})\|_2^2/2$ is just (8.1). The proof is now complete.

So, in view of Theorem 8.1, with joint limits $\theta^{\pm}$ and joint velocity limits $\dot{\theta}^{\pm}$ included, the state-adjustment scheme is formulated as (8.1)–(8.3). Furthermore, by using the constraint-combining technique [78, 100, 162, 163], joint physical limits (8.2) and (8.3) can be combined and handled via the following bound constraint:

$$\eta^- \leqslant \dot{\theta} \leqslant \eta^+, \tag{8.10}$$

where the ith elements of $\eta^{\pm}$ are defined as $\eta_i^+ = \min\{\dot{\theta}_i^+, \mu(\theta_i^+ - \theta_i)\}$ and $\eta_i^- = \max\{\dot{\theta}_i^-, \mu(\theta_i^- - \theta_i)\}$, $i = 1, 2, \cdots, n$, with design-parameter $\mu > 0$ used to scale the feasible region of $\dot{\theta}$ and set as 4 in the computer simulations, and the experiment performed on the six-DOF planar robot manipulator.

With decision variable vector $\mathbf{x} = \dot{\theta} \in \mathbb{R}^n$ [100, 162], the presented state-adjustment scheme (8.1)–(8.3) is now reformulated as the following QP:

$$\text{minimize} \quad \frac{1}{2}\mathbf{x}^{\mathrm{T}} W \mathbf{x} + \mathbf{z}^{\mathrm{T}}\mathbf{x} \tag{8.11}$$

$$\text{subject to} \quad \eta^{-} \leqslant \mathbf{x} \leqslant \eta^{+}, \tag{8.12}$$

where coefficients $W = I$ and $\mathbf{z} = \gamma(\theta - \theta_{\mathrm{d}})$. It is worth noting that, according to Theorem 8.1, the initial joint velocity $\dot{\theta}(0) = -\gamma(\theta(0) - \theta_{\mathrm{d}})$ could be very large (even reach the values of joint velocity limits sometimes), which is difficult or impossible in practice, or even causes mechanical damage to the robot. To prevent the occurrence of such a large initial joint velocity, we exploit the continuation technique by imposing a zero-initial-velocity (ZIV) constraint $\mathbf{x}^{-}(t) \leqslant \mathbf{x} \leqslant \mathbf{x}^{+}(t)$ (i.e., $\mathbf{x}^{-}(t) \leqslant \dot{\theta} \leqslant \mathbf{x}^{+}(t)$) with

$$\mathbf{x}^{-}(t) = \sin(\pi t/2T)\eta^{-}, \quad \mathbf{x}^{+}(t) = \sin(\pi t/2T)\eta^{+},$$

where $T > 0$ is the task duration, that is, time $t \in [0, T]$. So, this QP problem (8.11)–(8.12) is finally modified as the following time-varying QP (with $\mathbf{z}(t)$, $\mathbf{x}^{\pm}(t)$ and $\mathbf{x}(t)$ all time-varying):

$$\text{minimize} \quad \frac{1}{2}\mathbf{x}^{\mathrm{T}}(t) W \mathbf{x}(t) + \mathbf{z}^{\mathrm{T}}(t)\mathbf{x}(t) \tag{8.13}$$

$$\text{subject to} \quad \mathbf{x}^{-}(t) \leqslant \mathbf{x}(t) \leqslant \mathbf{x}^{+}(t). \tag{8.14}$$

Note that, by imposing the zero-initial-velocity constraint depicted in (8.14), the zero initial velocity of the motion is guaranteed and is now acceptable in practical applications.

8.3 QP Solution and Control of Robot Manipulator

In this section, numerical algorithm E47 presented in Chapter 5 is employed to solve the corresponding bound-constraint QP problem (8.13)–(8.14). In addition, the motors of the robot manipulator are driven by the pulse signals, and thus the sampling period is set to be $\Delta t = 0.01$ s for the solution of the QP and the generation of the pulse signals, according to the authors' experimental experience and the precision requirement.

As mentioned in the previous chapters, the motors of the robot joints are driven by the pulse signals transmitted from the host computer, and the resultant joint variables (i.e., joint-angle θ and joint-velocity $\dot{\theta}$) should thus be converted into pulses per second (PPS) to control the manipulator. For the first joint driven by a servomotor, the number of PPS is

$$\mathrm{PPS}_1 = \gamma\dot{\theta}_1/(2\pi), \tag{8.15}$$

where $\gamma = 3.2 \times 10^5$ is the parameter related to the six-DOF planar robot manipulator. For the second joint through the sixth joint driven by stepping motors, following the law of cosines, we have the rotation rate of the $(i+1)$th motor as

$$v_{i+1} = \frac{a_{i+1}b_{i+1}\dot{\theta}_{i+1}\cos\theta_{i+1}}{s_{i+1}\sqrt{a_{i+1}^2 + b_{i+1}^2 + 2a_{i+1}b_{i+1}\sin\theta_{i+1}}},$$

with $i = 1, 2, \cdots, 5$. As the stepping angles of the second joint through the sixth joint are all 0.01π rad and the subdividing multiples are 32, the number of pulses per second for the $(i+1)$th joint $(i = 1, 2, \cdots, 5)$ is

$$\mathrm{PPS}_{i+1} = \left(2\pi/(0.01\pi/32)\right)v_{i+1} = \frac{6400a_{i+1}b_{i+1}\dot{\theta}_{i+1}\cos\theta_{i+1}}{s_{i+1}\sqrt{a_{i+1}^2 + b_{i+1}^2 + 2a_{i+1}b_{i+1}\sin\theta_{i+1}}} \quad (8.16)$$

By exploiting Equations (8.15) and (8.16), we can obtain the PPS signals for controlling the robot manipulator.

8.4 Computer Simulations and Comparisons

In this section, the presented state-adjustment scheme (8.1)–(8.3) (i.e., correspondingly the QP (8.11)–(8.12) or (8.13)–(8.14)) and numerical algorithm E47 are simulated for the state adjustment of the presented six-DOF planar robot manipulator. Specifically, in the simulations, without loss of generality, the task of the six-DOF planar robot manipulator is to adjust its configuration from the initial state $\theta(0) = [1.2219, 0.3491, 0.5062, 0.0611, 0.0349, 0.0349]^{\mathrm{T}}$ to the desired state $\theta_{\mathrm{d}} = [0.3491, 0.3491, \pi/12, \pi/12, \pi/12, \pi/12]^{\mathrm{T}}$ in radians (both of which can be set as any values within the joint limits). In addition, the simulation studies are divided into two subsections. The first one discusses the convergence property of the state-adjustment scheme (8.11)–(8.12), which does not impose the zero-initial-velocity constraint. On the other hand, the efficacy, flexibility and accuracy of the presented state-adjustment scheme (8.13)–(8.14) are investigated in the second subsection in detail, which imposes the zero-initial-velocity constraint for velocity-continuation purposes.

8.4.1 State Adjustment without ZIV Constraint

In this subsection, the state-adjustment scheme (8.1)–(8.3) without imposing the zero-initial-velocity constraint, that is, correspondingly the QP (8.11)–(8.12), is simulated based on the six-DOF planar robot manipulator. The simulation results are shown in Figure 8.1, Table 8.1, and Figure 8.2.

Figure 8.1 illustrates the transient behaviors of joint-angle and joint-velocity variables of the six-DOF planar robot manipulator in the situation of $\gamma = 1$ and $T = 30$ s. As seen from the figure, the values of the joint angles and joint velocities exponentially converge to their desired values. In addition, Table 8.1 shows the joint-angle values at different time instants (i.e., $t = 0.5, 1, 2, 3, 6, 10, 15$ s). With respect to Table 8.1, calculating by (8.8), we obtain the quantified convergence results that the θ_{d}-difference ratio $\delta(0.5)$ (i.e., at time instant $t = 0.5$ s) is about 60.50% (which approximates $\exp(-0.5)$), $\delta(1)$ is about 36.60% (which approximates $\exp(-1)$), $\delta(2)$ is about 13.40% (which approximates $\exp(-2)$), $\delta(3)$ is about 4.90% (which approximates $\exp(-3)$), $\delta(6)$ is about 0.24% (which approximates $\exp(-6)$), and $\delta(10)$ is about 0.00%, which implies that $\theta(t)$ actually achieves the desired state θ_{d} from a practical viewpoint. These confirm the exponential convergence property of (8.1) analyzed in Theorem 8.1. In addition, the initial joint-velocity $\dot{\theta}(0) = [-0.8728, 0.0000, -0.2444, 0.2007, 0.2269, 0.2269]^{\mathrm{T}}$ rad/s is the maximal joint-velocity during the task, which can be seen from the second subfigure of Figure 8.1.

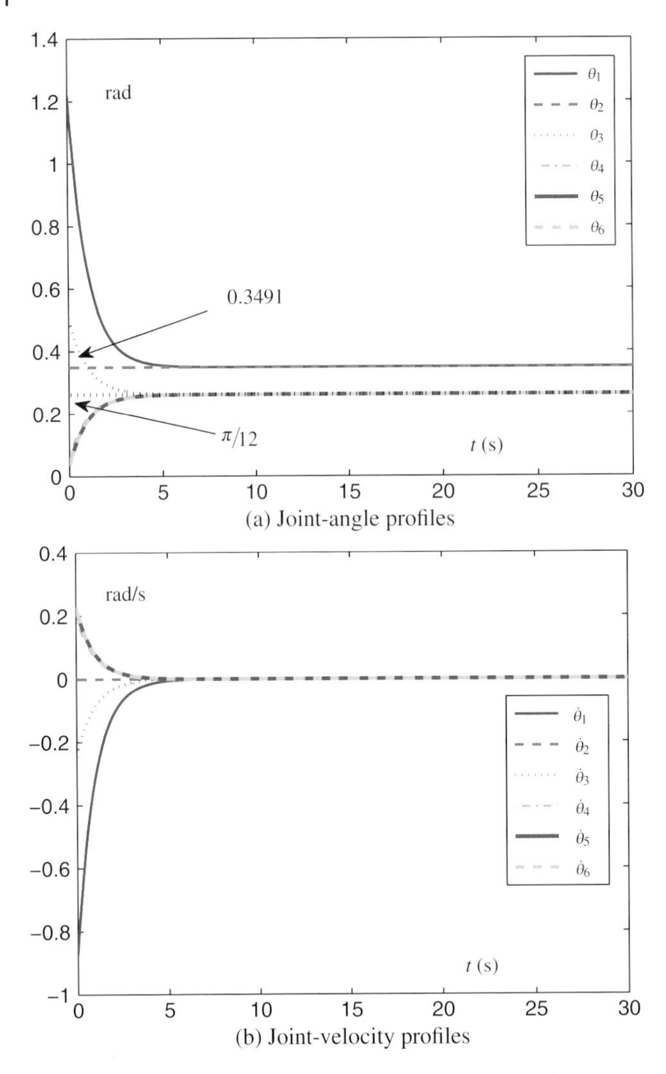

Figure 8.1 (a) Joint-angle and (b) joint-velocity profiles of a six-DOF planar robot manipulator synthesized by state-adjustment scheme (8.1)–(8.3) without imposing a zero-initial-velocity constraint and with parameters $\gamma = 1$ and $T = 30$ s.

Table 8.1 Joint-angle values with $\gamma = 1$ and $T = 30$ s at different time t instants.

t (s)	θ_1 (rad)	θ_2 (rad)	θ_3 (rad)	θ_4 (rad)	θ_5 (rad)	θ_6 (rad)
0.5	0.87712	0.34910	0.40966	0.14037	0.12453	0.12453
1	0.66856	0.34910	0.35126	0.18833	0.17875	0.17875
2	0.46603	0.34910	0.29454	0.23491	0.23140	0.23140
3	0.39190	0.34910	0.27379	0.25196	0.25067	0.25067
6	0.35120	0.34910	0.26240	0.26132	0.26125	0.26125
10	0.34914	0.34910	0.26181	0.26180	0.26180	0.26180
15	0.34910	0.34910	0.26180	0.26180	0.26180	0.26180

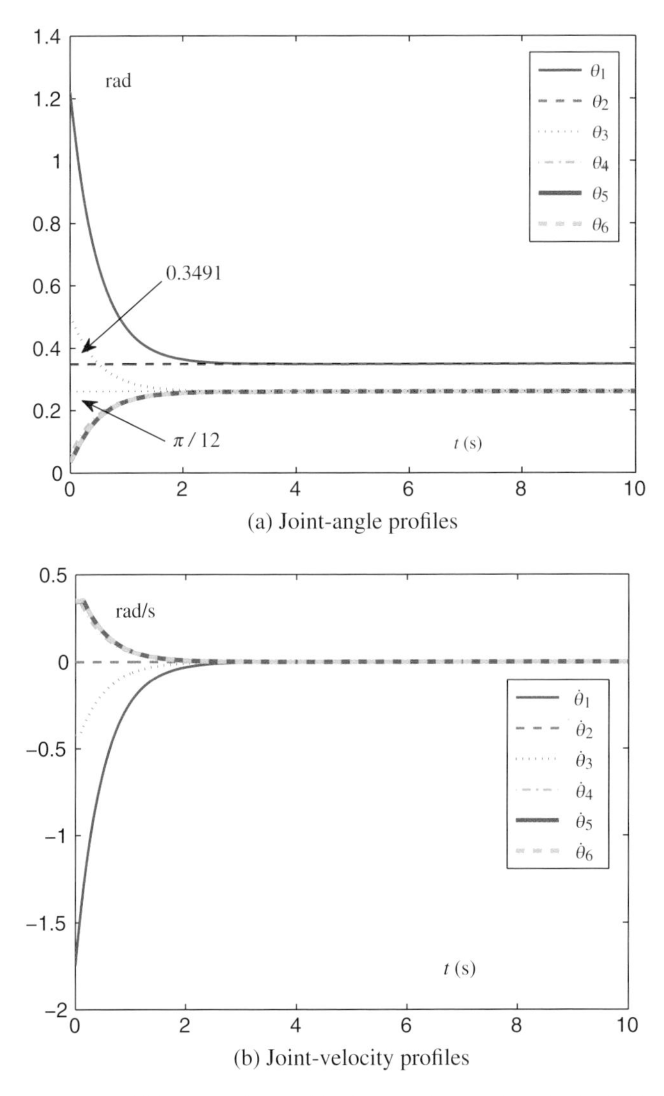

Figure 8.2 (a) Joint-angle and (b) joint-velocity profiles of a six-DOF planar robot manipulator synthesized by state-adjustment scheme (8.1)–(8.3) without imposing a zero-initial-velocity constraint and with parameters $\gamma = 2$ and $T = 10$ s.

For further investigation, a different value of γ (i.e., $\gamma = 2$) is set, and the simulation results are shown in Figure 8.2. As seen from the figure, as the value of γ increases from 1 to 2, the initial velocity $\dot{\theta}(0)$ increases accordingly (as compared to the second subfigure of Figure 8.1), which verifies well $\dot{\theta}(0) = -\gamma(\theta(0) - \theta_{\mathrm{d}})$ (i.e., (8.6) with $t = 0$ s). Note that, as shown in the second subfigure of Figure 8.2, some of the initial joint-velocities, that is, $\dot{\theta}_4(0)$, $\dot{\theta}_5(0)$ and $\dot{\theta}_6(0)$, meet the manipulator's physical limits, that is, $\dot{\theta}_4^+(0)$, $\dot{\theta}_5^+(0)$ and $\dot{\theta}_6^+(0)$, respectively. This is because $\dot{\theta}_4(0) = 0.4014$ rad/s, $\dot{\theta}_5(0) = 0.4538$ rad/s and $\dot{\theta}_6(0) = 0.4538$ rad/s as computed by (8.6) are larger than the joint-velocity limits $\dot{\theta}_4^+(0) = 0.3470$ rad/s, $\dot{\theta}_5^+(0) = 0.3450$ rad/s and $\dot{\theta}_6^+(0) = 0.3487$ rad/s, respectively. These simulations have also substantiated the state-adjustment scheme's avoidance feature of joint physical limits.

8.4.2 State Adjustment with ZIV Constraint

As seen from these simulation results, the initial joint-velocities are nonzero, which may be less desirable for practical applications. As presented in Section 8.2, the zero-initial-velocity constraint can be imposed to make the initial joint-velocity be zero (which is more acceptable in practice). In this subsection, we simulate the state-adjustment scheme (8.1)–(8.3) with the zero-initial-velocity constraint imposed, that is, QP (8.13)–(8.14), for the state adjustment of the six-DOF planar robot manipulator. The computer simulation results are illustrated in Figure 8.3 and Figure 8.4, Table 8.2 (where the final error of the ith joint $|e_i(T)| = |\theta_i(T) - \theta_{\mathrm{d}i}|(i = 1, 2, \cdots, 6)$ with $|\cdot|$ denoting the absolute value of a scalar), Figure 8.5 and Figure 8.6.

As synthesized by the state-adjustment scheme (8.1)–(8.3) with the zero-initial-velocity constraint imposed, Figure 8.3 shows the simulated motion process of the

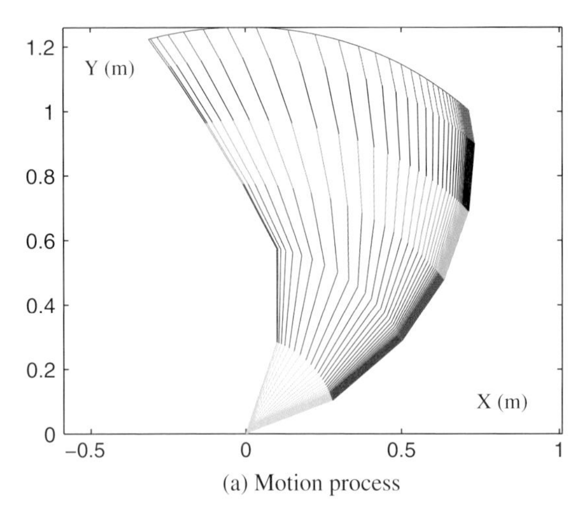

(a) Motion process

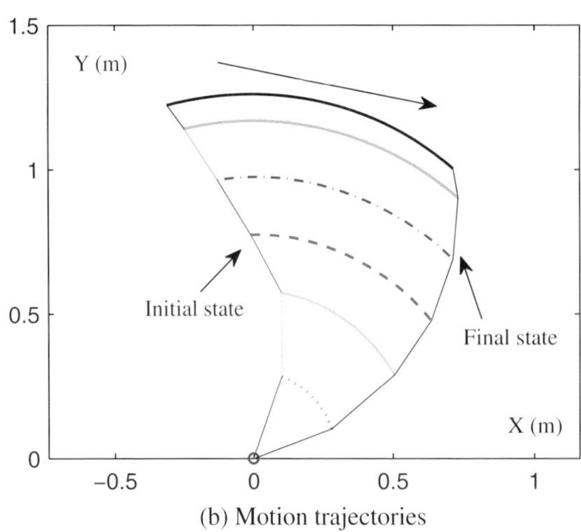

(b) Motion trajectories

Figure 8.3 (a) Motion process and (b) trajectories of a six-DOF planar robot manipulator synthesized by state-adjustment scheme (8.1)–(8.3) with a zero-initial-velocity constraint imposed (i.e., QP (8.13)–(8.14)) and with $\gamma = 1$ and $T = 10$ s.

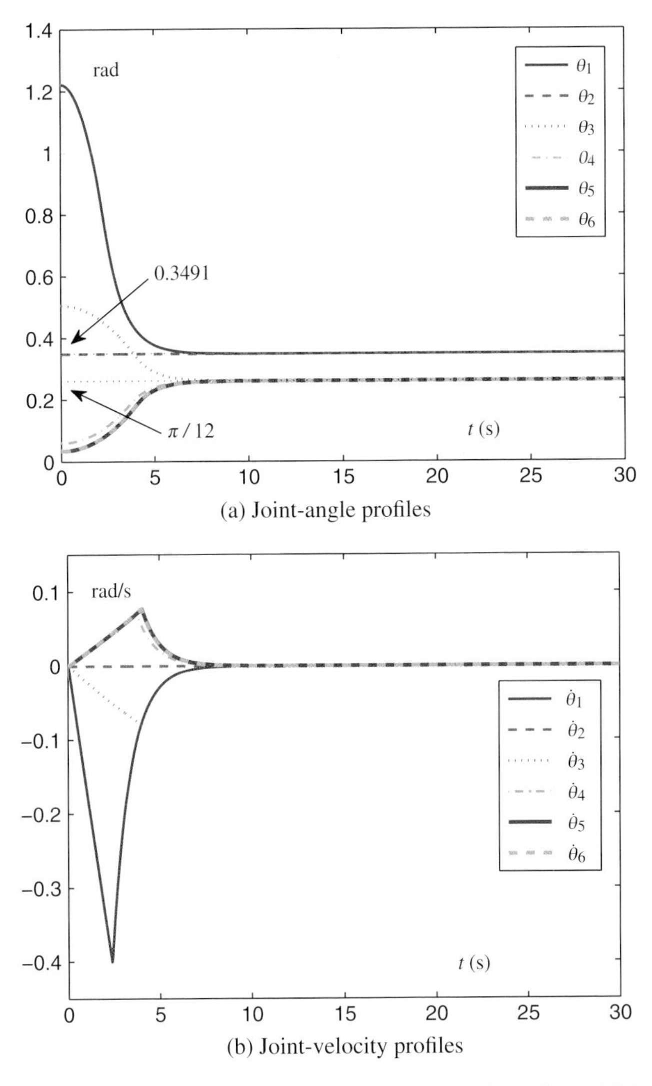

Figure 8.4 (a) Joint-angle and (b) joint-velocity profiles of a six-DOF planar robot manipulator synthesized by state-adjustment scheme (8.1)–(8.3) with zero-initial-velocity constraint imposed (i.e., QP (8.13)–(8.14)) and with $\gamma = 1$ and $T = 30$ s.

Table 8.2 Joint-angle errors $\{|e_i(T)|\}_{i=1}^{6}$ of state-adjustment task for different γ and T.

Test #	Joint 1	Joint 2	Joint 3	Joint 4	Joint 5	Joint 6
Situation 1 (10^{-7} rad)	3.32930	3.32929	2.45206	2.45206	2.45206	2.45206
Situation 2 (10^{-5} rad)	7.40662	0.03329	3.84190	2.93454	3.76557	3.72127
Situation 3 (10^{-7} rad)	1.77145	1.66465	1.34676	1.10760	1.05763	1.06175

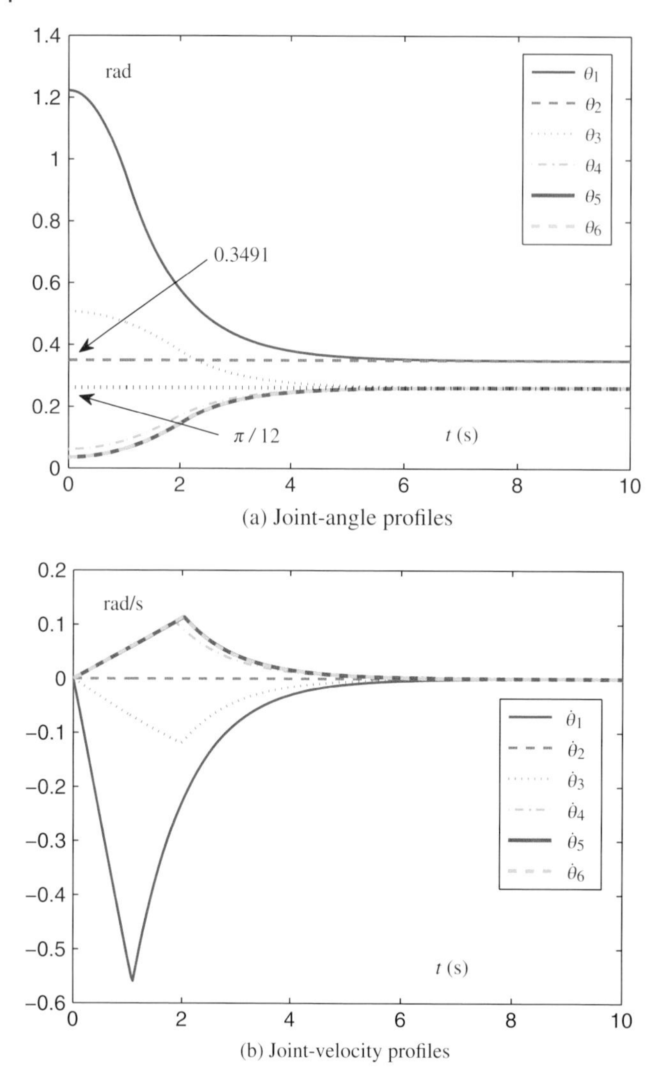

Figure 8.5 (a) Joint-angle and (b) joint-velocity profiles of a six-DOF planar robot manipulator synthesized by state-adjustment scheme (8.1)–(8.3) with a zero-initial-velocity constraint imposed (i.e., QP (8.13)–(8.14)) and with $\gamma = 1$ and $T = 10$ s.

six-DOF planar robot manipulator when it executes a state-adjustment task. The arrow appearing in the figure shows the motion direction. More specifically, the first subfigure of Figure 8.3 reflects the motion process and the change of joint-velocity (i.e., from slowness to fastness and then to slowness), while the second subfigure of Figure 8.3 shows the motion trajectories of joints during the task execution. From Figure 8.3, we can see that the motion velocities are conformable to the zero-initial-velocity constraint (8.14), and that the joint trajectories are smooth, which are suitable for practical applications. For further investigation, comparison, and illustration, we simulate the state-adjustment scheme (8.1)–(8.3) in the following different situations (i.e., with different values of T and γ).

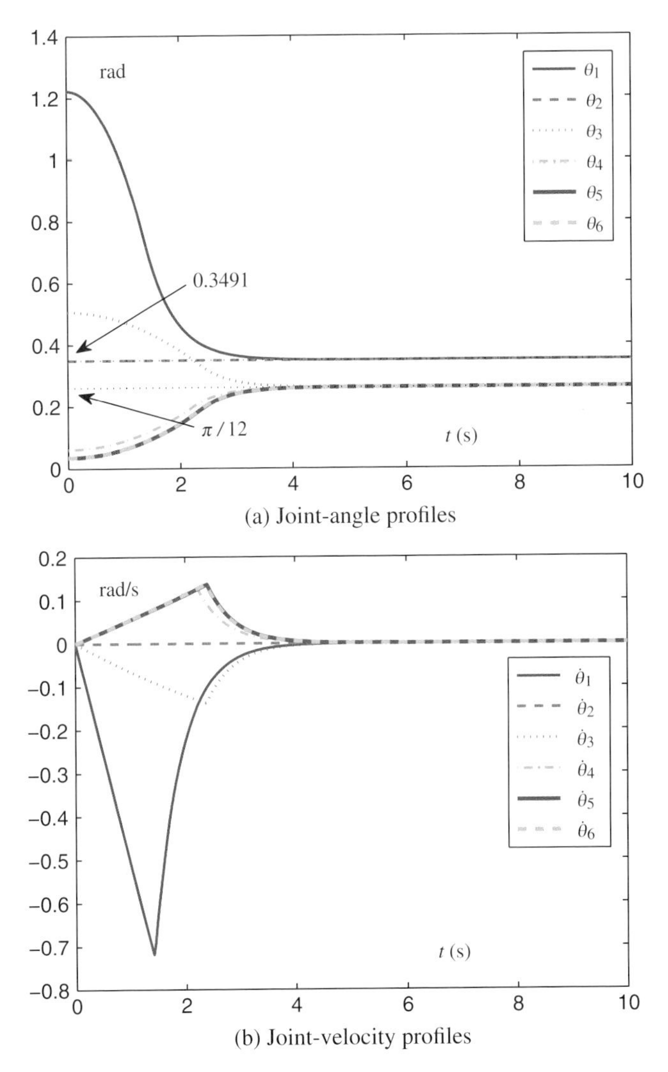

Figure 8.6 (a) Joint-angle and (b) joint-velocity profiles of a six-DOF planar robot manipulator synthesized by state-adjustment scheme (8.1)–(8.3) with a zero-initial-velocity constraint imposed (i.e., QP (8.13)–(8.14)) and with $\gamma = 2$ and $T = 10$ s.

Situation 1. $\gamma = 1$ **and** $T = 30$ **s**

Figure 8.4 illustrates the transient behaviors of joint-angle and joint-velocity variables with $\gamma = 1$ and $T = 30$ s. From the first subfigure of Figure 8.4, we see that all joints rapidly converge to their desired state (i.e., θ_{d}). Specifically speaking, $\theta_1(t)$ converges to 0.3491 rad, and $\theta_3(t)$ through $\theta_6(t)$ all converge to $\pi/12$ rad. Note that $\theta_2(t)$ keeps unchanged during the task execution because of $\theta_2(0) = \theta_{\mathrm{d}2}$. Correspondingly, as seen from the second subfigure of Figure 8.4, the joint-velocity starts evidently from zero and finally approaches zero within 10 s. It is worth mentioning that, as shown in Table 8.2, the maximal joint-angle error in this situation is 3.32930×10^{-7} rad, which is very close to zero when the execution time t arrives at 30 s. These illustrate the efficacy of the presented state-adjustment scheme QP (8.13)–(8.14) on robots' motion control.

Situation 2. $\gamma = 1$ **and** $T = 10$ **s**

As illustrated in the previous situation, the state-adjustment task can be completed in 10 s. So, 30 s may far too "time-consuming" for the state-adjustment task. Thus, we decrease the task duration T from 30 s to 10 s to further investigate the presented state-adjustment scheme (8.13)–(8.14). Figure 8.5 illustrates the profiles of joint variables synthesized by (8.13)–(8.14) with $\gamma = 1$ and $T = 10$ s. From the first subfigure of Figure 8.5, we see that the joint-angle variables converge to the desired values within 6.5 s, shorter than 10 s shown in Figure 8.4. Correspondingly, as seen from the second subfigure of Figure 8.5, the joint-velocity $\dot{\theta}(t)$ starts from zero, then becomes larger to meet the short time requirement, and finally returns to zero, where the maximal joint-velocity magnitude is about 0.58 rad/s. These substantiate that the presented scheme is flexible for the state-adjustment task, especially under the short time requirement (i.e., $T = 10$ s).

Situation 3. $\gamma = 2$ **and** $T = 10$ s

As seen from Table 8.2, the maximal joint-angle error of Situation 2 is 7.40662×10^{-5} rad, which is larger than that of Situation 1. In other words, if we simply decrease the task duration T to satisfy a shorter execution time requirement, the joint-angle error may increase a little bit. In order to improve the precision while using the short task-execution time, we can increase the value of design parameter γ readily. Figure 8.6 shows the profiles of joint variables synthesized by the presented state-adjustment scheme (8.13)–(8.14) with $\gamma = 2$ and $T = 10$ s. From the first subfigure of Figure 8.6, we see that the joint-angle variables converge to the desired values within 5.0 s, which is less than 6.5 s shown in Figure 8.5. In addition, as seen from the second subfigure of Figure 8.6, starting from zero, the joint-velocity variables are larger in magnitude than those in Figure 8.5 to meet the short time requirement (i.e., $\dot{\theta}$ converges to zero in shorter time). Furthermore, as shown in Table 8.2, the presented state-adjustment scheme with $\gamma = 2$ generates a considerably smaller joint-angle error of the state-adjustment task; for example, the maximal joint-angle error is 1.77145×10^{-7} rad.

In summary, these simulation results (as well as the corresponding analysis) have shown the exponential convergence property of the minimization (8.1) and related joint variables. In addition, by imposing the zero-initial-velocity constraint, the initial joint-velocity of the motion can be guaranteed to be zero. Moreover, by choosing γ and T appropriately, higher precision and a shorter execution time are achieved. These have shown in general the efficacy, flexibility, and accuracy of the presented state-adjustment scheme (8.13)–(8.14) for the configuration control of redundant robot manipulators.

8.5 Physical Experiments

In this section, the experimental test of the presented state-adjustment scheme (8.13)–(8.14) and numerical algorithm E47 are performed on the actual six-DOF planar robot manipulator. The initial joint-angle vector is set the same as before; that is, $\theta(0) = [1.222, 0.349, 0.506, 0.061, 0.035, 0.035]^{\mathrm{T}}$ rad. To avoid too large a motion scope and to take pictures conveniently, the desired value of θ_1 is set to be a more appropriate one that can retain a smaller motion scope of the robot manipulator. Thus, the desired joint-angle vector θ_{d} of the physical robot manipulator is set as $[1.135, 0.349, \pi/12, \pi/12, \pi/12, \pi/12]^{\mathrm{T}}$ in radians. In addition, the task duration $T = 10$ s and design parameter $\gamma = 1$.

By exploiting numerical algorithm E47 to solve the presented state-adjustment scheme (8.13)–(8.14), the joint variables (i.e., θ and $\dot{\theta}$) are obtained. In addition, by exploiting Equations (8.15) and (8.16), the joint variables are converted into PPS for controlling the six-DOF planar robot manipulator, which are shown in Figures 8.7

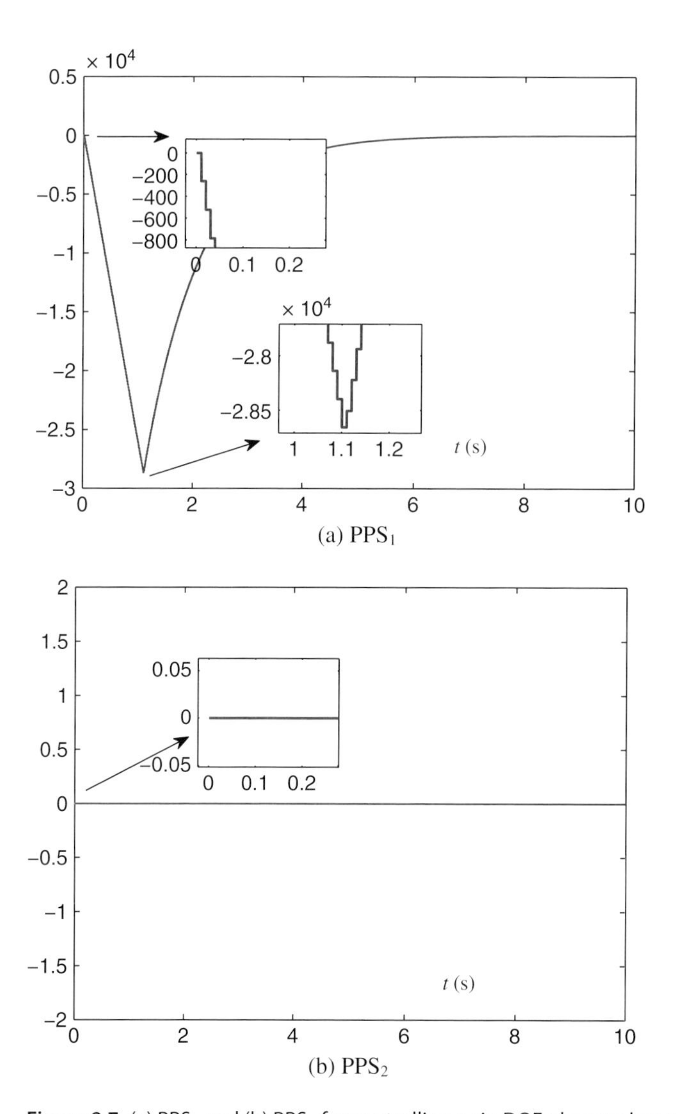

Figure 8.7 (a) PPS_1 and (b) PPS_2 for controlling a six-DOF planar robot manipulator.

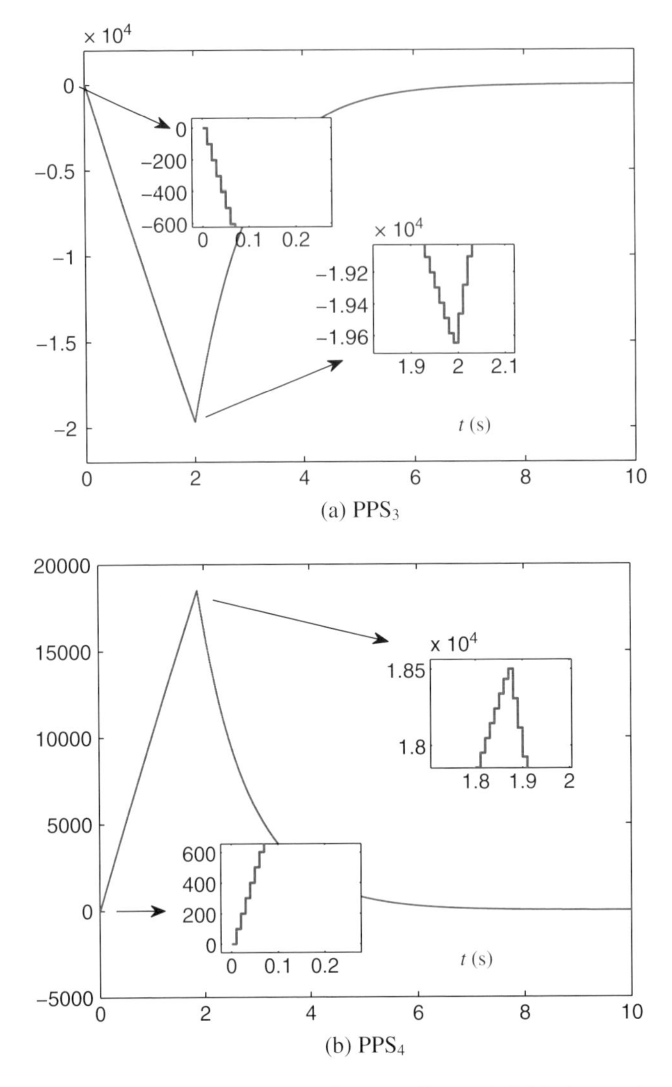

Figure 8.8 (a) PPS_3 and (b) PPS_4 for controlling a six-DOF planar robot manipulator.

through 8.9. As seen from Figures 8.7 through 8.9, with the zero-initial-velocity constraint imposed, the initial PPS of each joint is 0. Note that the plots of PPS are step-like because numerical algorithm E47 is a discrete-time QP solver and the sampling period $\Delta t = 0.01$ s. The state adjustment of the robot manipulator can be achieved by sending such PPS to drive the motors. Figure 8.10 shows the motion transients of the physical six-DOF planar robot manipulator during the state-adjustment task execution from the initial state $\theta(0)$ to the final state $\theta(T)$ with $T = 10$ s. As seen from the figure, all joints of the robot manipulator move towards their desired states (i.e., the element values of θ_{d}). Note that Joint 2 of the robot manipulator remains still during the state-adjustment task execution, which fits well with the fact that its desired state θ_{d2} equals the initial one $\theta_2(0)$. It is worth pointing out here that,

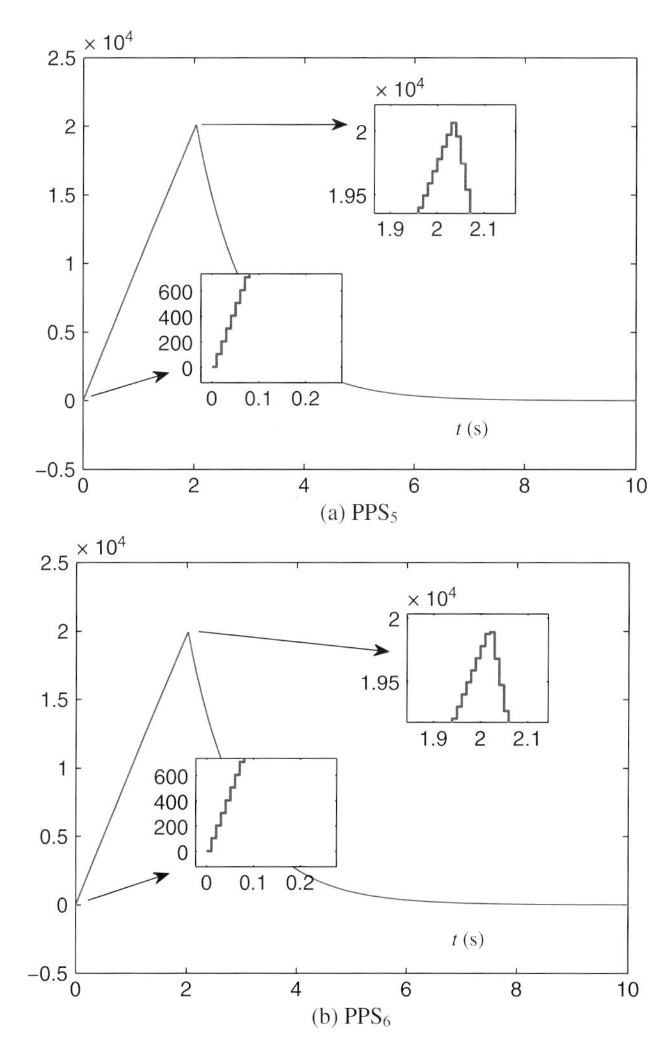

Figure 8.9 (a) PPS_5 and (b) PPS_6 for controlling a six-DOF planar robot manipulator.

corresponding to Figures 8.7 through 8.9, the change of joint-velocity is from slowness to fastness and then to slowness (i.e., the dynamic effects of the robot manipulator during the state adjustment), which is similar to the reflection of the first subfigure of Figure 8.3 (i.e., the sparser the motion process of the manipulator is, the faster the manipulator moves, because all the sampling periods between every two adjacent trajectories are the same). These substantiate further the realizability and effectiveness of the presented state-adjustment scheme (8.13)–(8.14). Similar to those in Table 8.2, as the corresponding computer-simulation shows, the final error $[|e_1(T)|, \cdots, |e_6(T)|]^{\mathrm{T}}$ between $\theta(T)$ and θ_{d} is $10^{-5} \times [0.46534, 0.01664, 3.82939, 2.94700, 3.77802, 3.73372]^{\mathrm{T}}$ in radians, and the maximal joint-angle error in this situation (i.e., with $\gamma = 1$, $T = 10$ s and the previous $\theta(0)$ and θ_{d}) is 3.82939×10^{-5} rad, both of which are small and acceptable in practice.

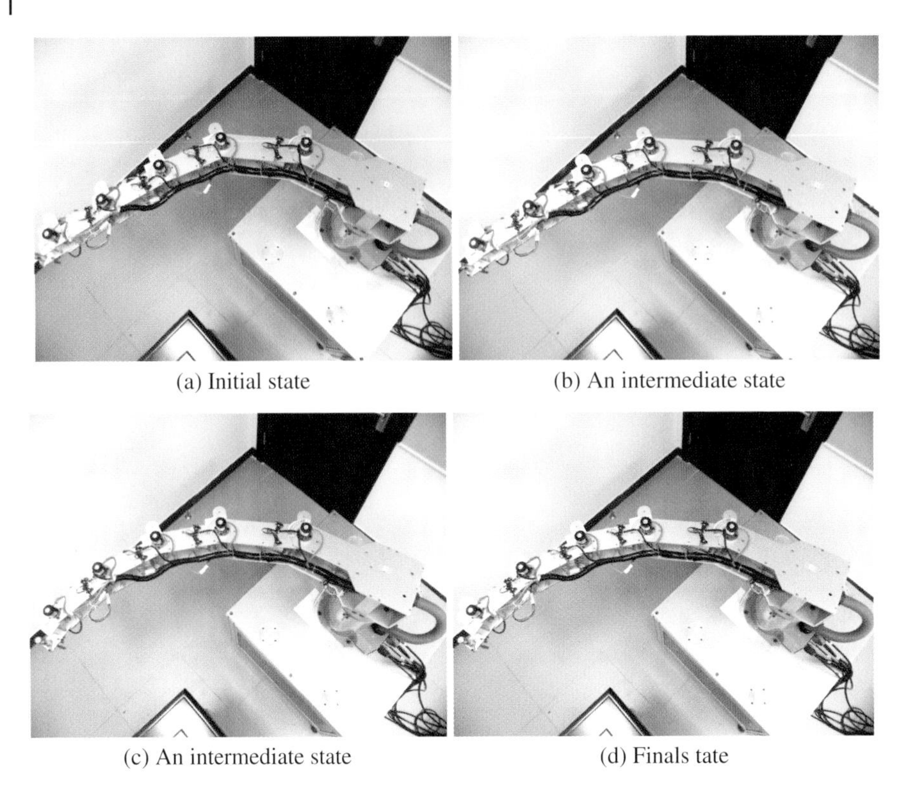

(a) Initial state (b) An intermediate state

(c) An intermediate state (d) Finals tate

Figure 8.10 Motion transients of a physical six-DOF planar robot manipulator from initial state $\theta(0)$ to desired state θ_d, which is synthesized by state-adjustment scheme (8.13)–(8.14) with $T = 10$ s. *Source:* Li 2012. Reproduced with permission of Cambridge University Press.

8.6 Chapter Summary

In this chapter, a QP-based state-adjustment scheme (8.13)–(8.14) with no end-effector task explicitly assigned has been presented and investigated to achieve a desired configuration for robot manipulators. Such a scheme has incorporated the joint physical limits, and has finally been formulated as a solvable QP. In addition, being a velocity-continuation technique, the zero-initial-velocity constraint has been discussed and employed for the presented state-adjustment scheme. Moreover, the state-adjustment scheme and the resultant QP problem (8.13)–(8.14) have been solved effectively via numerical algorithm E47 at the joint-velocity level. Both computer-simulation and experiment results based on a six-DOF planar robot manipulator have shown the realizability, effectiveness, flexibility, and accuracy of the presented state-adjustment scheme (8.1)–(8.3) [or correspondingly QP (8.13)–(8.14)].

Part V

Self-Motion Planning

9 QP-Based Self-Motion Planning

9.1 Introduction

Kinematically redundant manipulators are those that have more DOF than required for (specific) end-effectors' position-and-orientation tasks [62]. This implies that redundancy can also be established with respect to some particular tasks; for example, positioning only. For non-redundant manipulators, the joint motion is usually determined uniquely by a prescribed end-effector trajectory and, thus, there is no freedom left to handle secondary tasks, such as the handling of joint physical limits and environmental constraints (e.g., obstacles). In comparison, redundant manipulators have wider operational space and extra DOF to meet a number of functional constraints and performance criteria because there exist an infinite number of joint configurations as feasible solutions (corresponding to a prescribed end-effector task).

A fundamental issue in operating redundant mechanical systems is the inverse-kinematics problem (or to say, the redundancy-resolution problem) [1]. It can generally be described; "given the desired Cartesian trajectory $\mathbf{r}_{\mathrm{d}}(t) \in \mathbb{R}^m$ at the manipulator end-effector, how can we generate the corresponding joint trajectory $\theta(t) \in \mathbb{R}^n$ in real time t?". By resolving the redundancy properly, the robots can avoid joint physical limits, obstacles, as well as to optimize various secondary criteria [1, 61, 100]. However, for a redundant manipulator performing a specified end-effector task, multiple solutions (or even an infinite number of solutions) exist. In this sense, the redundancy of joint motion may complicate the manipulator control problem significantly, in addition to the kinematic and dynamic nonlinearities. To take full advantage of the redundancy, various computational schemes have been developed. The conventional solution of such an inverse-kinematics problem is the pseudoinverse-based approach [173–175]; that is, one minimum-norm particular solution plus a homogeneous solution. The research of recent years [1, 61, 100], however, shows that the redundancy-resolution problem can be solved in a more favorable manner via online optimization (specifically, QP) techniques.

In this chapter, our attention is focused on the self-motion trajectory planning of redundant PUMA560 and PA10 manipulators. That is, keeping the end-effector at a certain position (and/or orientation) (i.e., $\dot{\mathbf{r}}_{\mathrm{d}}(t) \equiv 0 \in \mathbb{R}^m$), we need to adjust the manipulator configuration (in terms of joint space) from one state (i.e., an initial joint state, $\theta(0) \in \mathbb{R}^n$) to another state (i.e., a desired joint state, $\theta_{\mathrm{d}} \in \mathbb{R}^n$). Such self-motion

Robot Manipulator Redundancy Resolution, First Edition. Yunong Zhang and Long Jin.

planning can be used for avoiding joint physical limits, obstacles, and singularities, as well as for the situation where the manipulator's end-effector is locked. Many researchers have exploited and investigated deeply such a self-motion-planning problem of redundant manipulators [176–179]. For example, in [176], the concept of self-motion manifold was presented, in addition to a proposition that, in the case of robots with revolute joints, the number of disjoint self-motion manifolds cannot exceed the maximum number of inverse-kinematics solutions of a non-redundant manipulator of the same class. In [177], some potential applications of self-motion in redundant robots were discussed. In [178], a discontinuous switching control scheme was proposed to stabilize all the joints to the desired positions while keeping the end-effector unmoved. A bi-directional self-motion planning scheme was proposed for redundant manipulators in [179], which together with [178], however, both didn't consider the avoidance of joint physical limits (e.g., joint limits and joint velocity limits). As we know, (1) almost all robots have joint physical limits, and (2) if the limits are violated, then the primary end-effector task can not be fulfilled successfully, not to mention additional physical damages possibly caused by commanded joints hitting their mechanical bounds. This motivates this work as well.

In this chapter, a quadratic performance index is presented for self-motion planning of robot manipulators, which is then reformulated as a QP. To solve this QP problem online, an LVI-aided primal-dual neural network is developed. As bound constraints, the joint limits and joint velocity limits are both incorporated into the QP formulation and the QP neural-solver. Computer-simulation results based on the PUMA560 and PA10 manipulators substantiate further the efficacy of such a neural-QP-based self-motion-planning scheme of robots.

9.2 Preliminaries and QP Formulation

Consider a redundant robot manipulator, of which the end-effector's position and orientation vector $\mathbf{r}_{\mathrm{d}}(t) \in \mathbb{R}^m$ in Cartesian space is related to the joint-space vector $\theta(t) \in \mathbb{R}^n$ through the forward kinematic equation $\mathbf{f}(\theta) \rightarrow \mathbf{r}_{\mathrm{d}}$, where $\mathbf{f}(\cdot)$ is defined as before in the previous chapter. For the purpose of the previously discussed inverse-kinematics, however, the forward kinematic equation is usually difficult to solve due to its nonlinearity and redundancy. Such problem solving is then, usually, considered at the joint-velocity level by differentiating the forward kinematic equation as $J(\theta)\dot{\theta} = \dot{\mathbf{r}}_{\mathrm{d}}$, where $\dot{\mathbf{r}}_{\mathrm{d}}$ and $\dot{\theta}$ defined as before. Note that, for redundant manipulators, because $m < n$, equations $\mathbf{r}_{\mathrm{e}} = \mathbf{f}(\theta)$ and $J(\theta)\dot{\theta} = \dot{\mathbf{r}}_{\mathrm{d}}$ are (usually) both under-determined and thus admit an infinite number of feasible solutions.

9.2.1 Self-Motion Criterion

For repetitive motion planning (RMP) (also called, drift-free redundancy resolution), papers [36, 90] presented a minimization scheme in terms of a quadratic function defined as the joint displacement between the current joint state and initial state. The performance index therein to be minimized is

$$(\dot{\theta} + \mathbf{c})^{\mathrm{T}}(\dot{\theta} + \mathbf{c})/2 \text{ with } \mathbf{c} = \lambda(\theta - \theta(0)), \tag{9.1}$$

where λ is a positive design-parameter used to scale the magnitude of the manipulator response to such joint-drifts.

Extending this RMP idea to the self-motion-planning (SMP) situation for redundant robot manipulators, we can present the following quadratic performance index used to achieve any suitable eventual state $\theta_{\mathrm{d}} \in \mathbb{R}^n$ from $\theta(0) \in \mathbb{R}^n$:

$$(\dot{\theta} + \mathbf{c})^{\mathrm{T}}(\dot{\theta} + \mathbf{c})/2 \text{ with } \mathbf{c} = \lambda(\theta - \theta_{\mathrm{d}}),$$

where θ_{d} can also be termed the desired joint configuration, and $\lambda > 0$ is defined the same as Equation (9.1). Because almost all manipulators are physically constrained by their joint limits and joint velocity limits, it could be more realistic and useful to consider the avoidance of such joint physical limits. Hence, we can have the following problem formulation for self-motion-planning of redundant robot manipulators:

$$\text{minimize} \quad (\dot{\theta} + \mathbf{c})^{\mathrm{T}}(\dot{\theta} + \mathbf{c})/2 \tag{9.2}$$

$$\text{subject to} \quad \mathbf{c} = \lambda(\theta - \theta_{\mathrm{d}}),$$

$$J(\theta)\dot{\theta} = 0, \tag{9.3}$$

$$\theta^- \leqslant \theta \leqslant \theta^+, \tag{9.4}$$

$$\dot{\theta}^- \leqslant \dot{\theta} \leqslant \dot{\theta}^+. \tag{9.5}$$

As we are considering the self-motion of manipulators, the end-effector should stay unchanged (i.e., $\dot{\mathbf{r}}_{\mathrm{d}} \equiv 0$), which together with $J(\theta)\dot{\theta} = \dot{\mathbf{r}}_{\mathrm{d}}$ yields the equality constraint (9.3). Besides, in bound constraints (9.4) and (9.5), superscripts $^+$ and $^-$ denote, respectively, the upper and lower limits of a joint variable vector (e.g., joint variable θ or joint velocity $\dot{\theta}$).

9.2.2 QP Formulation

As we discussed in the previous chapters, bound constraints (9.4) and (9.5) can be combined and unified into one dynamic bound-constraint, $\xi^- \leqslant \dot{\theta} \leqslant \xi^+$, where the ith elements of ξ^- and ξ^+ are defined, respectively:

$$\xi_i^- = \max\{\dot{\theta}_i^-, \mu(\theta_i^- - \theta_i)\}, \ \xi_i^+ = \min\{\dot{\theta}_i^+, \mu(\theta_i^+ - \theta_i)\}.$$

It follows from Subsection 9.2.1 that the self-motion-planning scheme (9.2)–(9.5) of physically-constrained robot manipulators can be reformulated finally as the QP:

$$\text{minimize} \quad \mathbf{x}^{\mathrm{T}} W \mathbf{x}/2 + \mathbf{c}^{\mathrm{T}}\mathbf{x}, \tag{9.6}$$

$$\text{subject to} \quad J\mathbf{x} = \mathbf{b}, \tag{9.7}$$

$$\xi^- \leqslant \mathbf{x} \leqslant \xi^+, \tag{9.8}$$

where decision variable vector $\mathbf{x} = \dot{\theta}$, and coefficients $W = I$, $\mathbf{c} = \lambda(\theta - \theta_{\mathrm{d}})$ and $\mathbf{b} = 0$, with $\xi^\pm$ defined as before.

9.3 LVIAPDNN Assisted QP Solution

In this section, the linear variational inequality (LVI) aided primal-dual neural network (LVIAPDNN) presented in Chapter 5 is employed to solve the time-varying QP problem (9.6)–(9.8) as well as the robot self-motion-planning problem (9.2)–(9.5) in real time. For readers' convenience, as well as the completeness of this chapter, the design procedure of the LVIAPDNN solver is shown as follows.

Firstly, we can convert the QP problem (9.6)–(9.8) to a set of linear variational inequalities (LVI). That is, to find a primal-dual equilibrium vector $\mathbf{y}^* \in \Omega = \{\mathbf{y}|\mathbf{y}^- \leqslant \mathbf{y} \leqslant \mathbf{y}^+\} \subset \mathbb{R}^{n+m}$ such that

$$(\mathbf{y} - \mathbf{y}^*)^{\mathrm{T}}(Q\mathbf{y}^* + \mathbf{p}) \geqslant 0, \ \forall \mathbf{y} \in \Omega, \tag{9.9}$$

where the primal-dual decision variable vector $\mathbf{y}$ and its upper/lower bounds are defined, respectively, as (with $\varpi \gg 0 \in \mathbb{R}^m$ being sufficiently large to replace the m-dimensional $+\infty$ numerically):

$$\mathbf{y} = \begin{bmatrix} \mathbf{x} \\ \mathbf{u} \end{bmatrix}, \ \mathbf{y}^+ = \begin{bmatrix} \xi^+ \\ \varpi \end{bmatrix}, \ \mathbf{y}^- = \begin{bmatrix} \xi^- \\ -\varpi \end{bmatrix} \in \mathbb{R}^{n+m}. \tag{9.10}$$

In (9.9), $\mathbf{u} \in \mathbb{R}^m$ is the dual decision vector corresponding to equality constraint (9.7), and the coefficients Q and $\mathbf{p}$ are augmented and defined as

$$Q = \begin{bmatrix} W & -J^{\mathrm{T}} \\ J & 0 \end{bmatrix} \in \mathbb{R}^{(n+m)\times(n+m)}, \ \mathbf{p} = \begin{bmatrix} \mathbf{c} \\ -\mathbf{b} \end{bmatrix} \in \mathbb{R}^{n+m}.$$

In summary, we have the following theoretical results [61, 92].

Theorem 9.1 With the existence of (at least) one optimal solution $\mathbf{x}^*$, the QP problem (9.6)–(9.8) can be converted to the LVI problem (9.9).

Secondly, it is known that LVI problem (9.9) is equivalent to the following system of piecewise-linear equations [92, 180]:

$$\mathbf{P}_\Omega(\mathbf{y} - (Q\mathbf{y} + \mathbf{p})) - \mathbf{y} = 0, \tag{9.11}$$

where $\mathbf{P}_\Omega(\cdot) : \mathbb{R}^{n+m} \to \Omega$ is a piecewise-linear projection operator from $\mathbb{R}^{n+m}$ onto set Ω, with the ith element of $\mathbf{P}_\Omega(\mathbf{y})$ defined as [note that $y_i^\pm$ are the ith elements of $\mathbf{y}^\pm$ in (10.10)]:

$$\begin{cases} y_i^-, & \text{if } y_i < y_i^- \\ y_i, & \text{if } y_i^- \leqslant y_i \leqslant y_i^+ \\ y_i^+, & \text{if } y_i > y_i^+ \end{cases}, \ \forall i \in \{1, 2, \cdots, n+m\}.$$

Thirdly, to solve (9.11), guided by dynamic-system-solver design experience [15, 78], we can adopt the following dynamics [for the real-time LVIAPDNN solver for QP (9.6)–(9.8)]:

$$\dot{\mathbf{y}} = \gamma(I + Q^{\mathrm{T}})(\mathbf{P}_\Omega(\mathbf{y} - (Q\mathbf{y} + \mathbf{p})) - \mathbf{y}), \tag{9.12}$$

where design parameter $\gamma > 0$ is used to scale the convergence rate of the LVIAPDNN. It can be obtained from Theorem 5.2 that the presented LVIAPDNN possesses the global convergence property [61, 92].

9.4 PUMA560 Based Computer Simulations

In this section, we would apply the QP-based redundancy-resolution scheme to the self-motion planning of physically constrained PUMA560 robot manipulator [182].

Without loss of generality, we only consider the positioning of the end-effector and the degrees of redundancy thus: $n - m = 3$. For illustration purposes, three different simulations are carried out to validate the general applicability and efficiency of our SMP scheme. That is, from the initial configuration A (i.e., some $\theta(0)$) to the desired configuration B (i.e., some θ_d); from the same initial configuration A (i.e., $\theta(0)$) to a different desired configuration C (i.e., another θ_d); from a different initial configuration E (i.e., another $\theta(0)$) to a different desired configuration F (i.e., another θ_d). For illustration and better readability, the values of initial configurations and desired configurations are all shown in Table 9.1. In the simulations of this chapter, the limited joint ranges of PUMA560 robot manipulator we take into account in the SMP redundancy-resolution scheme are assumed to be $[-2.984, 2.984]$, $[-3.378, 0.807]$, $[-0.974, 3.378]$, $[-2.064, 3.190]$, $[-1.877, 0.038]$, $[-3.378, 3.378]$ rad, respectively, for θ_1 through θ_6. The limited range of each joint velocity $\dot{\theta}_i$ is $[-1.5, 1.5]$ rad-per-second,

Table 9.1 Initial configurations and desired configurations (rad) of PUMA560 robot manipulator involved in the simulations.

Joint #	A used as $\theta(0)$	B used as θ_d
θ_1	0.00000000000	+0.224234026947615
θ_2	0.00000000000	+0.050660994984405
θ_3	0.00000000000	−0.037325009949644
θ_4	0.00000000000	−1.920055856170361
θ_5	0.00000000000	−0.413604750702359
θ_6	0.00000000000	+0.000000000000000

Joint #	A used as $\theta(0)$	C used as θ_d
θ_1	0.00000000000	-0.31819107703883
θ_2	0.00000000000	+0.14585270971836
θ_3	0.00000000000	−0.31942626042259
θ_4	0.00000000000	+2.06307001144010
θ_5	0.00000000000	−0.72039386298613
θ_6	0.00000000000	+0.00000000000000

Joint #	E used as $\theta(0)$	F used as θ_d
θ_1	0.00000000000	+0.241644973239444
θ_2	$-\pi/4$	−0.983411131771378
θ_3	0.00000000000	+0.187576110192631
θ_4	$\pi/2$	−0.007321039981582
θ_5	$-\pi/4$	+0.034910106475825
θ_6	0.00000000000	+0.000000000000000

$i = 1, 2, \cdots, 6$. In addition, for the original robotic scheme-formulation (9.2)–(9.5) and its transformed QP scheme-formulation (9.6)–(9.8), we can set the parameter $\mu = 20$ and the SMP-task duration $T = 5$ s. Moreover, in order to ensure that the simulation results have a certain precision, we set the design parameter $\gamma = 1 \times 10^8$.

9.4.1 From Initial Configuration A to Desired Configuration B

In this simulation, we firstly assume the initial configuration to be A (i.e., $\theta(0)$) and the desired configuration to be B (i.e., θ_d). Synthesized with $\lambda = 4$, Figure 9.1 and Figure 9.2 show the computer-simulation results. Figure 9.1 illustrates the three-dimensional

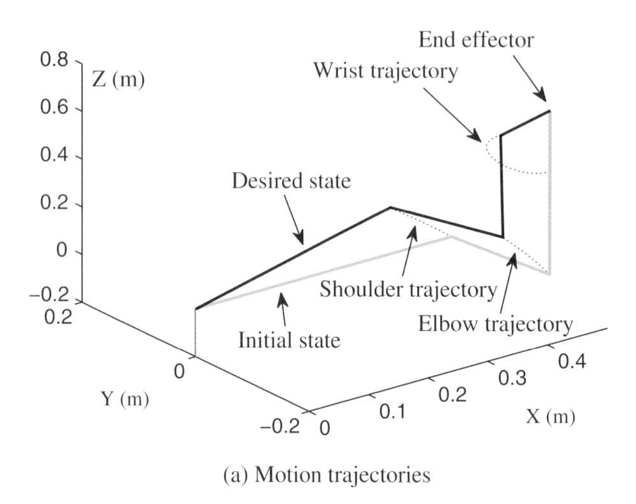

(a) Motion trajectories

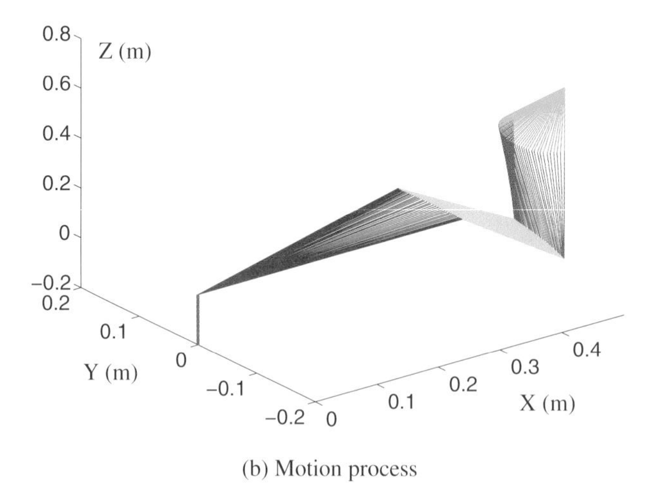

(b) Motion process

Figure 9.1 Three-dimensional motion trajectories of a PUMA560 robot manipulator performing self-motion from configurations A to B with $\lambda = 4$.

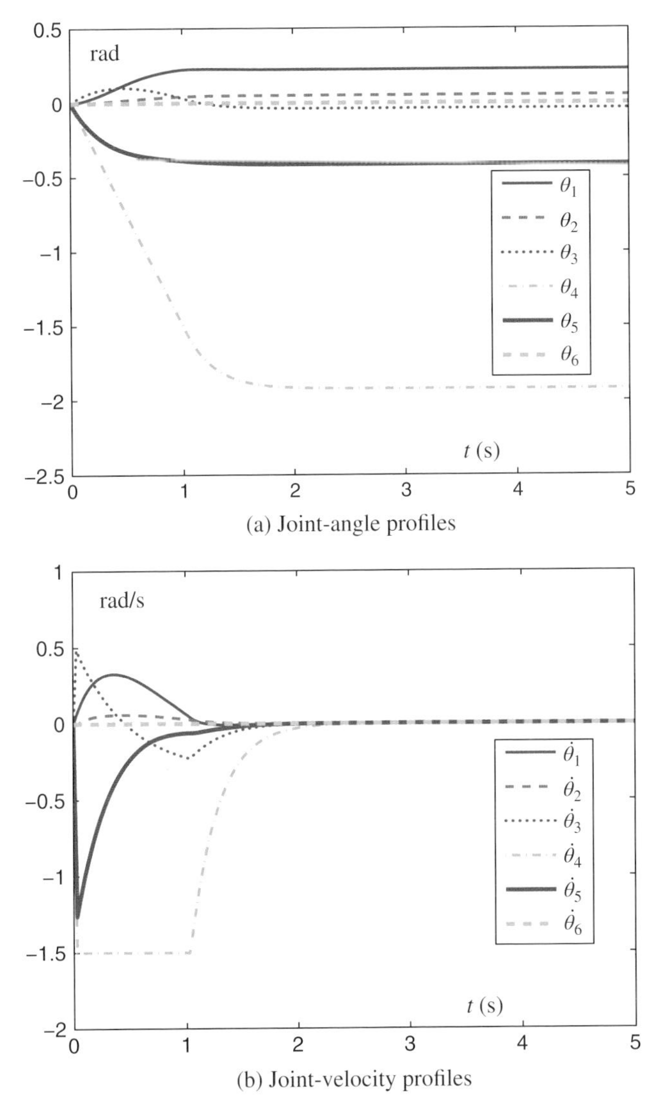

Figure 9.2 (a) Joint-angle and (b) joint-velocity profiles of PUMA560 robot manipulator performing self-motion from configurations A to B with $\lambda = 4$.

motion trajectories of PUMA560 robot manipulator. We can see that the manipulator configuration successfully move to the desired joint state B (i.e., $\theta_{\rm d}$) from the initial joint state A (i.e., $\theta(0)$) with the end-effector remaining at the same position. From Figure 9.2, we can see that all joint variables and joint velocity variables are convergent to their desired values and never exceed their limited ranges. In addition, as seen from Figure 9.2, the adjusting time from the initial configuration to the desired configuration

Table 9.2 Joint values and errors of PUMA560 robot manipulator before and after self-motion from configurations A to B (rad).

Joint #	Initial configuration $\theta(0)$	Desired configuration θ_d
θ_1	+0.000000000000000	+0.224234026947615
θ_2	+0.000000000000000	+0.050660994984405
θ_3	+0.000000000000000	−0.037325009949644
θ_4	+0.000000000000000	−1.920055856170361
θ_5	+0.000000000000000	−0.413604750702359
θ_6	+0.000000000000000	+0.000000000000000

Joint #	Final configuration $\theta(5)$	Error $\theta(5) - \theta_d$ ($\times 10^{-8}$)
θ_1	+0.224234033361507	+0.64138915567558
θ_2	+0.050660989165984	−0.58184214113410
θ_3	−0.037324993475390	+1.64742542455021
θ_4	−1.920055810421939	+4.57484219396065
θ_5	−0.413604747015443	+0.36869158837050
θ_6	+0.000000000000000	+0.00000000000000

is about 3 s. Furthermore, the errors between the actual joint-values after self-motion and the desired joint-values are shown in Table 9.2, which appear to be very tiny (of order 1×10^{-8} rad).

9.4.2 From Initial Configuration A to Desired Configuration C

In this simulation, we assume the initial configuration $\theta(0)$ the same as before (i.e., A), and we hope the joints could move to a different desired configuration C (i.e., used as θ_d). The design parameter λ is also set to be 4. Computer-simulation results are illustrated in Figure 9.3 and Figure 9.4. We can see that the manipulator carries out the self-motion from the initial state A to the desired state C successfully as expected. In addition, the adjusting time from the initial configuration A to the desired configuration C is about 3 s. For illustrative and comparative purposes, the errors between the actual joint-values after self-motion and the desired joint-values are shown in Table 9.3, which is very tiny as well (i.e., of order 1×10^{-8} rad) and should be acceptable in practice.

As a simple summary, these two simulations based on PUMA560 manipulator both illustrate the efficacy of our self-motion planning scheme for redundant robot manipulators.

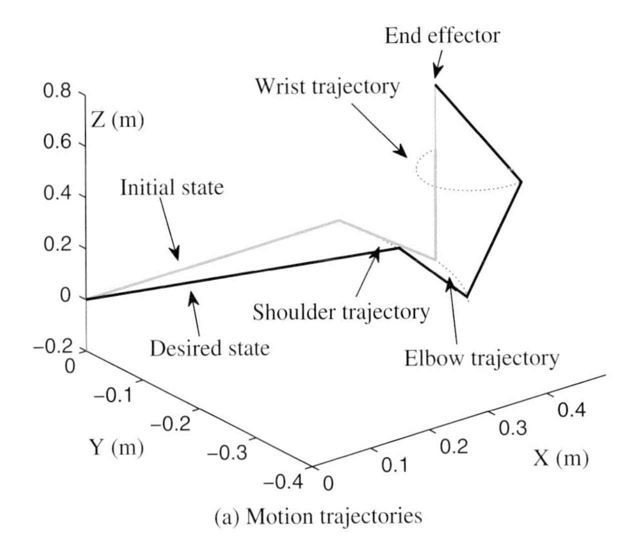

(a) Motion trajectories

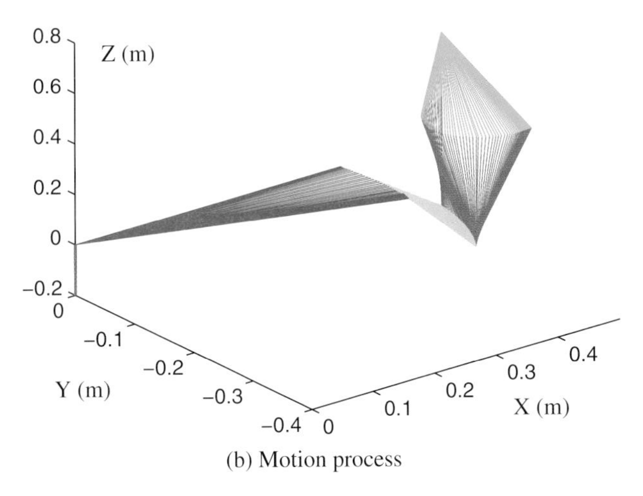

(b) Motion process

Figure 9.3 Three-dimensional motion trajectories of a PUMA560 robot manipulator performing self-motion from configurations A to C with $\lambda = 4$.

9.4.3 From Initial Configuration E to Desired Configuration F

In this simulation, for more comparison and illustration, we investigate a different initial configuration E (i.e., used as $\theta(0)$) and a different desired configuration F (i.e., used as θ_{d}). The design parameter λ is set to be 4 at first (and later 20 for comparison). Computer-simulation results are shown in Figure 9.5 and Figure 9.6. As seen from the figures, all joint variables and joint velocity variables are successfully convergent to their

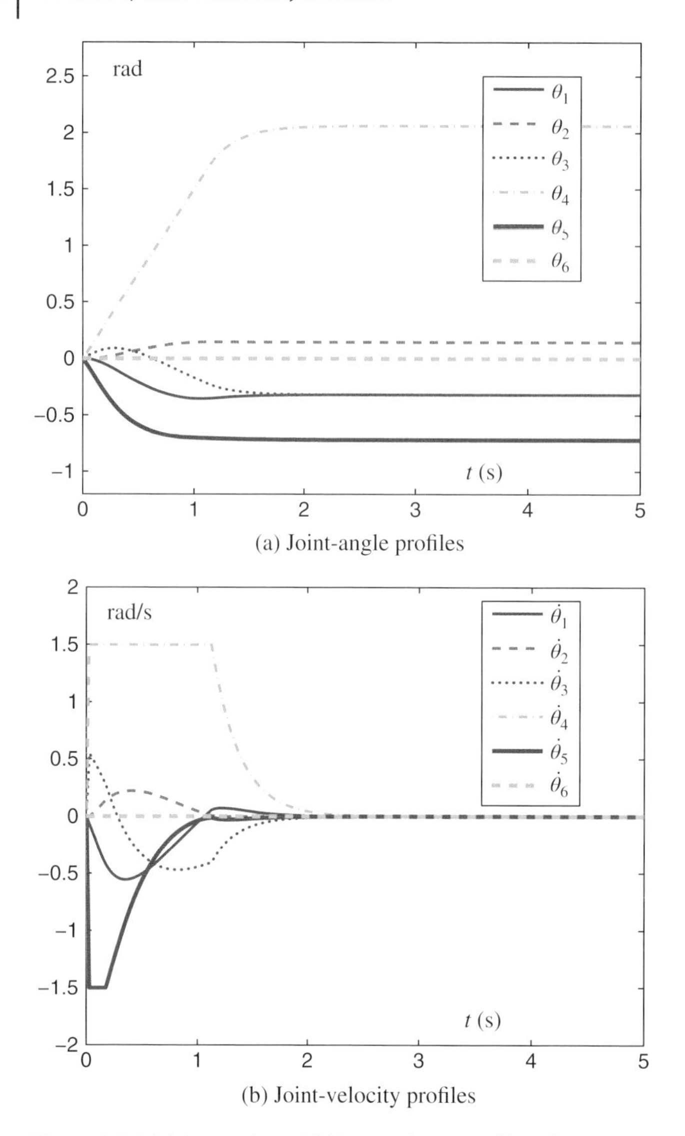

Figure 9.4 (a) Joint-angle and (b) joint-velocity profiles of a PUMA560 robot manipulator performing self-motion from configurations A to C with $\lambda = 4$.

desired values, while keeping within their limited ranges throughout the SMP task. The second subgraph of Figure 9.5 shows that the maximal end-effector position error during the self-motion process is less than 1.2×10^{-8} m, where ϵ_X, ϵ_Y, and ϵ_Z denote the components of position error $\epsilon(t) = \mathbf{r}(t) - \mathbf{f}(\theta(t))$, respectively, along the x-, y-, and z-axes in terms of the base frame of the robot system. In addition, as seen from Figure 9.5,

Table 9.3 Joint-angle values and errors of PUMA560 robot manipulator before and after self-motion from configurations A to C (rad).

Joint #	Initial configuration $\theta(0)$	Desired configuration θ_d
θ_1	+0.000000000000000	−0.31819107703883
θ_2	+0.000000000000000	+0.14585270971836
θ_3	+0.000000000000000	−0.31942626042259
θ_4	+0.000000000000000	+2.06307001144010
θ_5	+0.000000000000000	−0.72039386298613
θ_6	+0.000000000000000	+0.00000000000000

Joint #	Final configuration $\theta(5)$	Error $\theta(5) - \theta_d$ ($\times 10^{-8}$)
θ_1	−0.318191088634618	−1.15957884960771
θ_2	+0.145852709695678	−0.00226820229265
θ_3	−0.319426243633992	−1.67885978452631
θ_4	+2.063069942180092	−6.92600083951334
θ_5	−0.720393857828048	+0.51580818505315
θ_6	+0.000000000000000	+0.00000000000000

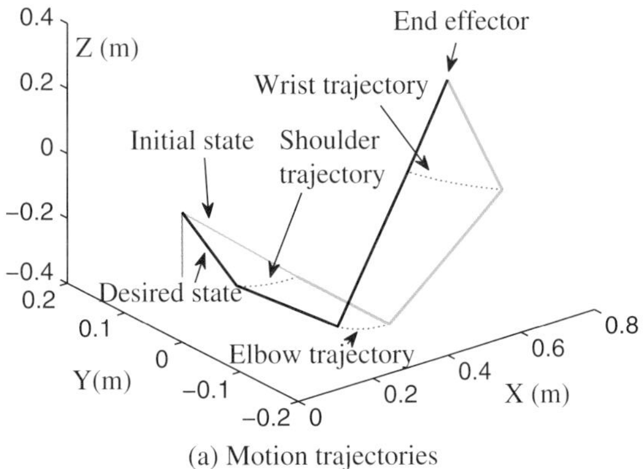

(a) Motion trajectories

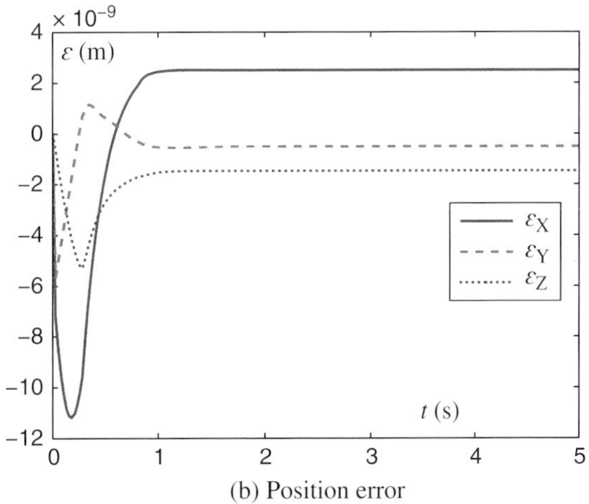

(b) Position error

Figure 9.5 (a) Three-dimensional motion trajectories and (b) maximal end-effector position error of a PUMA560 robot manipulator from configurations E to F with $\lambda = 4$.

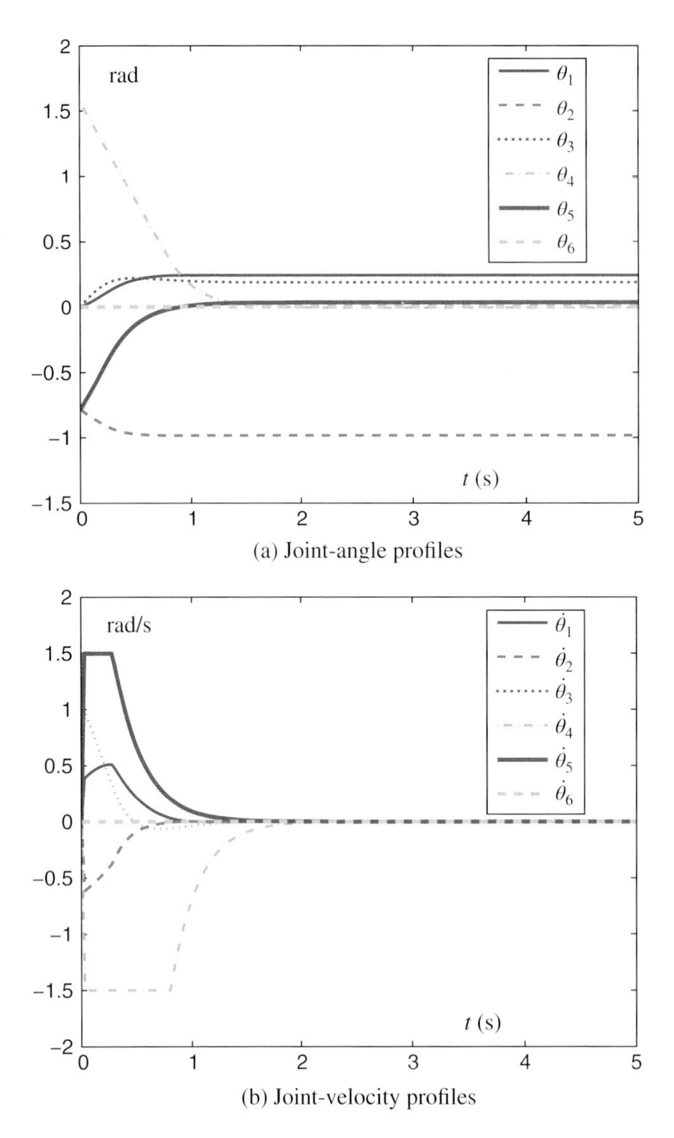

Figure 9.6 (a) Joint-angle and (b) joint-velocity profiles of a PUMA560 robot manipulator from configurations E to F with $\lambda = 4$.

Table 9.4 Joint-angle values and errors of PUMA560 robot manipulator before and after self-motion from configurations E to F (rad).

Joint #	Initial configuration $\theta(0)$	Desired configuration θ_d
θ_1	+0.000000000000000	+0.241644973239444
θ_2	−0.785398163397440	−0.983411131771378
θ_3	+0.000000000000000	+0.187576110192631
θ_4	+1.570796326794890	−0.007321039981582
θ_5	−0.785398163397440	+0.034910106475825
θ_6	+0.000000000000000	+0.000000000000000

Joint #	Final configuration $\theta(5)$	Error $\theta(5) - \theta_d$ ($\times 10^{-8}$)
θ_1	+0.241644973456330	+0.02168860924190
θ_2	−0.983411132056172	−0.02847940772099
θ_3	+0.187576111348927	+0.11562958091016
θ_4	−0.007321020849323	+1.91322592863960
θ_5	+0.034910103935376	−0.25404485667901
θ_6	+0.000000000000000	+0.00000000000000

the adjusting time from the initial state to the desired state is about 2.5 s (which can be decreased further by increasing the design parameter λ, as pointed out and compared in the next paragraph). Moreover, the errors between the desired and actual joint-values after self-motion are shown in Table 9.4.

For comparison, in order to decrease the adjusting time (i.e., 2.5 s), as another simulation we can set the design parameter λ to be a larger value; that is, $\lambda = 20$ (but with other parameters being the same as before). Figure 9.7 illustrates the computer-simulation result. From this figure, we can see that the adjusting time has successfully decreased to a smaller value; that is, about 1.5 s, and that all joint variables and joint velocity variables are still convergent to their desired values and kept within their limited ranges. Similarly, the adjusting time of the first two SMP simulations (i.e., from configurations A to B and to C) can also be decreased in this way (but with figures omitted here due to space limitation).

In summary, these computer-simulation results based on a PUMA560 robot manipulator (should be able to) substantiate well the efficacy of the original robotic scheme-formulation (9.2)–(9.5) and its transformed QP formulation (9.6)–(9.8) on the self-motion planning of redundant robot manipulators.

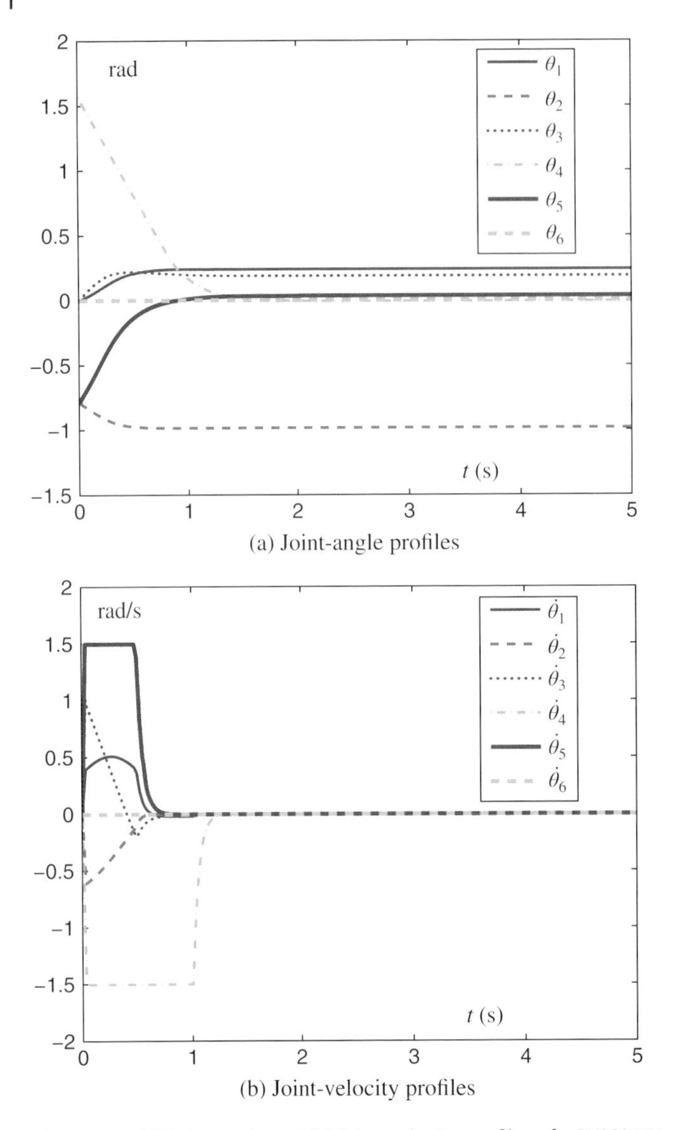

Figure 9.7 (a) Joint-angle and (b) joint-velocity profiles of a PUMA560 robot manipulator performing self-motion from configurations E to F with $\lambda = 20$.

9.5 PA10 Based Computer Simulations

The PA10 robotic manipulator has seven DOF (three rotating axes and four pivot axes) [181]. Its joint limits and joint velocity limits are shown in Table 9.5 [100]. In this section, we apply the QP-based redundancy-resolution scheme (9.12) to the self-motion planning of physically constrained PA10 manipulator. In this study, for

Table 9.5 Joint-angle limits and joint-velocity limits used in this chapter for a PA10 manipulator.

#	Joint	Axis	$\theta^{\pm}$ (rad)	$\dot{\theta}^{\pm}$ (rad/s)
1	Shoulder 1	Rotating	$\pm\pi$	± 1
2	Shoulder 2	Pivot	± 1.7637	± 1
3	Shoulder 3	Rotating	$\pm\pi$	± 2
4	Elbow 1	Pivot	± 2.6831	± 2
5	Elbow 2	Rotating	$\pm 3\pi/2$	$\pm 2\pi$
6	Wrist 1	Pivot	$\pm\pi$	$\pm 2\pi$
7	Wrist 2	Rotating	$\pm 2\pi$	$\pm 2\pi$

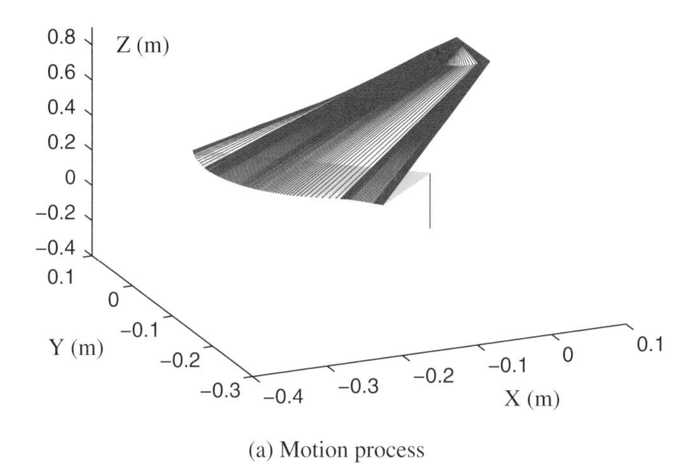

(a) Motion process

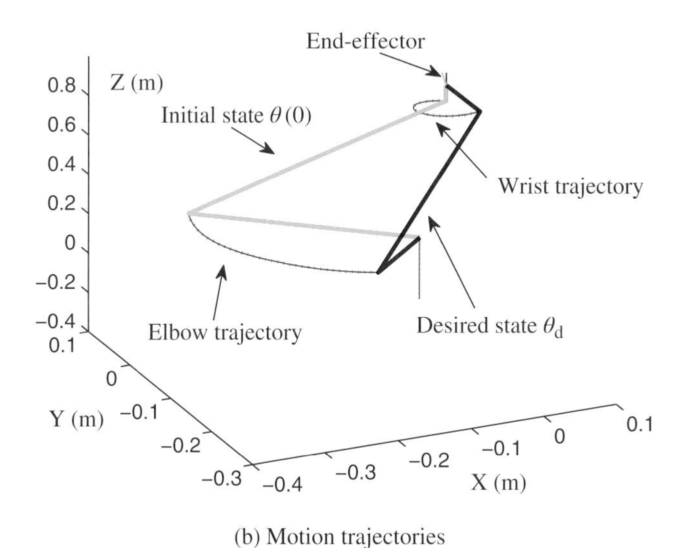

(b) Motion trajectories

Figure 9.8 Three-dimensional motion trajectories of a PA10 manipulator performing self-motion (from $\theta(0)$ to θ_d without moving the end-effector) with $\lambda = 3$.

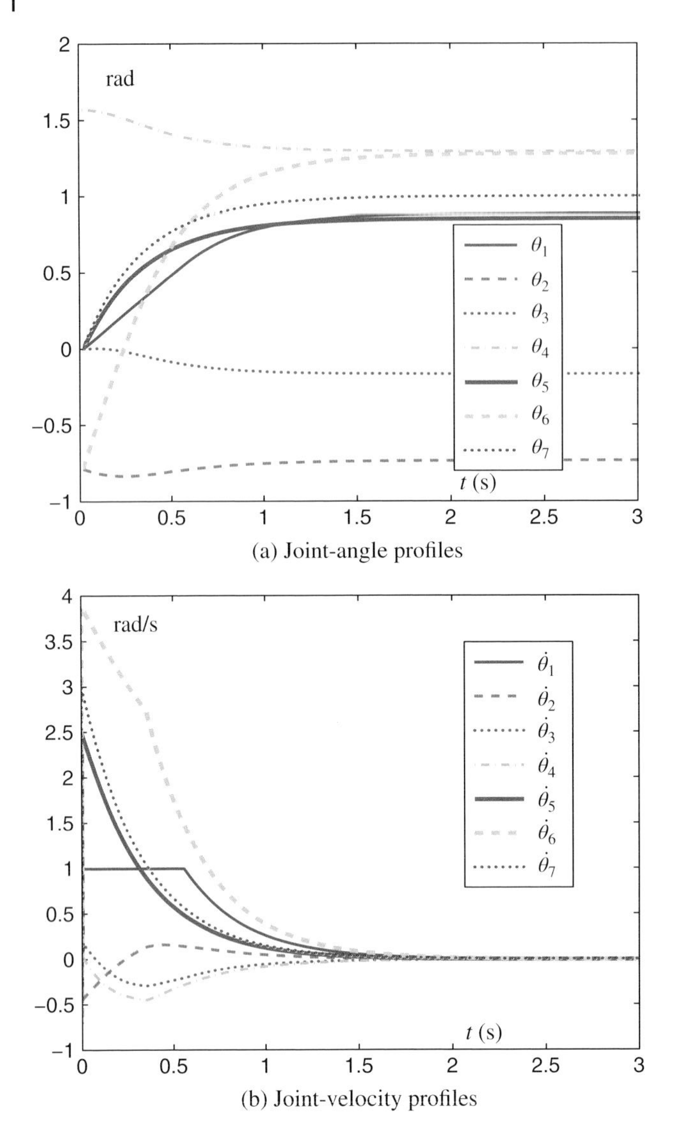

Figure 9.9 (a) Joint-angle and (b) joint-velocity profiles of a PA10 manipulator performing self-motion (from $\theta(0)$ to θ_d without moving end-effector) with $\lambda = 3$.

illustrative purposes and without loss of generality, only the end-effector positioning is considered (i.e., to be unchanged), and thus the Jacobian matrix is 3×7 in dimension (with $m = 3$ and the degrees of redundancy being $7 - 3 = 4$). The initial state $\theta(0) = [0, \pi/4, 0, \pi/2, 0, -\pi/4, 0]^\mathrm{T}$ rad, whereas, without loss of generality, we can assume the desired final state $\theta_\mathrm{d} = [0.888796, -0.732464, -0.164280, 1.294830, 0.854875, 1.281256, 1.000000]^\mathrm{T}$ rad for illustration and verification purposes. For better readability and visual effects, please refer to Figure 9.8. In addition, for the original scheme formulation (9.2)–(9.5), QP-based scheme formulation (9.6)–(9.8) and

Table 9.6 Joint-angle values and errors of a PA10 manipulator before and after self-motion with $\lambda = 3$ (rad).

Joint #	$\theta(0)$	$\theta(3)$	θ_d	$\theta(3) - \theta_d$
θ_1	+0.000000	+0.888578	+0.888796	−0.000218
θ_2	−0.785398	−0.732513	−0.732464	−0.000049
θ_3	+0.000000	−0.164242	−0.164280	+0.000038
θ_4	+1.570796	+1.294898	+1.294830	+0.000068
θ_5	+0.000000	+0.854768	+0.854875	−0.000107
θ_6	−0.785398	+1.280936	+1.281256	−0.000320
θ_7	+0.000000	+0.999876	+1.000000	−0.000124

neural-solver (9.12), we can set the design parameters $\mu = 1$ and $\gamma = 10^7$, as well as the simulation time $T = 3$ s.

Firstly, with $\lambda = 3$, the LVI-aided primal-dual neural network (9.12) is applied to the PA10 manipulator. Figure 9.8 and Figure 9.9 show the simulation results. From the left graph of Figure 9.9, we can see that all joint variables are convergent to their desired values while never exceeding their mechanical ranges (as listed in Table 9.5). Moreover, from the right graph of Figure 9.9, we can see that all joint velocities have been kept within their physical limits as well. This is evidently because we have taken them into account in the redundancy-resolution scheme. The errors between the actual joint-values after self-motion and desired joint-values are shown in Table 9.6, which appear to be very small (i.e., of order 1×10^{-4} rad). Figure 9.10 shows that the

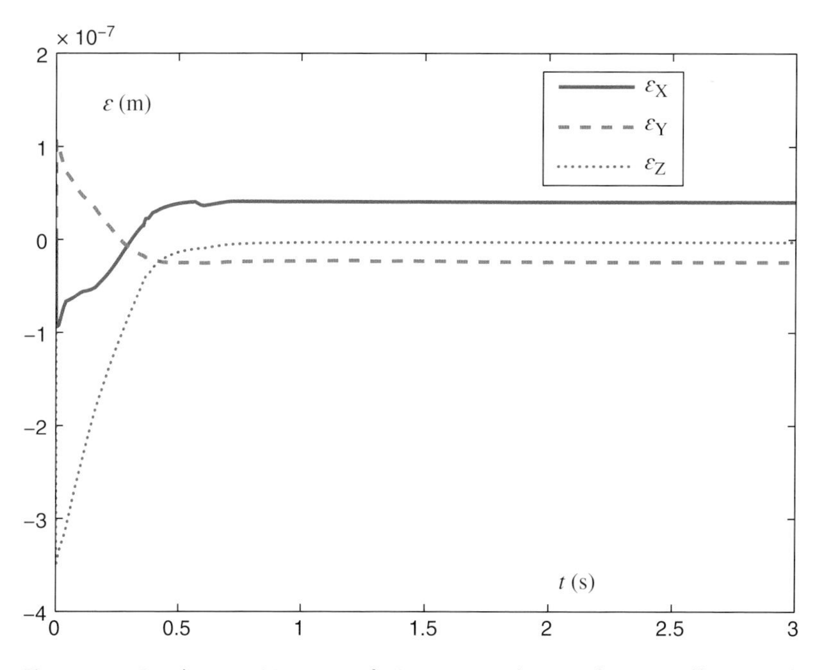

Figure 9.10 Resultant position error of a PA10 manipulator performing self-motion (from $\theta(0)$ to θ_d theoretically without moving the end-effector) with $\lambda = 3$.

maximal end-effector error is less than 4×10^{-4} mm, where ϵ_X, ϵ_Y, and ϵ_Z denote the components of position error $\epsilon(t) = \mathbf{f}(\theta(0)) - \mathbf{f}(\theta(t))$, respectively, along the x-, y-, and z-axes in terms of the base frame of the robot system.

Evidently, the computer-simulation based on a PA10 manipulator illustrates the effectiveness of our self-motion-planning scheme for redundant robot manipulators. As seen from Figure 9.9 and Figure 9.10, it is worth mentioning that the adjusting time from $\theta(0)$ to θ_d is only 2.0 s or less, which can be decreased further by increasing the design parameter λ.

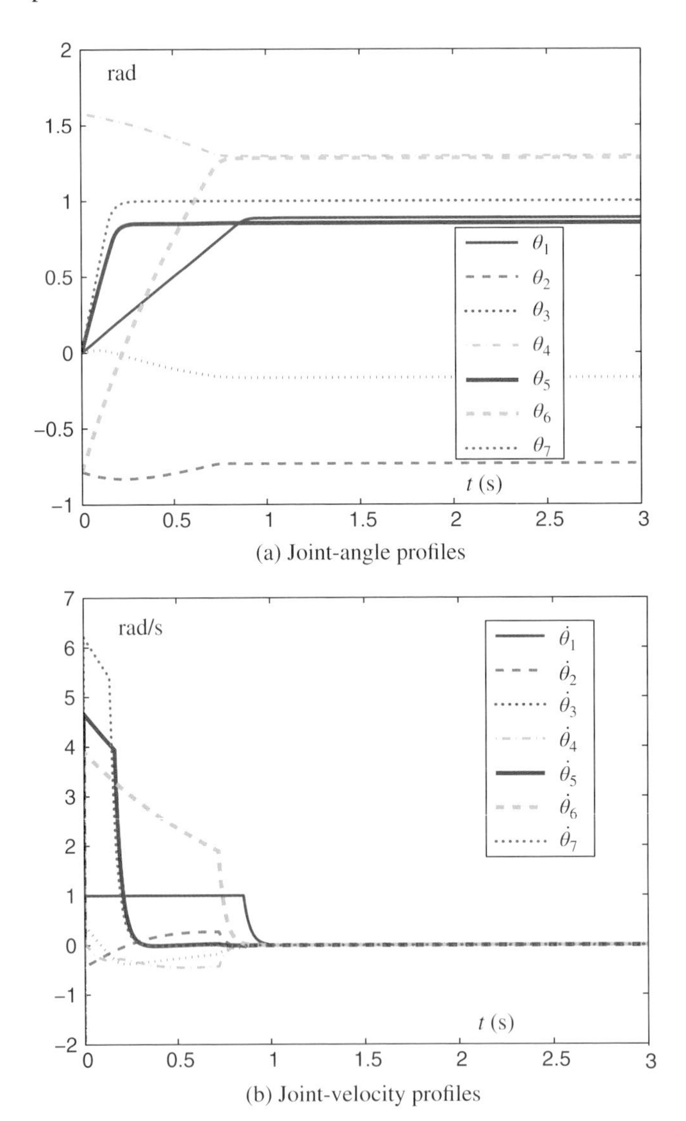

Figure 9.11 (a) Joint-angle and (b) joint-velocity profiles of a PA10 manipulator performing self-motion (from $\theta(0)$ to θ_d without moving the end-effector) with $\lambda = 30$.

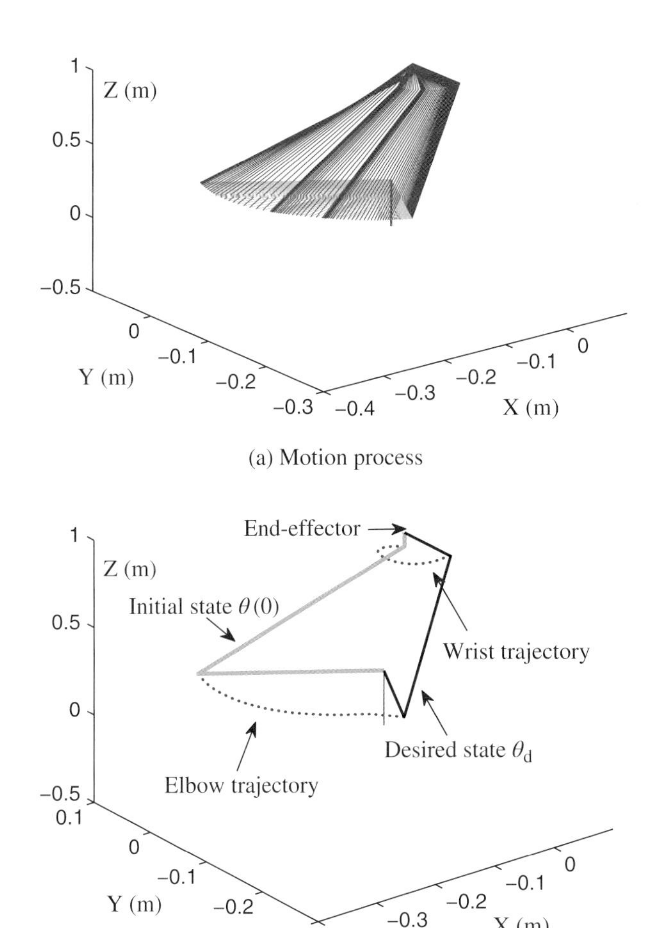

Figure 9.12 Three-dimensional motion trajectories of a PA10 manipulator performing self-motion (from $\theta(0)$ to θ_d without moving the end-effector) with $\lambda = 30$.

Secondly, for comparison as well as for illustration, the LVI-aided primal-dual neural network (9.12) is applied again to the PA10 manipulator with $\lambda = 30$. As shown in Figure 9.11, all joint variables are convergent to the desired values while staying within their mechanical ranges. Figure 9.12 illustrates the three-dimensional motion trajectories of PA10 manipulator, of which the two graphs correspond, respectively, to the performing or having-performed situation of such a self-motion task. Figure 9.13 gives the actual position error profile while the joint errors are shown in Table 9.7, both of which are very small. We also see from Figure 9.11 that the adjusting time from $\theta(0)$ to θ_d is now only 1.0 s or less. In addition, for the PA10-manipulator example, such an adjusting time can not be less than 0.888796 s, as the maximum velocity of the first joint is 1 rad/s (while the initial and desired values of such a joint are respectively 0.000000 and +0.888796 rad).

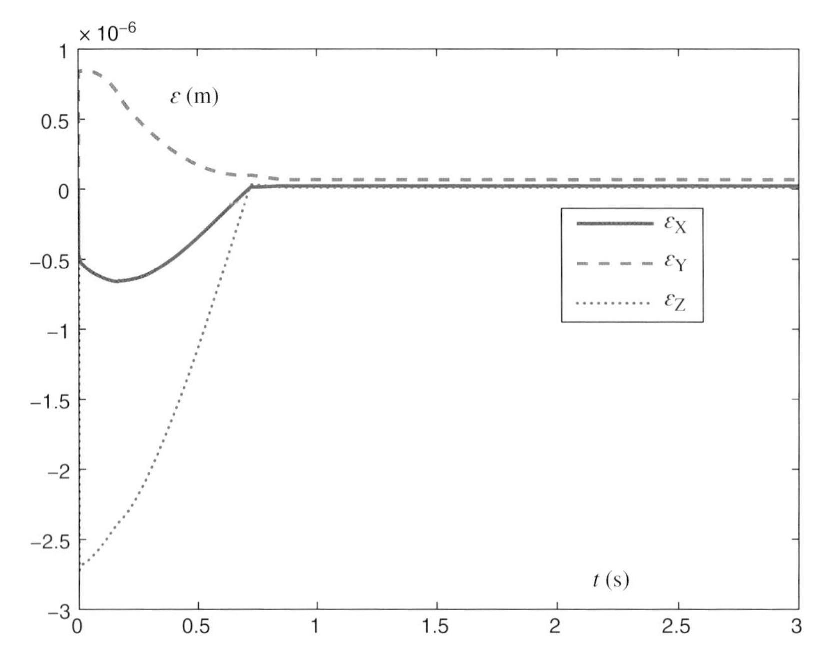

Figure 9.13 Resultant position error of a PA10 manipulator performing self-motion (from $\theta(0)$ to θ_d theoretically without moving the end-effector) with $\lambda = 30$.

Table 9.7 Joint-angle values and errors of a PA10 manipulator before and after self-motion with $\lambda = 30$ (rad).

Joint #	$\theta(0)$	$\theta(3)$	θ_d	$\theta(3) - \theta_d$
θ_1	+0.000000	+0.888796	+0.888796	+0.000000
θ_2	−0.785398	−0.732468	−0.732464	−0.000004
θ_3	+0.000000	−0.164282	−0.164280	−0.000002
θ_4	+1.570796	+1.294835	+1.294830	+0.000005
θ_5	+0.000000	+0.854875	+0.854875	+0.000000
θ_6	−0.785398	+1.281257	+1.281256	+0.000001
θ_7	+0.000000	+1.000000	+1.000000	+0.000000

In summary, these PA10-based simulation results have substantiated the efficacy of the original problem formulation (9.6)–(9.8), QP formulation (9.6)–(9.8), and neural-solver (9.12) on the self-motion-planning of redundant robot manipulators.

9.6 Chapter Summary

In this chapter, as illustrated based on PUMA560 and PA10 robot manipulators, we have roughly solved the self-motion-planning problem of physical-constrained redundant manipulators. The online solution, based on a QP problem formulation, can be achieved by the LVI-aided primal-dual neural network. The neural-network's architectural and computational complexities are lower than those of other existing recurrent neural networks, in addition to its piecewise-linear dynamics. The LVI-aided primal-dual neural network can also be proved with global exponential convergence to optimal solutions of convex QPs used here or in other contexts. Computer-simulation results based on PUMA560 and PA10 manipulators have shown further the efficacy of the self-motion-planning scheme-formulation (9.6)–(9.8) and its neural-network solver (9.12) on real-time kinematic control of joint-constrained redundant manipulators.

10

Pseudoinverse Method and Singularities Discussed

10.1 Introduction

A manipulator is said to be redundant when more DOF are available than the minimum number required to perform a specific end-effector primary task [1]. For redundant manipulators, as discussed in previous chapters, the inverse kinematic resolution is a fundamental issue in operating the robot systems [1, 3, 63, 65, 78, 183–188]. The general description of this problem is that, given the desired Cartesian path $\mathbf{r}_{\mathrm{d}}(t) \in \mathbb{R}^m$ for the manipulator's end-effector to track, how can we generate the corresponding joint trajectory $\theta(t) \in \mathbb{R}^n$ in real time t? Note that $n > m$ and, thus, there exist multiple or an infinite number of feasible solutions to the inverse-kinematic problem. In this sense, the redundancy of joint motion, together with kinematic and dynamic nonlinearities, complicate the manipulator control problem significantly.

However, by resolving the redundancy problem properly, the manipulators can avoid joint physical limits and obstacles, and optimize various secondary criteria [1, 78, 183, 184]. To take full advantage of the redundancy, various computational schemes have been developed. The conventional solution to such an inverse-kinematic problem is the pseudoinverse-based formulation; that is, one minimum-norm particular solution plus a homogeneous solution [55, 173]. The research of recent years [15, 36, 90, 189] shows that the redundancy-resolution problem can be solved more favorably via online optimization techniques in a different way.

The block diagram of the dynamic configuration of SMP for redundant manipulators is shown in Figure 10.1, which describes the main problems under study. Our research in this chapter is focused on the redundancy resolution of QP-based SMP scheme for redundant manipulators with the pseudoinverse method compared and singularities discussed. Note that such an SMP can be used for avoiding joint physical limits, obstacles, and singularities, and also for the situation where the manipulator's end-effector is locked [184, 186]. Consider a person who is doing push-up exercise with his/her palms still and we can see an application of the self-motion. Thus, self-motion research has practical value and potential biological background. This is one of the motivations of the research. It is worth noting that many methods and applications are presented for SMP of redundant manipulators [176, 177, 184, 186, 190–192]. For example, in [184], a method using self-motion is presented for online collision avoidance of redundant robots with obstacles. In [176], the concept of self-motion manifold is

Robot Manipulator Redundancy Resolution, First Edition. Yunong Zhang and Long Jin.

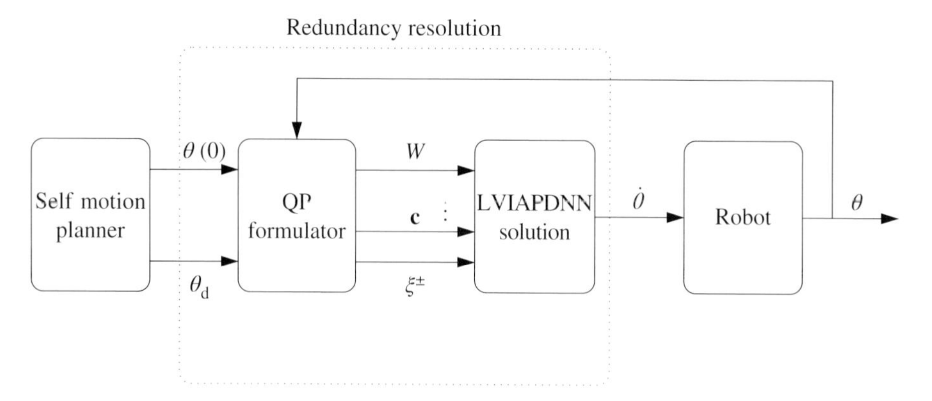

Figure 10.1 Block diagram of dynamic configuration of SMP for a redundant manipulator.

presented, together with a proposition that, in the case of robots with revolute joints, the number of disjoint self-motion manifolds cannot exceed the maximum number of inverse-kinematic solutions of a non-redundant manipulator of the same class. In addition, some potential applications of self-motion in redundant robots are discussed in [177].

In this chapter, a modified performance index is presented for SMP of redundant manipulators. The feasibility of such an index is proved through the theoretical derivation. The modified index, together with joint physical limits, is further formulated into a QP-based SMP scheme subject to equality and bound constraints. As an efficient real-time QP solver, the primal-dual neural network based on linear variational inequality (LVIAPDNN) presented in previous chapters, which has a simple piecewise-linear dynamic structure and global exponential convergence to optimal solution, is generalized from the authors' previous work and applied to the QP-based SMP scheme. For comparison purposes, the classical pseudoinverse-based method is presented for SMP of redundant manipulators. Computer-simulation results based on a three-link planar manipulator, PUMA560 manipulator, and PA10 manipulator further substantiate the efficacy and superiority of the LVIAPDNN solver and the QP-based SMP scheme. Besides, the effect of design parameters λ and γ is investigated, and the singularities of the self-motion for redundant manipulators are discussed.

10.2 Preliminaries and Scheme Formulation

For readers' convenience and also for this chapter's completeness, the general description of the redundancy-resolution problem is given again as follows. Considering a redundant manipulator, of which the end-effector pose vector $\mathbf{r}_{\mathrm{e}}(t) \in \mathbb{R}^m$ in Cartesian space is related to the joint-space vector $\theta(t) \in \mathbb{R}^n$ through the forward kinematic equation: $\mathbf{f}(\theta) = \mathbf{r}_{\mathrm{e}} \rightarrow \mathbf{r}_{\mathrm{d}}$, where $\mathbf{f}(\cdot)$ is a differentiable nonlinear function with a known structure and parameters for a given manipulator, and $\mathbf{r}_{\mathrm{d}} \in \mathbb{R}^m$ denotes the desired end-effector pose vector. However, this equation is usually difficult to solve due to its nonlinearity and redundancy. Such problem solving is then considered at the joint-velocity level as $J(\theta)\dot{\theta} = \dot{\mathbf{r}}_{\mathrm{e}}$, where $\dot{\mathbf{r}}_{\mathrm{e}}$ and $\dot{\theta}$ denote the m-dimensional end-effector velocity vector and the n-dimensional joint velocity vector, respectively. Jacobian matrix $J(\theta)$ is defined as $J(\theta) = \partial \mathbf{f}(\theta)/\partial\theta \in \mathbb{R}^{m\times n}$.

10.2.1 Modified Performance Index for SMP

For the repetitive motion planning (RMP) [104], based on the previous two chapters, the following quadratic performance index to be minimized is presented directly:

$$\frac{1}{2}(\dot{\theta} + \mathbf{c})^{\mathrm{T}}(\dot{\theta} + \mathbf{c}) \quad \text{with } \mathbf{c} = \lambda(\theta - \theta_{\mathrm{d}}), \tag{10.1}$$

where θ_{d} is the desired joint configuration (or called the desired joint state), and $\lambda > 0$ is a positive design parameter used to scale the magnitude of the joint displacements. The correctness and effectiveness of such an SMP performance index can be verified via the following GD method, which is presented in Chapter 1.

Firstly, to achieve the self-motion, we define the following energy function: $\mathcal{E} = \|\theta(t) - \theta_{\mathrm{d}}\|_2^2/2$. Evidently, the minimum value of the function $\mathcal{E}$ is obtained when $\theta(t) - \theta_{\mathrm{d}} = 0$ (i.e., the joint configuration achieves the desired one).

Secondly, using the GD method presented in Chapter 1, we set

$$\dot{\theta} = -\lambda(\partial\mathcal{E}/\partial\theta) = -\lambda(\theta - \theta_{\mathrm{d}}),$$

which can further be written as $\dot{\theta} + \lambda(\theta - \theta_{\mathrm{d}}) = 0$.

Thirdly, considering that joint physical limits always exist in robot manipulators, the dynamic equation $\dot{\theta} + \lambda(\theta - \theta_{\mathrm{d}}) = 0$ can thus be achieved only theoretically. Instead, minimizing $\|\dot{\theta} + \lambda(\theta - \theta_{\mathrm{d}})\|_2^2/2$ appears to be more reasonable in practice for SMP of redundant manipulators. It follows that expanding $\|\dot{\theta} + \lambda(\theta - \theta_{\mathrm{d}})\|_2^2/2$ yields $(\dot{\theta} + \lambda(\theta - \theta_{\mathrm{d}}))^{\mathrm{T}}(\dot{\theta} + \lambda(\theta - \theta_{\mathrm{d}}))/2$. By defining $\mathbf{c} = \lambda(\theta - \theta_{\mathrm{d}})$, it is proved readily that minimizing this expression is equivalent to minimizing the performance index $(\dot{\theta} + \mathbf{c})^{\mathrm{T}}(\dot{\theta} + \mathbf{c})/2$.

As we know, almost all manipulators have joint physical limits, and if the limits are violated, the primary end-effector task can not be fulfilled successfully, not to mention additional physical damages possibly caused. It is thus more realistic and useful for robotic practitioners to consider simultaneously the avoidance of such joint physical limits. Hence, we have the following scheme formulation for SMP of redundant manipulators (note that, for the self-motion, the end-effector remains motionless, i.e., $\dot{\mathbf{r}}_{\mathrm{d}} = 0$):

$$\text{minimize} \quad \frac{1}{2}(\dot{\theta} + \mathbf{c})^{\mathrm{T}}(\dot{\theta} + \mathbf{c}) \text{ with } \mathbf{c} = \lambda(\theta - \theta_{\mathrm{d}}) \tag{10.2}$$

$$\text{subject to} \quad J(\theta)\dot{\theta} = \dot{\mathbf{r}}_{\mathrm{d}} = 0, \tag{10.3}$$

$$\theta^- \leqslant \theta \leqslant \theta^+, \tag{10.4}$$

$$\dot{\theta}^- \leqslant \dot{\theta} \leqslant \dot{\theta}^+, \tag{10.5}$$

where superscripts $^+$ and $^-$ denote the upper and lower limits of the joint variable vector (e.g., θ or $\dot{\theta}$), respectively.

10.2.2 QP-Based SMP Scheme Formulation

As SMP is resolved at the joint-velocity level (i.e., in terms of $\dot{\theta}$), in view of [36, 104], the SMP scheme (10.2)–(10.5) can be reformulated and unified as the following QP (termed the QP-based SMP scheme) with duration T (i.e., $t \in [0, T]$):

$$\text{minimize} \quad \frac{1}{2}\dot{\theta}^{\mathrm{T}} W \dot{\theta} + \mathbf{c}^{\mathrm{T}}\dot{\theta} \tag{10.6}$$

$$\text{subject to} \quad J\dot{\theta} = \mathbf{b}, \tag{10.7}$$

$$\xi^{-} \leqslant \dot{\theta} \leqslant \xi^{+}, \tag{10.8}$$

where coefficients $W = I_n$ (with I_n denoting the $n \times n$ identity matrix) and $\mathbf{b} = \dot{\mathbf{r}}_{\mathrm{d}} = 0$, and the ith elements of ξ^{-} and ξ^{+} are defined, respectively, as

$$\xi_i^{-} = \max\{\dot{\theta}_i^{-}, \mu(\theta_i^{-} - \theta_i)\}, \;\; \xi_i^{+} = \min\{\dot{\theta}_i^{+}, \mu(\theta_i^{+} - \theta_i)\},$$

with intensity coefficient $\mu > 0$ being used to scale the feasible region of $\dot{\theta}$ caused by the above transformation.

10.3 LVIAPDNN Assisted QP Solution with Discussion

In view of high-performance parallel-computation ability and hardware-realization convenience, the neural-network approach is regarded as a powerful alternative to online optimization problems solving [46, 137, 193]. In this section, for readers' convenience and also for this chapter's completeness, the LVIAPDNN solver is presented again. LVIAPDNN has a simple piecewise-linear dynamic structure, global exponential convergence to the optimal solution, and capability of handling general QP problems in an inverse-free manner. The design procedure of LVIAPDNN can be generalized from the previous chapters.

Here, the LVIAPDNN solver is presented directly by the following dynamic equation:

$$\dot{\mathbf{y}} = \gamma(I + Q^{\mathrm{T}})(\mathbf{P}_{\Omega}(\mathbf{y} - (Q\mathbf{y} + \mathbf{p})) - \mathbf{y}), \tag{10.9}$$

where design parameter $\gamma > 0$, and the primal-dual decision variable vector $\mathbf{y} \in \mathbb{R}^N$ (with $N = n + m$) and its upper/lower bounds $\mathbf{y}^{\pm}$ are defined, respectively, as

$$\mathbf{y} = \begin{bmatrix} \dot{\theta} \\ \mathbf{u} \end{bmatrix}, \mathbf{y}^{+} = \begin{bmatrix} \xi^{+} \\ \varpi \end{bmatrix} \in \mathbb{R}^N, \mathbf{y}^{-} = \begin{bmatrix} \xi^{-} \\ -\varpi \end{bmatrix} \in \mathbb{R}^N,$$

with $\mathbf{y} \in \mathbb{R}^m$ being the dual decision vector defined corresponding to equality constraint (10.7), and with constant $\varpi \gg 0 \in \mathbb{R}^m$ defined to replace the m-dimensional $+\infty$ numerically. Matrix Q and vector $\mathbf{p}$ are augmented as

$$Q = \begin{bmatrix} W & -J^{\mathrm{T}} \\ J & 0 \end{bmatrix} \in \mathbb{R}^{N \times N}, \quad \mathbf{p} = \begin{bmatrix} \mathbf{c} \\ -\mathbf{b} \end{bmatrix} \in \mathbb{R}^N.$$

Remark 10.1 If constraint (10.8) is always satisfied or there is no constraint (10.8), the QP-based SMP scheme (10.6)–(10.8) reduces to the following equality-constrained optimization problem:

$$\text{minimize} \quad \frac{1}{2}\dot{\theta}^{\mathrm{T}} W \dot{\theta} + \mathbf{c}^{\mathrm{T}}\dot{\theta} \tag{10.10}$$

$$\text{subject to} \quad J\dot{\theta} = \mathbf{b}, \tag{10.11}$$

which can be solved by using the pseudoinverse-based method analytically as

$$\dot{\theta} = -(I_n - J^{\mathrm{T}}(JJ^{\mathrm{T}})^{-1}J)\mathbf{c}, \tag{10.12}$$

It is proved that, in the situation with bound constraint (10.8) satisfied (i.e., $\mathbf{P}_{\Omega}(\mathbf{y}) - \mathbf{y}$), the upper part of (10.9) in the limit tends to (10.12). The detailed derivation is addressed in the Appendix of this chapter.

Remark 10.2 It is worth noting that the pseudoinverse-based method (10.12) requires a Jacobian pseudoinverse solving at each sampling point, which would increase the computational complexity. In contrast, LVIAPDNN solver (10.9) can avoid calculating the Jacobian pseudoinverse, and thus it takes much less time than the pseudoinverse-based method at each sampling point. It is observed from numerical tests that the LVIAPDNN computing time at a sampling point is about 5.3×10^{-5} s and the pseudoinverse-method computing time at a sampling point is about 3.1×10^{-4} s, about 5.85 times larger than the former. In addition, the pseudoinverse-based method would be difficult in handling the inequality-based physical limits. That is, when constraint (10.8) is not taken into account, the pseudoinverse-based method can generate a self-motion solution; and, because constraint (10.8) does exist and is then imposed, the pseudoinverse-based method actually loses its efficacy. This will be shown in Section 10.4. In the same situation, LVIAPDNN solver (10.9) is still effective and efficient, thus preferred.

Moreover, it is worth discussing the reason why dynamic equation (10.9) corresponds to an LVIAPDNN solver. To the authors' knowledge, a recurrent neural network (RNN) can be viewed as the implementation of a dynamic system in physical devices [1]. That is, an RNN appears in the form of a networked hardware-realizable dynamic system [e.g., dynamic system (10.9)]. So, in this sense, dynamic equation (10.9) corresponds to a recurrent neural network. Because dynamic equation (10.9) is designed based on the conversion of QP (10.6)–(10.8) to an LVI problem and utilizes both primal and dual decision variables, it is thus termed LVIAPDNN. For more details, please refer to the authors' previous works (e.g., [137, 193]). In addition, for a better understanding of dynamic equation (10.9) implemented as an LVIAPDNN solver, its structure diagram is provided in Figure 10.2 and explained as follows.

Corresponding to Figure 10.2, dynamic equation (10.9) is expressed in the ith neuron form (with $i = 1, 2, \cdots, N$):

$$\dot{y}_i = \sum_{k=1}^{N} g_{ik}\left(P_{\Omega k}\left(\sum_{j=1}^{N} h_{kj}y_j - p_k\right) - y_k\right),$$

where h_{kj} denotes the kjth entry of matrix $H = I - Q$, g_{ik} denotes the ikth entry of combined scaling matrix $G = \gamma(I + Q^{\mathrm{T}})$, and $P_{\Omega k}(\cdot)$ denotes the kth processing element of activation function array $\mathbf{P}_{\Omega}(\cdot)$. From Figure 10.2, which consists of three layers with each layer having N neurons, we can see that, in the first layer, the piecewise linear activation function $P_{\Omega k}(\cdot)$ is employed in the kth neuron and $-p_k$ is the bias for the kth neuron, with $k = 1, 2, \cdots, N$. In addition, the piecewise linear activation function $P_{\Omega k}(\cdot)$ can be realized by using operational amplifiers known as limiters [1]. The second layer is constructed by using linear neurons, each of which sums up correspondingly

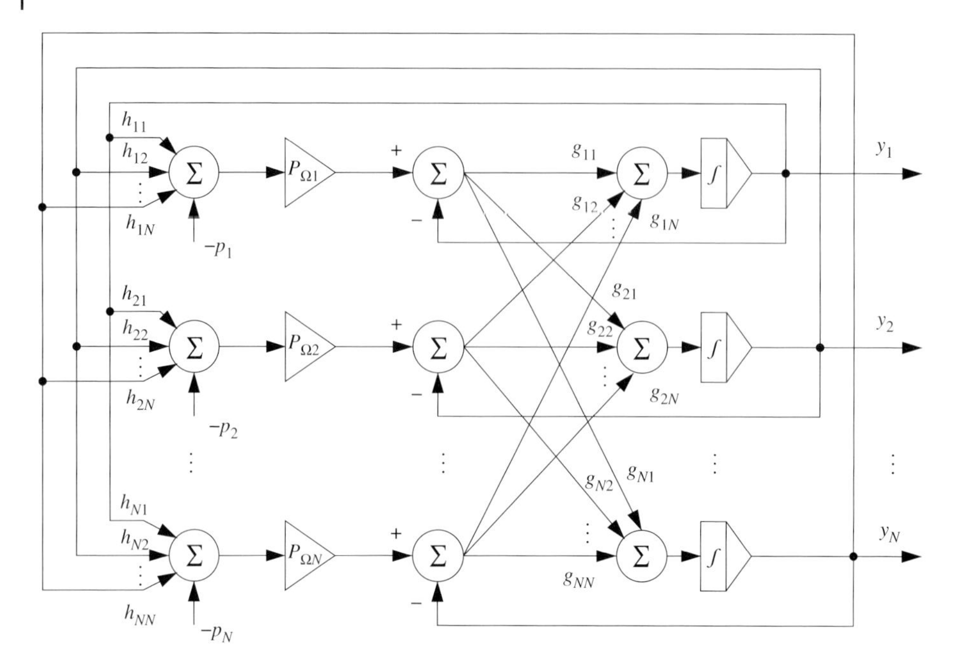

Figure 10.2 LVIAPDNN structure corresponding to dynamic equation (10.9).

the first-layer neuron output and the negative feedback of the last-layer neuron output. In the third layer, integrating neurons are employed to complete the neural network structure and generate the network outputs $u_i, i = 1, 2, \cdots N$. When such an LVI-aided primal-dual neural network (10.9) works, the solution to QP-based SMP scheme (10.6)–(10.8) can be established online. Furthermore, we have the following theoretical results on global exponential convergence of LVIAPDNN solver (10.9) [137, 193], as well as a remark about the initialization of LVIAPDNN solver (10.9).

According to the theoretical analysis of the previous chapter, we know that LVI-APDNN solver (10.9) possesses a global exponential convergence property, with the exponential convergence rate being proportional to $\gamma\rho$. Also, we have the following remark about the initialization of LVIAPDNN solver (10.9).

Remark 10.3 The variable vector $\mathbf{y}$ of (10.9) is composed of two parts: primal decision variable vector $\dot{\theta}$ and dual decision variable vector $\mathbf{y}$ [corresponding to equality constraint (10.7)]. As stated in the previous chapter, the initial condition $\mathbf{y}(0)$ can be randomly generated. However, because the initial velocity $\dot{\theta}(0)$ of redundant manipulators is generally zero at the beginning in practice, we set the initial primal decision variable $\dot{\theta}(0)$ to zero, while initial dual decision variable $\mathbf{y}(0)$ is randomly generated. In addition, without loss of generality, initial configuration $\theta(0)$ is prescribed by users for a redundant manipulator. Thus, with these initial conditions, MATLAB routine "ode15s" can be used to simulate such a dynamic system (10.9).

In summary, these analyses show the efficacy and superiority of LVIAPDNN solver (10.9), as compared with the classical pseudoinverse-based method (10.12). Therefore, it can be seen as a new method presented to solve such a self-motion QP in a different way.

10.4 Computer Simulations

In this section, the presented QP-based SMP scheme and LVIAPDNN are tested on a three-link planar manipulator, PUMA560 manipulator, and PA10 manipulator. Because QP (10.6)–(10.8) has a zero solution at least (i.e., $\dot{\theta} = 0$), we have no special assumption about kinematic singularities in this chapter. Generally speaking, initial state $\theta(0)$ and desired state θ_d can be selected quite arbitrarily by a user for performing a self-motion. However, due to the existence of various kinds of physical limits in those manipulators, from an initial state of a robot manipulator (illustrated in Figure 10.3 using a three-link planar robot as an example) and with the robot end-effector required to be kept motionless, generally, the following three situations are possibly encountered in practice.

1) The initial state $\theta(0)$ can change to the desired state θ_d. Evidently, in this situation, $\theta(0)$ and θ_d satisfy the relationship $\mathbf{f}(\theta(0)) = \mathbf{f}(\theta_d)$ and θ_d lies in the self-motion manifold passing through $\theta(0)$. This, as illustrated via Figure 10.3(a), is a usual situation mainly investigated in the chapter.
2) The initial state $\theta(0)$ can not change to another state; or to say, QP (10.6)–(10.8) has only a zero solution, that is, $\dot{\theta} = 0$. For example, if the three-link planar manipulator is at a straight-line configuration as shown in Figure 10.3(b), it cannot and will not change to another configuration because its end-effector is required to be motionless [and, in this situation, LVIAPDNN solver (10.9) generates $\dot{\theta} = 0$]. This can be viewed as a kinematic singularity problem related to self-motion planning, and the presented method solves the problem correctly by generating $\dot{\theta} = 0$.
3) The initial state $\theta(0)$ can change to another state, but not the desired one (i.e., θ_d). This is generally because the θ_d provided by a user does not lie in the self-motion manifold passing through $\theta(0)$; in mathematics, $\mathbf{f}(\theta_d) \neq \mathbf{f}(\theta(0))$. For example, please see Figure 10.3(c). This can also be viewed as a singularity problem related to self-motion planning. However, in this situation, LVIAPDNN solver (10.9) still works well by generating an optimal solution θ_f which minimizes $\|\theta(t) - \theta_d\|_2^2$ with $\mathbf{f}(\theta(t)) = \mathbf{f}(\theta(0))$ satisfied. This interesting simulation result of this situation (using PA10 as an example) is discussed at the end of this section.

Without loss of generality, we set design parameter $\gamma = 10^9$ and intensity coefficient $\mu = 4$ for those robot manipulators.

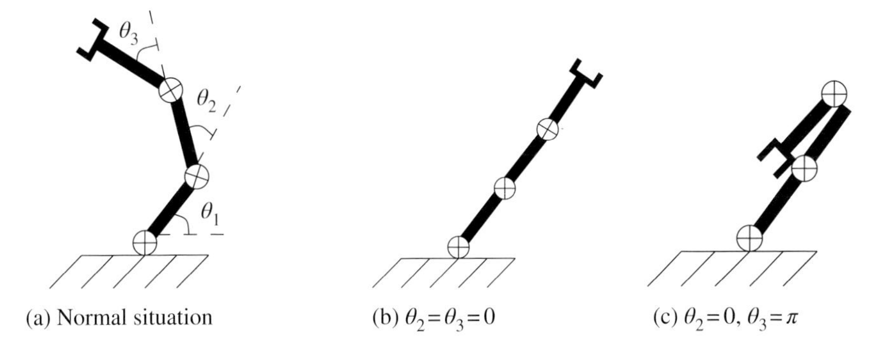

Figure 10.3 Examples of three initial states of self-motion for a three-link redundant planar manipulator.

10.4.1 Three-Link Redundant Planar Manipulator

The three-link redundant planar manipulator is of three DOF, and has one redundant degree with respect to a two-dimensional workspace (because we only consider the end-effector position). The joint angle limits and joint velocity limits are set $\theta^{\pm} = \pm\pi$ rad and $\dot{\theta}^{\pm} = \pm\pi$ rad/s, respectively. Without loss of generality, we assume initial configuration $\theta(0) = [\pi/4, 0, -\pi/4]^{\mathrm{T}}$ rad and desired configuration $\theta_{\mathrm{d}} = [\pi/4, -\pi/4, \pi/4]^{\mathrm{T}}$ rad.

10.4.1.1 Verifications

In this simulation test, LVIAPDNN solver (10.9) is applied to the three-link redundant planar manipulator with design parameter $\lambda = 2$. Figure 10.4 shows the self-motion trajectories and end-effector position error of the three-link planar manipulator. It is seen from Figure 10.4(a) that the three-link manipulator moves from the initial configuration to the desired configuration successfully. The position error (also termed the position change) of the end-effector shown in Figure 10.4(b) is less than 3.4×10^{-8} m. It is very

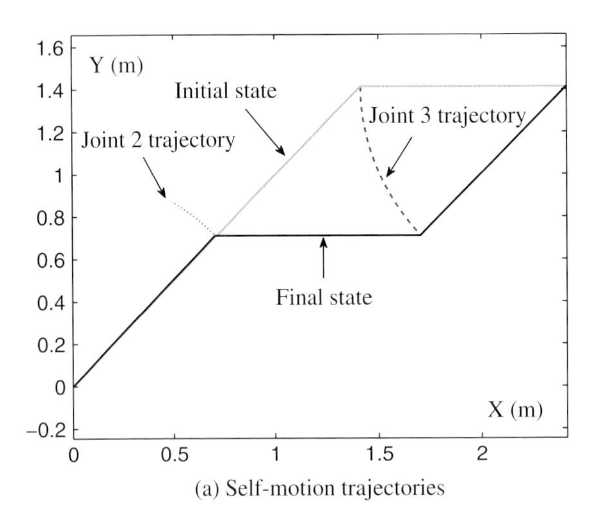

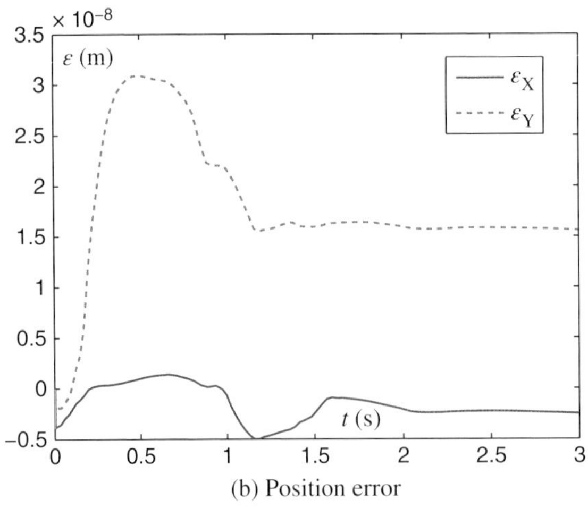

Figure 10.4 (a) Motion trajectories and (b) end-effector position error of a three-link robot performing self-motion with $\lambda = 2$.

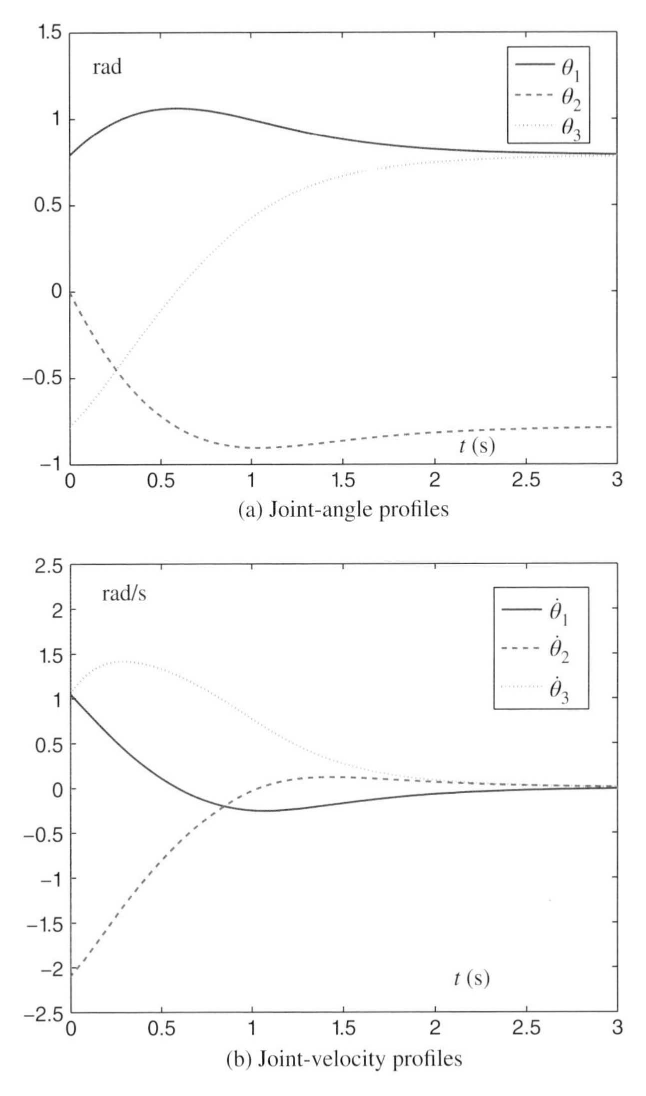

Figure 10.5 (a) Joint-angle and (b) joint-velocity profiles of a three-link robot performing self-motion with $\lambda = 2$.

acceptable in practice for self-motion planning of redundant manipulators. Figure 10.5 shows the joint-angle and joint velocity profiles of the three-link planar manipulator performing self-motion. As seen from Figure 10.5(a), the joint variables (i.e., θ) converge to the desired values within around 3 s. In Figure 10.5(b), all joint velocity variables (i.e., $\dot{\theta}$) approach 0 within around 3 s too. That is to say, 3 s are needed for joint variables coming to the desired values. It is worth mentioning that, due to the consideration of bound constraint (10.8), all joint variables and joint velocity variables are kept within their limits (i.e., $\pm\pi$), as shown in Figure 10.5. Furthermore, the errors between the actual joint values (after self-motion) and the desired joint values are shown in Table 10.1, which appear to be less than 0.0051 rad. It means that the final configuration is close enough

Table 10.1 Joint values and errors (rad) of three-link robot before and after self-motion with $\lambda = 2$.

Joint #	Initial configuration $\theta(0)$	Desired configuration θ_d
1	+0.78539816339745 $(+\pi/4)$	+0.78539816339745 $(+\pi/4)$
2	+0.00000000000000	−0.78539816339745 $(-\pi/4)$
3	−0.78539816339745 $(-\pi/4)$	+0.78539816339745 $(+\pi/4)$
Joint #	**Final configuration θ_f**	**Joint error $\theta_f - \theta_d$**
1	+0.79041450189865	+0.00501633850120
2	−0.79037874780731	−0.00498058440986
3	+0.78032077528588	−0.00507738811157

Table 10.2 Final joint values and errors (rad) of a three-link robot before and after self-motion with $\lambda = 4$.

Joint #	Final configuration θ_f	Joint error $\theta_f - \theta_d$
1	+0.78541143492341	$+1.327152596 \times 10^{-5}$
2	−0.78541129581962	$-1.313242217 \times 10^{-5}$
3	+0.78538452711143	$-1.363628602 \times 10^{-5}$

Table 10.3 Final joint values and errors (rad) of a three-link robot before and after self-motion with $\lambda = 8$.

Joint #	Final configuration θ_f	Joint error $\theta_f - \theta_d$
1	+0.78539807344451	$-0.8995294 \times 10^{-7}$
2	−0.78539788134918	$+2.8204827 \times 10^{-7}$
3	+0.78539779050265	$-3.7289480 \times 10^{-7}$

to the desired configuration. In summary, the self-motion planning is conducted well by the presented QP-based SMP scheme.

In addition, the QP-based SMP scheme performs much more efficiently and accurately by increasing the value of design parameter λ. For illustrative and comparative purposes, the errors between the actual joint values (after self-motion) and the desired joint values with $\lambda = 4$ are shown in Table 10.2, which are less than 1.364×10^{-5} rad. In addition, all joint and joint velocity variables converge faster than those in the situation with $\lambda = 2$, and the convergence time (about 1.5 s) is less than the convergence time of the previous simulation. Table 10.3 shows the errors between the actual joint values (after self-motion) and the desired joint values with $\lambda = 8$, which are less than 3.73×10^{-7} rad. These imply that the final configuration is much closer to the desired configuration than that in the previous simulations. Note that the convergence time (about 1.2 s) is just a little shorter than that in the previous simulation with $\lambda = 4$. In view of the authors' further simulation results with other large values of design parameter λ, the minimum convergence time exists, that is, about 1 s. This is because the joint velocity limits $\dot{\theta}^{\pm}$ exist and are considered in the QP-based SMP scheme.

Thus, we can conclude that the QP-based SMP scheme is effective on self-motion planning, and that design parameter λ plays an important role in the performance index. It is worth pointing out additionally that we can get similar results on the role of design parameter λ via other simulations based on different redundant manipulators.

10.4.1.2 Comparisons

In this simulation test, for illustration and comparison, the pseudoinverse-based method (10.12) is applied to the three-link planar manipulator. Firstly, when constraint (10.8) is not taken into account, Figure 10.6 shows that the pseudoinverse-based method can make the three-link planar manipulator achieve the self-motion task successfully

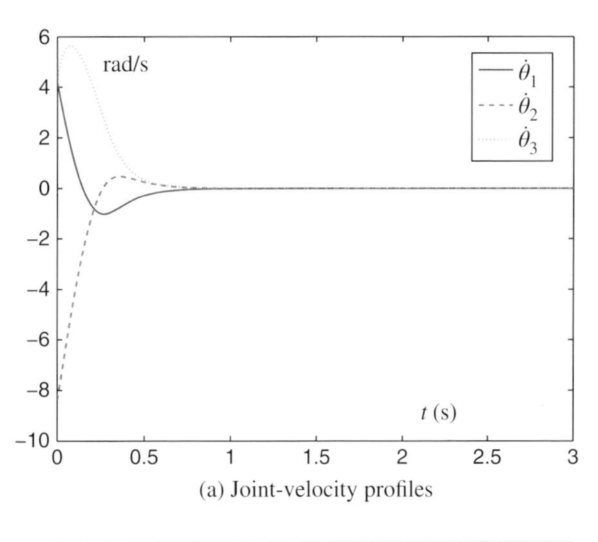

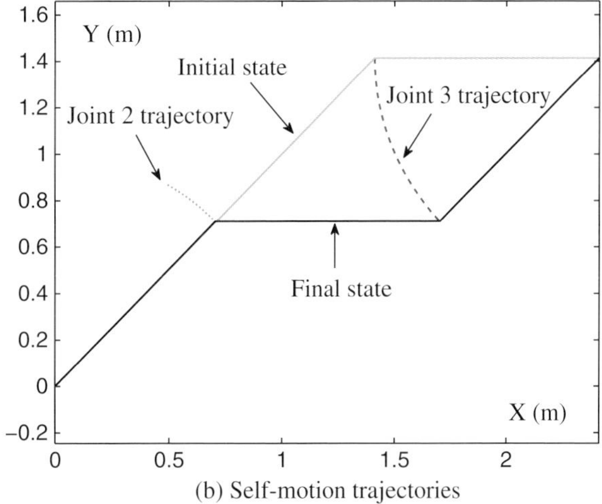

Figure 10.6 (a) Joint-velocity profiles and (b) motion trajectories of a three-link planar robot performing self-motion synthesized by the pseudoinverse-based method (10.12) with $\lambda = 8$ and without considering constraint (10.8).

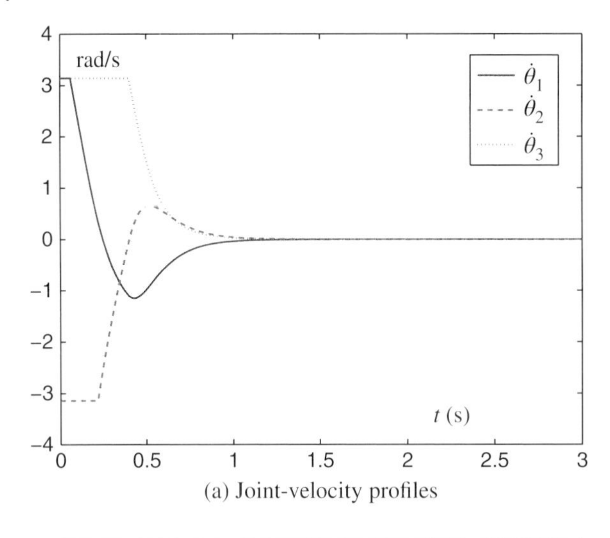

(a) Joint-velocity profiles

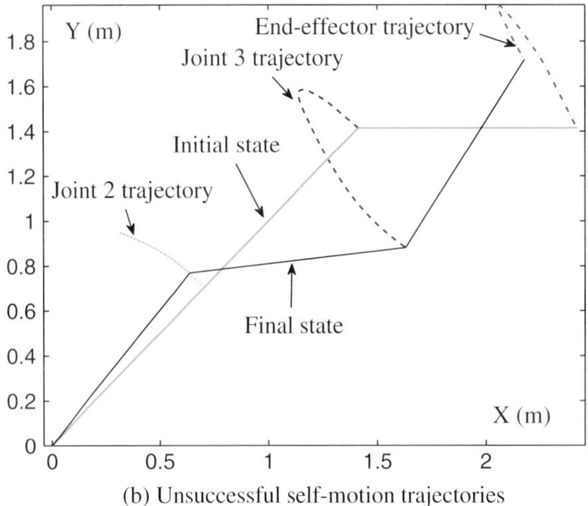

(b) Unsuccessful self-motion trajectories

Figure 10.7 (a) Joint-velocity profiles and (b) motion trajectories of a three-link planar robot performing self-motion synthesized by the pseudoinverse-based method (10.12) with $\lambda = 8$ and with constraint (10.8) imposed.

under the same conditions and with design parameter $\lambda = 8$. However, when joint physical constraint (10.8) is imposed, from Figure 10.7, we see that the end-effector of the three-link planar manipulator moves during the "self-motion" process, which implies the pseudoinverse-based method has lost its efficacy. In contrast, in this situation, the LVIAPDNN solver is still effective and efficient, which can be seen from Figure 10.8. This shows the efficacy and superiority of LVIAPDNN solver (10.9) as compared with the traditional pseudoinverse-based method (10.12). Therefore, (10.9) can be seen as a new method presented to solve the self-motion QP problem in a different way.

10.4.2 PUMA560 Robot Manipulator

The PUMA560 robot is a spatial manipulator with six joints [1, 194]. When we consider only the position of the end-effector, PUMA560 robot becomes a functionally

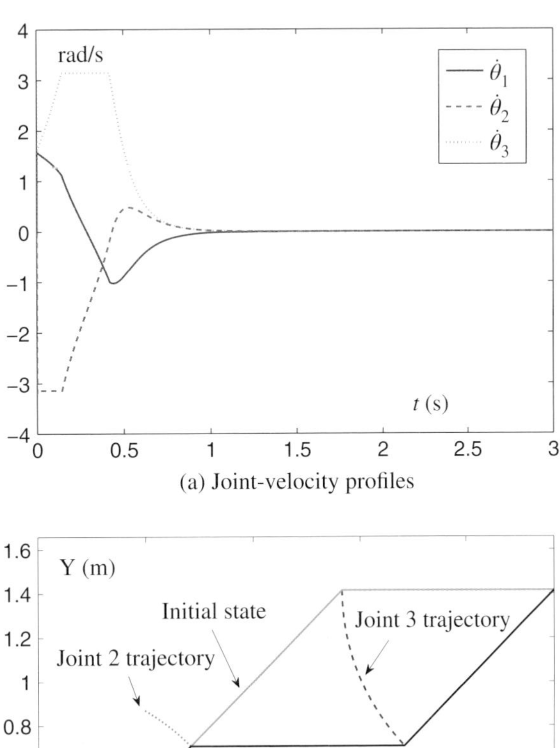

(a) Joint-velocity profiles

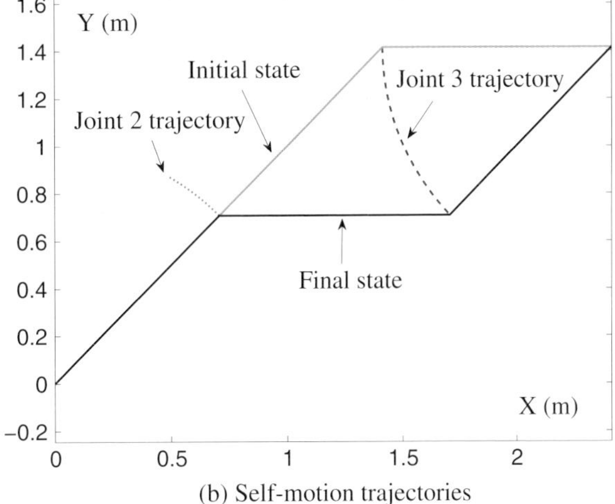

(b) Self-motion trajectories

Figure 10.8 (a) Joint-velocity profiles and (b) motion trajectories of a three-link planar robot performing self-motion synthesized by LVIAPDNN solver (10.9) with $\lambda = 8$ and with constraint (10.8) imposed as well.

redundant manipulator. The limited joint ranges are $[-2.775, 2.775]$, $[-3.892, 0.750]$, $[-0.785, 3.927]$, $[-1.919, 2.967]$, $[-1.745, 1.745]$, $[-4.625, 4.625]$ in radians, respectively, for θ_1 through θ_6, and the limited range of each joint velocity $\dot{\theta}_i$ is assumed to be $[-1.5, 1.5]$ rad/s, $i = 1, 2, \cdots, 6$. For illustration and for better readability, the values of initial and desired configurations are shown in Table 10.4. The computer-simulation results are shown in Figure 10.9 and Figure 10.10 as well as Table 10.4.

Figure 10.9(a) shows the three-dimensional motion trajectories of the PUMA560 robot manipulator. From it, we can see that the manipulator moves to the desired joint state (i.e., θ_{d}) from the initial joint state (i.e., $\theta(0)$) with the end-effector remaining at the same position. The position error of the end-effector depicted in Figure 10.9(b) is less than 5.4×10^{-9} m. As seen from Figure 10.10, all joint variables and joint velocity variables are convergent to their desired values and never exceeding their limited ranges. In addition, the errors between the actual joint-values (after self-motion)

Table 10.4 Joint values and errors (rad) of a PUMA560 robot before and after self-motion with $\lambda = 4$.

Joint #	Initial configuration $\theta(0)$	Desired configuration θ_d
1	+0.00000000000000	−0.32378107703583
2	+0.00000000000000	+0.14014270871832
3	+0.00000000000000	−0.28792626042232
4	+0.00000000000000	+1.97297001144022
5	+0.00000000000000	−0.69839386298614
6	+0.00000000000000	+0.00000000000000
Joint #	**Final configuration θ_f**	**Joint error $\theta_\mathrm{f} - \theta_\mathrm{d}$**
1	−0.32376069403331	$+0.2038300252105 \times 10^{-4}$
2	+0.14017617738045	$+0.3346866212825 \times 10^{-4}$
3	−0.28793486169026	$-0.0860126793734 \times 10^{-4}$
4	+1.97279199657110	$-1.7801486912217 \times 10^{-4}$
5	−0.69835615089934	$+0.3771208680281 \times 10^{-4}$
6	+0.00000000000000	$+0.0000000000000 \times 10^{-4}$

and the desired joint-values are shown in Table 10.4, which appear to be very small (i.e., of order 1×10^{-4} rad). Thus, the presented QP-based SMP scheme is also useful for the spatially-working PUMA560 robot manipulator.

It is worth discussing the reason why we select design parameter γ to be so large. As we may know [1], design parameter γ, being the reciprocal of a capacitance parameter in the hardware implementation, should be set as large as the hardware permits, or selected appropriately for experimental or simulative purposes. That is, in this situation, $\gamma = 10^9$ corresponds to 10^{-9} F (i.e., 1000 pF). This capacitance value can be acceptable in practice. In addition, as γ increases, the capacitance value will become smaller (such as 100pF, 10pF, and 1pF), which is also realistic and available in practical hardware implementation. That is why we select design parameter $\gamma = 10^9$. Moreover,

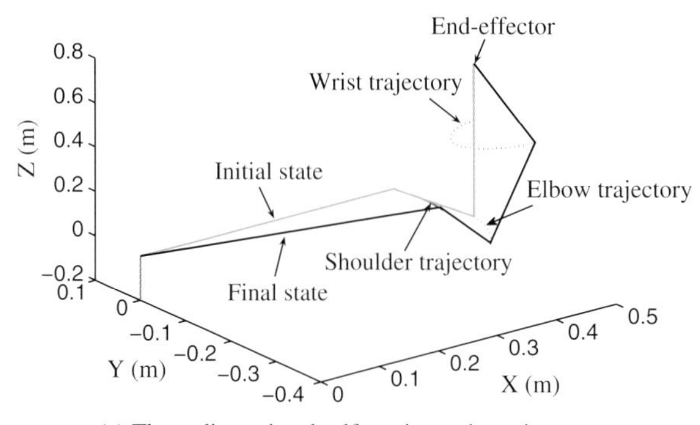

Figure 10.9 (a) Motion trajectories and (b) end-effector position error of PUMA560 robot performing self-motion.

Figure 10.9 *(Continued)*

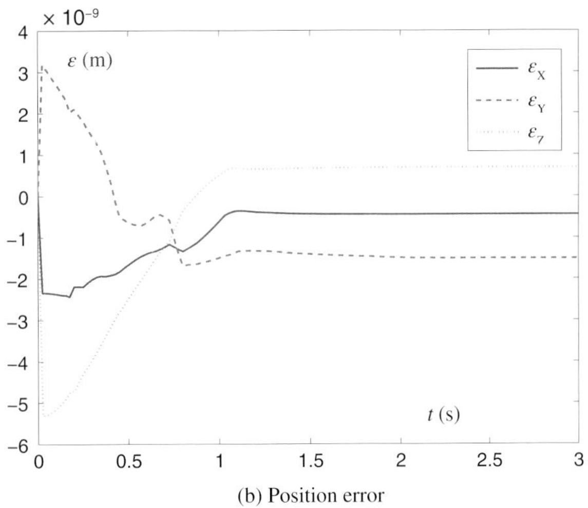

(b) Position error

Figure 10.10 (a) Joint-angle and (b) joint-velocity profiles of a PUMA560 robot performing self-motion with $\lambda = 4$.

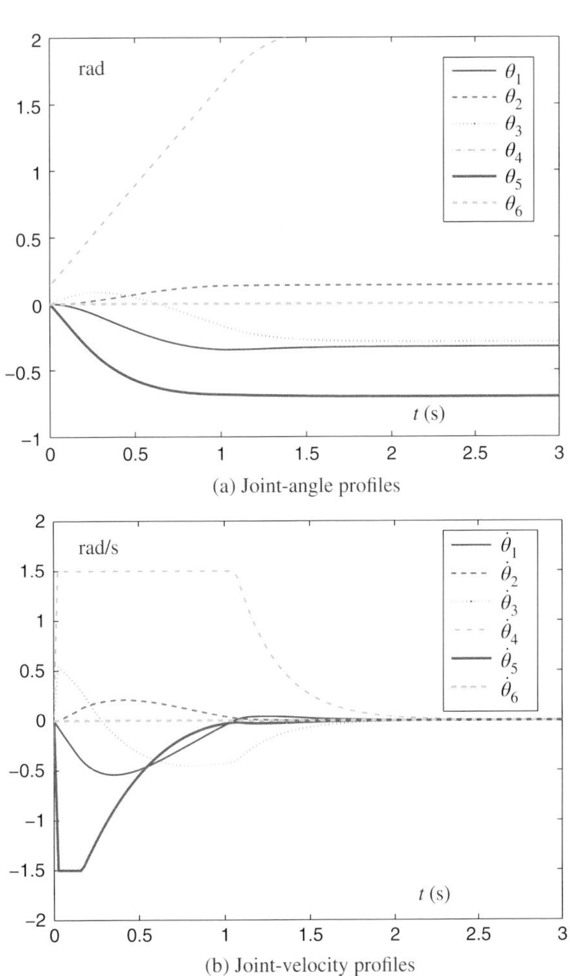

(b) Joint-velocity profiles

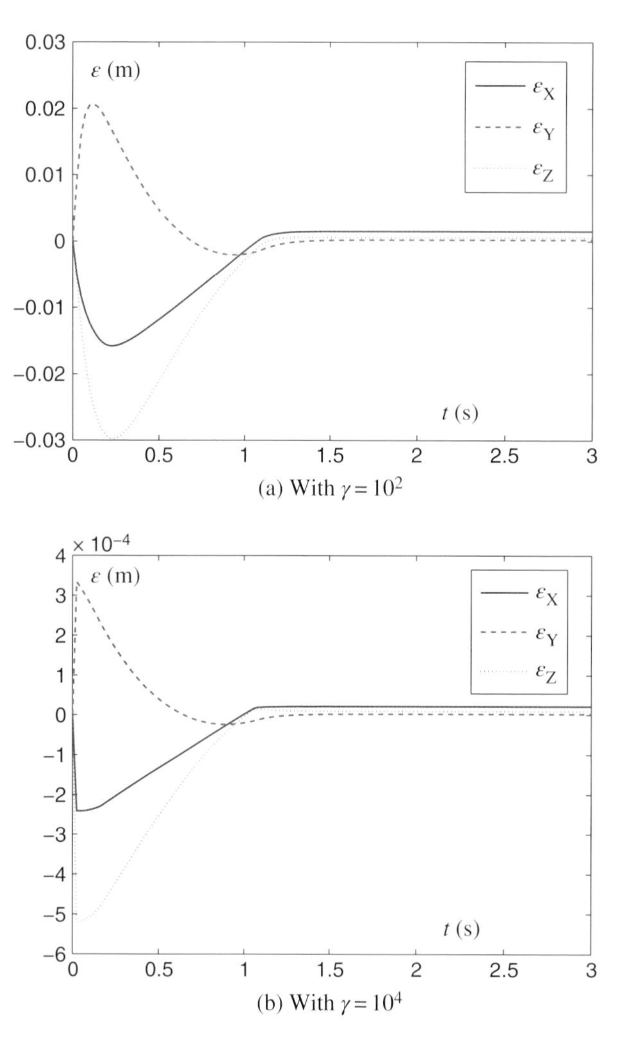

Figure 10.11 End-effector position errors of PUMA560 robot performing self-motion synthesized by LVIAPDNN solver (10.9) with different values of γ (i.e., (a) $\gamma = 10^2$ and (b) $\gamma = 10^4$).

by choosing different values of γ in the simulation tests, we see from Figure 10.11 and Figure 10.12 that the end-effector position error, resulting from the solution by the LVI-APDNN solver, can be decreased from the order of 10^{-2} m to the order of 10^{-8} m, as the value of γ is increased from $\gamma = 10^2$ to $\gamma = 10^8$. The simulation results illustrate that γ plays an important role in the convergence of state vector u of the LVIAPDNN solver to theoretical solution u^*.

10.4.3 PA10 Robot Manipulator

In this subsection, we consider only the end-effector position again, and thus PA10 robot manipulator has 4 degrees of redundancy, with $\theta^{\pm}$ and $\dot{\theta}^{\pm}$ given in [1]. Without loss of generality, the values of initial configurations and desired configurations are

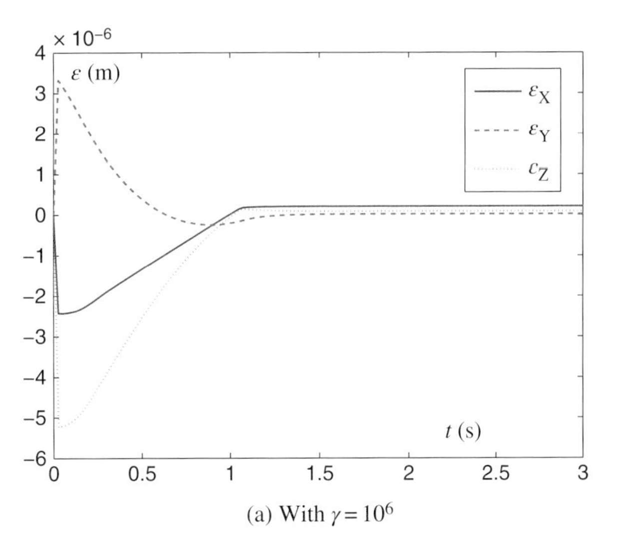

(a) With $\gamma = 10^6$

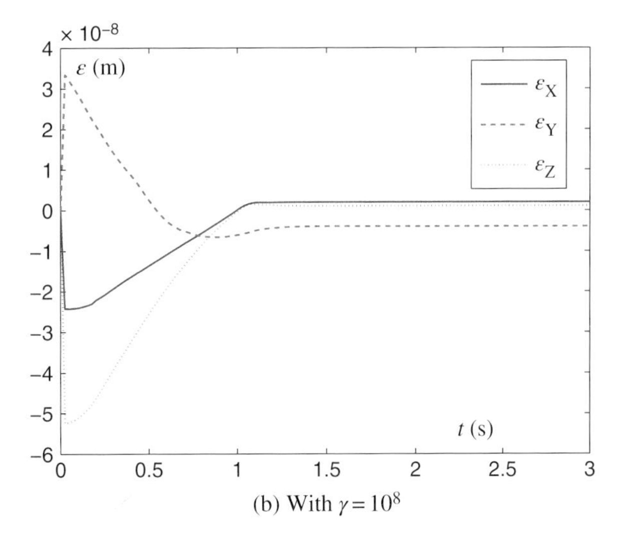

(b) With $\gamma = 10^8$

Figure 10.12 End-effector position errors of a PUMA560 robot performing self-motion synthesized by LVIAPDNN solver (10.9) with different values of γ (i.e., (a) $\gamma = 10^6$ and (b) $\gamma = 10^8$).

shown in Table 10.5 for illustration and verification purposes. As a further investigation, the LVIAPDNN solver is applied again to the PA10 robot manipulator. Figure 10.13, Figure 10.14, and Table 10.5 illustrate the computer-simulation results, which illustrate once more the effectiveness of the presented QP-based SMP scheme and the LVIAPDNN solver. That is to say, such a scheme can make the PA10 robot manipulator successfully realize self-motion.

As an interesting complement to previous simulations, in this simulation test, we use the same initial state $\theta(0)$ of PA10 as before, but use a "wrong" desired state $\theta_{\mathrm{d}} = [0, +\pi/4, 0, 0, \pi/3, -\pi/4, 0]$ rad, which does not lie in the self-motion manifold passing through $\theta(0)$, i.e., $\mathbf{f}(\theta_{\mathrm{d}}) \neq \mathbf{f}(\theta(0))$. The simulation result synthesized by the LVIAPDNN solver is shown in Figure 10.15. Specifically speaking, Figure 10.15(a) illustrates that the PA10 performs a self-motion successfully (i.e., with $\mathbf{f}(\theta(t)) = \mathbf{f}(\theta(0))$);

Table 10.5 Joint values and errors (rad) of a PA10 robot before and after self-motion with $\lambda = 4$.

Joint #	Initial configuration $\theta(0)$	Desired configuration θ_d
1	+0.00000000000000	+1.16008092352312
2	−0.78539816339744	−0.73646475061456
3	+0.00000000000000	−0.14572092931235
4	+1.57079632679489	+1.26018304599104
5	+0.00000000000000	+0.58487571470121
6	−0.78539816339744	+1.32105623270231
7	+0.00000000000000	+0.56135005321742
Joint #	**Final configuration θ_f**	**Joint error $\theta_f - \theta_d$**
1	+1.16002044686999	$-0.6047665312647 \times 10^{-4}$
2	−0.73647351686466	$-0.0876625009794 \times 10^{-4}$
3	−0.14577773286518	$-0.5680355282991 \times 10^{-4}$
4	+1.25995361137172	$-2.2943461932146 \times 10^{-4}$
5	+0.58488318585975	$+0.0747115854371 \times 10^{-4}$
6	+1.32097100512611	$-0.8522757619533 \times 10^{-4}$
7	+0.56134660440394	$-0.0344881348102 \times 10^{-4}$

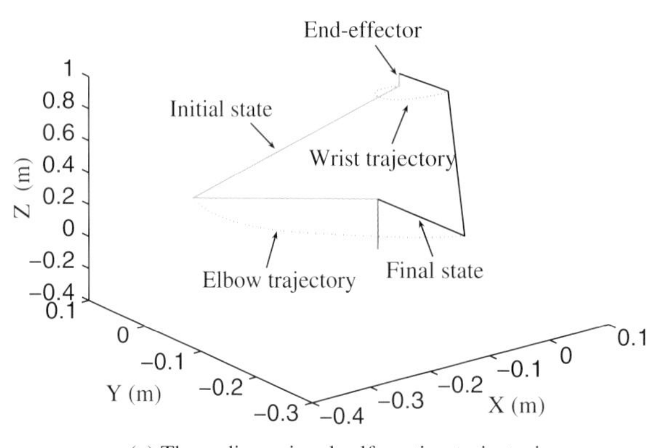

Figure 10.13 (a) Motion trajectories and (b) end-effector position error of a PA10 robot performing self-motion with $\lambda = 4$.

and, from Figure 10.15(b), we see further that θ_f minimizes $\|\theta(t) - \theta_d\|_2^2$ (though θ_f does not reach θ_d).

In summary, these computer-simulation results based on a three-link planar manipulator, PUMA560, and PA10 robot manipulators all have illustrated the efficacy and superiority of the presented QP-based SMP scheme. Furthermore, design parameter λ plays an important role in the quadratic performance index, which, together with γ, has to be selected appropriately to satisfy the solution precision we need in practice.

Figure 10.13 (*Continued*)

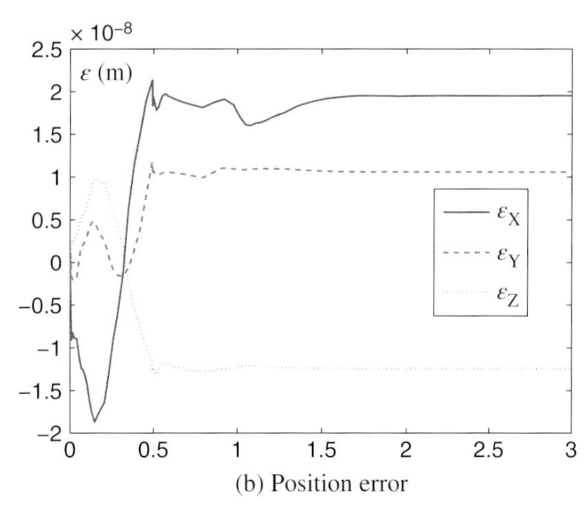

(b) Position error

Figure 10.14 (a) Joint-angle and (b) joint-velocity profiles of a PA10 robot performing self-motion with $\lambda = 4$.

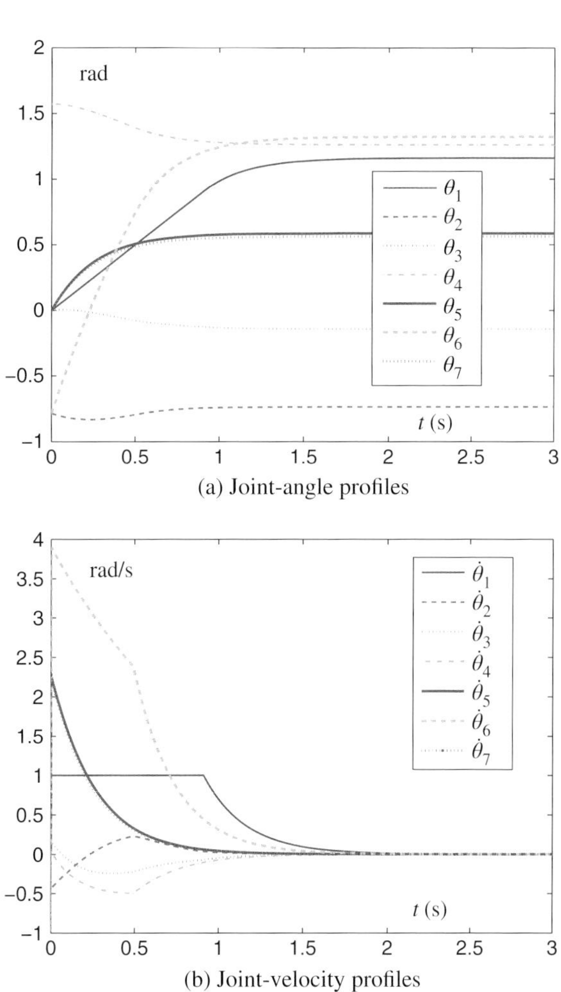

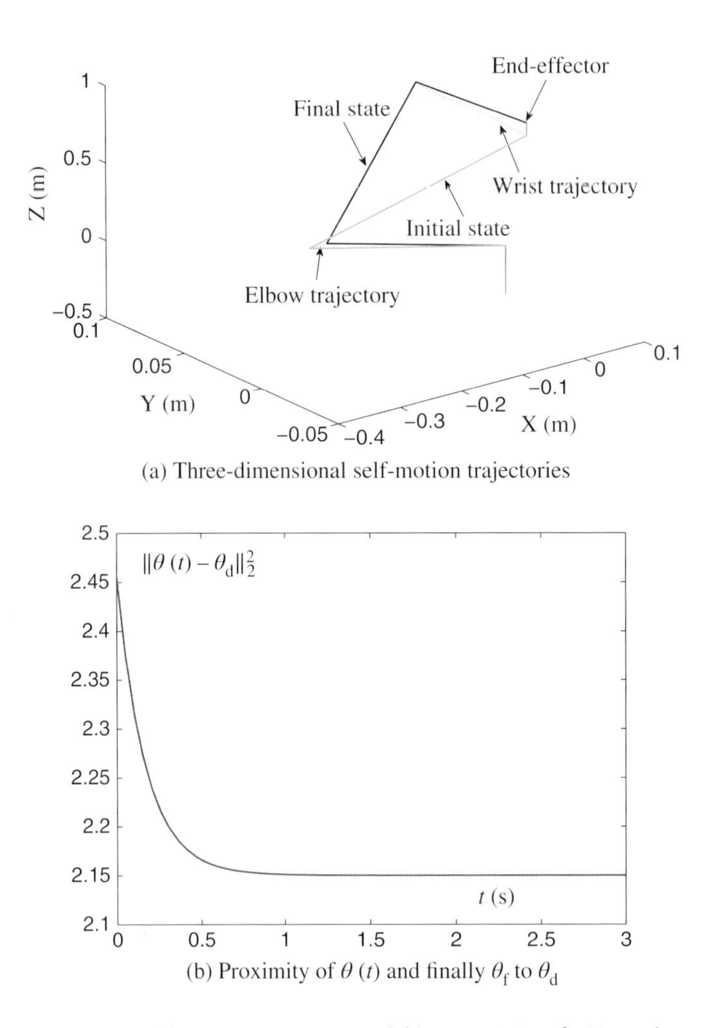

Figure 10.15 (a) Motion trajectories and (b) θ_d-proximity of a PA10 robot performing self-motion with $\mathbf{f}(\theta_d) \neq \mathbf{f}(\theta(0))$.

10.5 Chapter Summary

In this chapter, a modified performance index has been exploited for the self-motion planning of robot manipulators, and the corresponding theoretical analysis has been conducted to verify its correctness. Such a performance index, together with joint physical limits, has been formulated into the QP-based SMP scheme subject to equality and bound constraints. Moreover, for solving the presented scheme, we have generalized the LVIAPDNN solver, which has a simple piecewise-linear dynamic structure and global exponential convergence to the optimal solution. For comparison purposes, the classical pseudoinverse-based method has been presented for SMP of redundant robot manipulators. Computer-simulation results based on a three-link planar manipulator,

PUMA560, and PA10 manipulators have further substantiated the efficacy and superiority of the LVIAPDNN solver and the QP-based SMP scheme. Besides, the effect of design parameters λ and γ has been investigated, and the singularities of the self-motion for redundant manipulators have been discussed. Future work may lie in the applications and experiments of the presented QP-based SMP scheme onto actual redundant manipulators.

Appendix

Equivalence Analysis in Limit Situation

When constraint (10.8) is satisfied or there is no constraint (10.8), on one hand, in view of $P_\Omega(\mathbf{y} - (M\mathbf{y} + \mathbf{p})) = \mathbf{y} - (M\mathbf{y} + \mathbf{p})$, dynamic equation (10.9) reduces to

$$\dot{\mathbf{y}} = -\gamma(I_N + M^{\mathrm{T}})(M\mathbf{y} + \mathbf{p}),$$

where $\gamma(I_N + M^{\mathrm{T}})$ is a scaling matrix. So, if the dynamic equation reaches its equilibrium state (i.e., in the limit situation), we have $-\gamma(I_N + M^{\mathrm{T}})(M\mathbf{y} + \mathbf{p}) = 0$ and further $M\mathbf{y} + \mathbf{p} = 0$. Expanding $M\mathbf{y} + \mathbf{p} = 0$ yields

$$\begin{bmatrix} W & -J^{\mathrm{T}} \\ J & 0 \end{bmatrix} \begin{bmatrix} \dot{\theta} \\ \mathbf{y} \end{bmatrix} + \begin{bmatrix} \mathbf{c} \\ -\mathbf{b} \end{bmatrix} = \begin{bmatrix} \dot{\theta} - J^{\mathrm{T}}\mathbf{y} + \mathbf{c} \\ J\dot{\theta} - \mathbf{b} \end{bmatrix} = 0.$$

For the equation $\dot{\theta} - J^{\mathrm{T}}\mathbf{y} + \mathbf{c} = 0$, multiplying J yields

$$J\dot{\theta} - JJ^{\mathrm{T}}\mathbf{y} + J\mathbf{c} = 0. \tag{10.13}$$

In view of the self-motion requirement, i.e., $J\dot{\theta} = \mathbf{b} = 0$, it follows from (10.13) that

$$\mathbf{y} = (JJ^{\mathrm{T}})^{-1}J\mathbf{c}. \tag{10.14}$$

Then, substituting (10.14) into $\dot{\theta} - J^{\mathrm{T}}\mathbf{y} + \mathbf{c} = 0$ yields the pseudoinverse-based solution (10.12):

$$\dot{\theta} = -(I_n - J^{\mathrm{T}}(JJ^{\mathrm{T}})^{-1}J)\mathbf{c},$$

which completes the proof. That is, when constraint (10.8) is satisfied or there is no constraint (10.8), the upper part of LVIAPDNN solver (10.9) in the limit tends to (10.12) theoretically (i.e., being equivalent in the limit situation).

11

Self-Motion Planning with ZIV Constraint

11.1 Introduction

The number of DOF is very vital for a manipulator. A manipulator is kinematically and/or functionally redundant when more DOF are available than the minimum number required to perform the primary task of a specific manipulator [62]. Compared with a non-redundant manipulator, a redundant manipulator has wider operational space and extra degrees to configure the manipulator to meet additional functional constraints without affecting the position (and/or orientation) of the end-effector [62]. Therefore, much attention and effort in the robotic research community have been paid to studies on and applications of redundant robot manipulators [62, 94, 104, 159, 182, 195–199].

One of the important issues in controlling redundant robot manipulators is self-motion [159, 182, 200]; that is, going from one state to another state without moving the end-effector. By keeping the end-effector at a certain position/orientation (in mathematics, with the end-effector velocity $\dot{\mathbf{r}} \equiv 0 \in \mathbb{R}^m$), the manipulator can adjust its configuration $\theta(t) \in \mathbb{R}^m$ in the joint space from one state (e.g., an initial joint state, $\theta(0)$) to another state (e.g., a desired joint state, θ_{d}). By performing self-motion, the mechanism can obtain more maneuverability or achieve different secondary tasks, such as avoiding joint limits and achieving cyclicity [62, 104]. Therefore, self-motion planning is evidently very essential in redundant robot control, and it has significant physical and biological meanings [177, 179, 200].

In a long and extensive study, much effort has been devoted to the development of self-motion planning [177, 179, 182, 200]. Some computational methods and algorithms have been presented for self-motion planning in robotic applications, such as a reduction-type solution [177], a discontinuous-switching control scheme [200], and a bi-directional self-motion path-planning scheme [179]. However, in these techniques, physical constraints, such as joint-angle limits and joint-velocity limits, are usually not considered, which may lead to the saturation and infeasibility of the solution. Unlike the aforementioned techniques, a self-motion planning scheme based on quadratic programming, which can handle the joint physical limits, has been presented [182]. The feasibility of such a quadratic programming self-motion

Robot Manipulator Redundancy Resolution, First Edition. Yunong Zhang and Long Jin.

scheme has also been investigated and substantiated by computer simulation on a functionally redundant PUMA560 manipulator. Furthering the research of [182], this chapter mainly focuses on the realizability of the self-motion scheme. Thus, an actual six-DOF planar push-rod-joint robot manipulator is developed for the implementation of such a self-motion scheme. That is, a series of challenges related to the actual six-DOF planar robot manipulator is encountered during the implementation and correspondingly, some effective measures are presented and taken for the implementation. Thus, in this chapter, a zero-initial-velocity self-motion scheme for redundant robot manipulators, which can be formulated into a QP, is presented. Based on the conversion technique of QP to a linear variational inequality (LVI), a numerical-computing algorithm is finally developed and employed to solve the QP and the presented self-motion scheme. Note that, as stated in previous chapters, because the numerical-computing algorithm adopts the core equations [i.e., equations (4)–(7)] of [168], it can simply be called numerical algorithm E47 for presentation convenience. Furthermore, a pulse-conversion algorithm is given to convert the resultant joint angle and joint velocity to the motor-driving pulses for controlling the six-DOF planar robot manipulator.

11.2 Preliminaries and Scheme Formulation

In view of the robot forward-kinematic equation $\mathbf{r}_\mathrm{e} = \mathbf{f}(\theta) \rightarrow \mathbf{r}_\mathrm{d}$, to perform the redundant manipulator's self-motion, the minimization of the following quadratic performance index is presented to achieve a suitable desired state θ_d from an initial state $\theta(0)$: $(\dot{\theta} + \mathbf{z})^\mathrm{T}(\dot{\theta} + \mathbf{z})/2$ with $\mathbf{z} = \lambda(\theta - \theta_\mathrm{d})$, where superscript$^\mathrm{T}$ denotes the transpose of a matrix or vector, and $\lambda > 0 \in \mathbb{R}$ is used to scale the magnitude of the manipulator in response to the difference in these states $(\theta - \theta_\mathrm{d})$. As previously mentioned, the six-DOF planar robot manipulator is physically constrained by its joint-physical limits. Thus, we have the following scheme formulation for the self-motion planning of the redundant robot manipulators [182]:

$$\text{minimize} \quad (\dot{\theta} + \mathbf{z})^\mathrm{T}(\dot{\theta} + \mathbf{z})/2 \text{ with } \mathbf{z} = \lambda(\theta - \theta_\mathrm{d}), \tag{11.1}$$

$$\text{subject to} \quad J(\theta)\dot{\theta} = \dot{\mathbf{r}}_\mathrm{d} = 0, \tag{11.2}$$

$$\theta^- \leqslant \theta \leqslant \theta^+, \tag{11.3}$$

$$\dot{\theta}^- \leqslant \dot{\theta} \leqslant \dot{\theta}^+, \tag{11.4}$$

where $\dot{\mathbf{r}}_\mathrm{d} \in \mathbb{R}^m$ denotes the m-dimensional end-effector velocity vector (which should be zero for self-motion purposes); the Jacobian matrix $J(\theta)$ is defined as $J(\theta) = \partial f(\theta)/\partial\theta$; and $\theta^\pm$ and $\dot{\theta}^\pm$ denote the upper and lower limits of the joint-angle vector and joint-velocity vector, respectively.

11.2.1 Handling Joint Physical Limits

Figure 11.1 shows the safety device and limit-position indicators of the six-DOF planar robot manipulator. In normal situations, as shown in Figure 11.1(a), the LED indicators are "on" (i.e., they glow red). When a joint of the manipulator approaches

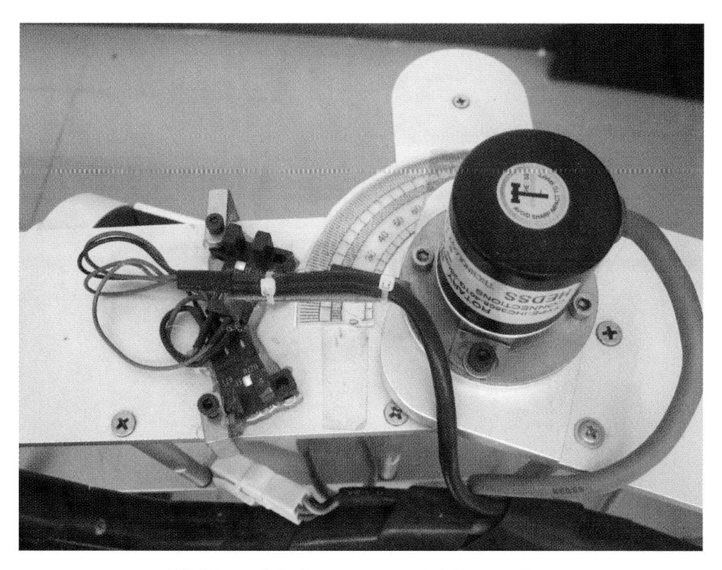

(a) Normal indicator state of joint motion

(b) Indicator state when a joint reaches its limit

Figure 11.1 Safety device and limit-position indicators of a six-DOF planar robot manipulator. *Source*: Li 2012. Reproduced with permission of Emerald Group Publishing Limited.

or reaches one of its physical limits, the corresponding LED indicator turns "off," as shown in Figure 11.1(b). However, when the limit position is reached, the joint would be locked by its physical safety devices until the end of the task, which would lead to the failure of the manipulator task execution. To prevent the actual physical limits

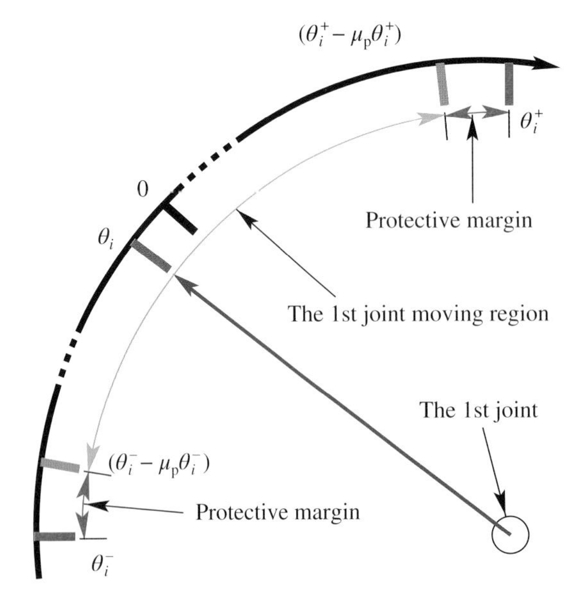

(a) First joint limits with protective margins

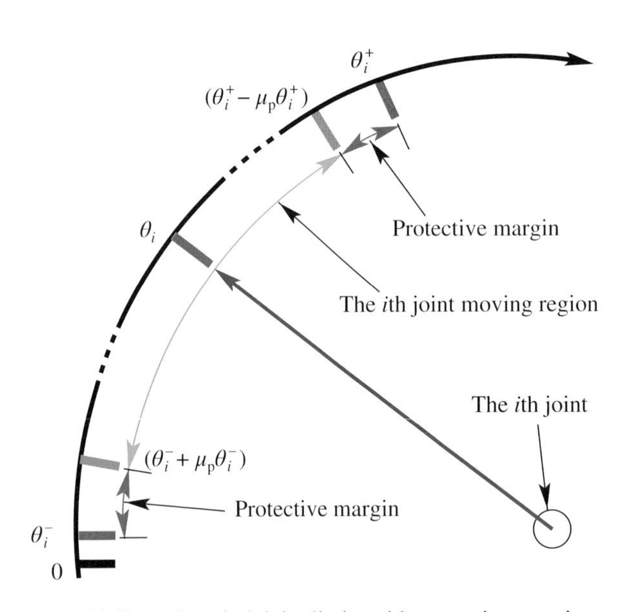

(b) Second to sixth joint limits with protective margins

Figure 11.2 Limits analysis of a six-DOF planar robot manipulator.

from being directly hit, protective margins are established as shown in Figure 11.2, where coefficient μ_p scales the lengths of the protective margins and is determined by experiments (here, $\mu_p = 0.1$). Note that Figure 11.2(a) corresponds to the first joint limits, and Figure 11.2(b) corresponds to the second up to the sixth joint limits. As synthesized through Figure 11.2, the joint angle constraints are unified as

$$\xi^- \leqslant \theta \leqslant \xi^+, \tag{11.5}$$

where the ith elements of $\xi^\pm$ are defined as $\xi_i^+ = \theta_i^+ - \mu_p\theta_i^+$ and $\xi_i^- = \theta_i^- + \mu_p|\theta_i^-|$ ($i \in \{1, 2, \cdots, 6\}$). By using the constraint-combining technique [104, 182], (11.4) and (11.5) can be combined at the joint-velocity level and handled through the following bound constraint:

$$\eta^- \leqslant \dot{\theta} \leqslant \eta^+, \tag{11.6}$$

where the ith elements of $\eta^\pm$ are defined as $\eta_i^+ = \min\{\dot{\theta}_i^+, \mu_v(\xi_i^+ - \theta_i)\}$, and $\eta_i^- = \max\{\dot{\theta}_i^-, \mu_v(\xi_i^- - \theta_i)\}$, with design parameter μ_v used to scale the feasible region of $\dot{\theta}$ being generally larger than $\max\{\dot{\theta}_i^+/(\xi_i^+ - \xi_i^-), |\dot{\theta}_i^-|/(\xi_i^+ - \xi_i^-)\}|_{i=1}^n$. In this chapter, μ_v is set to 4 in the computer simulations and the experiment is performed on the six-DOF planar robot manipulator.

11.2.2 QP Reformulation

In equation (11.1), $(\dot{\theta} + \mathbf{z})^T(\dot{\theta} + \mathbf{z})/2 = (\dot{\theta}^T\dot{\theta} + \dot{\theta}^T\mathbf{z} + \mathbf{z}^T\dot{\theta} + \mathbf{z}^T\mathbf{z})/2$. Since the self-motion scheme is resolved at the joint-velocity level and the decision variable vector is joint-velocity $\dot{\theta}$, the parameter $\mathbf{z}$ (i.e., $\mathbf{z} = \lambda(\theta - \theta_d)$) is viewed as a constant in the performance index. In this situation, $\mathbf{z}^T\mathbf{z}/2$ is also viewed as a constant (with respect to $\dot{\theta}$), and is thus set aside from the performance index. Therefore, equation (11.1) can be reformulated as "minimize $\dot{\theta}^T\dot{\theta}/2 + \mathbf{z}^T\dot{\theta}$" (note that $\mathbf{z}^T\dot{\theta} = \dot{\theta}^T\mathbf{z}$). By defining the decision variable vector $\mathbf{x} = \dot{\theta} \in \mathbb{R}^n$ [94, 104], the self-motion scheme (11.1)–(11.4) is then reformulated as the following QP:

$$\text{minimize} \quad \frac{1}{2}\mathbf{x}^T W\mathbf{x} + \mathbf{z}^T\mathbf{x}, \tag{11.7}$$

$$\text{subject to} \quad J\mathbf{x} = \mathbf{b}, \tag{11.8}$$

$$\eta^- \leqslant \mathbf{x} \leqslant \eta^+, \tag{11.9}$$

where coefficients $W = I$, $\mathbf{z} = \lambda(\theta - \theta_d)$ and $\mathbf{b} = \dot{\mathbf{r}}_d = 0$. Note that, as shown in Figure 2 as well as Figures 4 through 6 of [182], the magnitude of joint velocity $\dot{\theta}$ increases drastically near time instant $t = 0$, or even reaches the joint-velocity limits, which can be regarded as initial velocity $\dot{\theta}(0)$ being nonzero and may cause mechanical damage to the robot manipulator in practice. Thus, achieving the corresponding realization or implementation of the presented self-motion scheme in (11.1)–(11.4) will be difficult (or even impossible).

11.2.3 Design of ZIV Constraint

To prevent the occurrence of a large initial joint velocity, the continuation technique is exploited by imposing a zero-initial-velocity constraint on the joint-velocity limits (i.e., $\dot{\theta}^{\pm}$). A new bound constraint is then obtained as $\mathbf{x}^{-}(t) \leqslant \dot{\theta} \leqslant \mathbf{x}^{+}(t)$ (i.e., $\mathbf{x}^{-}(t) \leqslant \mathbf{x} \leqslant \mathbf{x}^{+}(t)$ in QP), where the ith elements are defined as

$$\mathbf{x}_i^{-}(t) = \max\{\dot{\theta}_i^{-} \sin(\pi t/2T), \mu_{\mathrm{v}}(\xi_i^{-} - \theta_i)\},$$
$$\mathbf{x}_i^{+}(t) = \min\{\dot{\theta}_i^{+} \sin(\pi t/2T), \mu_{\mathrm{v}}(\xi_i^{+} - \theta_i)\},$$

where $T > 0$ is the task duration and time $t \in [0, T]$. Thus, the QP problem (11.7)–(11.9) is finally modified as the following time-varying QP (with $\mathbf{z}$, $\mathbf{x}^{\pm}$ and $\mathbf{x}$ all time-varying):

$$\text{minimize} \quad \frac{1}{2}\mathbf{x}^{\mathrm{T}}\mathbf{x} + \mathbf{z}^{\mathrm{T}}\mathbf{x}, \tag{11.10}$$

$$\text{subject to} \quad J\mathbf{x} = \mathbf{b} = 0, \tag{11.11}$$

$$\mathbf{x}^{-} \leqslant \mathbf{x} \leqslant \mathbf{x}^{+}. \tag{11.12}$$

Note that, by imposing the zero-initial-velocity constraint depicted in (11.12), the zero initial velocity of the motion is guaranteed and can be acceptable in practical applications.

11.3 E47 Assisted QP Solution

In this section, we show the motion planning of the presented zero-initial-velocity self-motion scheme (11.10)–(11.12) and the control process of the six DOF planar robot manipulator, as presented in Figure 11.3. Note that symbol D serves as a time-delay unit in the figure. As mentioned previously, the six-DOF planar robot manipulator is driven by pulse signals, and thus a numerical-computing method (i.e., numerical algorithm E47) is presented to solve the QP problem and the presented self-motion scheme. Simply put, the QP problem is solved by numerical algorithm E47. The resultant joint variables (i.e., θ and $\dot{\theta}$) are then converted into pulses to control the six-DOF planar robot manipulator. Note that numerical algorithm E47 can be divided into two parts (i.e., the LVI conversion and numerical algorithm E47) to solve the QP. Using the LVI conversion design method [94, 104], the QP problem (11.10)–(11.12) can be converted first into an LVI. That is, a primal-dual equilibrium

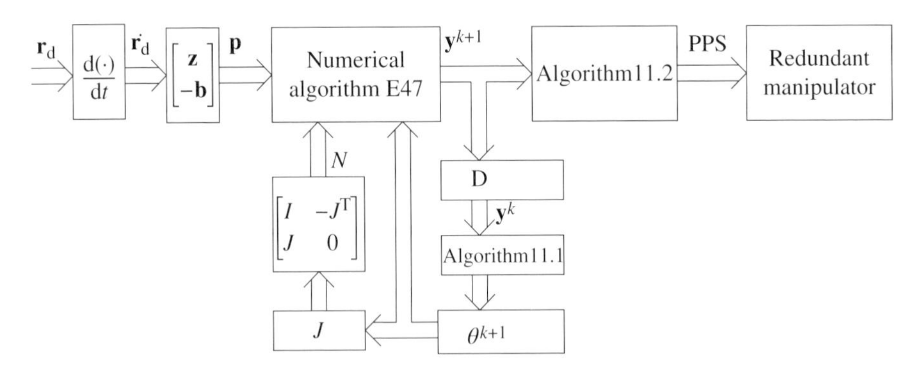

Figure 11.3 Block chart of zero-initial-velocity self-motion planning and control of a six-DOF planar robot manipulator.

vector $\mathbf{y}^* \in \Omega = \{\mathbf{y}|\mathbf{y}^- \leqslant \mathbf{y} \leqslant \mathbf{y}^+\} \subset \mathbb{R}^{n+m}$ should be obtained such that

$$(\mathbf{y} - \mathbf{y}^*)^{\mathrm{T}}(Q\mathbf{y}^* + \mathbf{p}) \geqslant 0, \ \forall \mathbf{y} \in \Omega, \tag{11.13}$$

where the primal-dual decision vector $\mathbf{y}$ and its upper/lower bounds $\mathbf{y}^{\pm}$ are defined, respectively, as

$$\mathbf{y} = \begin{bmatrix} \mathbf{x} \\ \mathbf{u} \end{bmatrix}, \ \mathbf{y}^+ = \begin{bmatrix} \mathbf{x}^+ \\ \varpi \mathbf{1}_{\mathrm{v}} \end{bmatrix}, \ \mathbf{y}^- = \begin{bmatrix} \mathbf{x}^- \\ -\varpi \mathbf{1}_{\mathrm{v}} \end{bmatrix},$$

where $\mathbf{1}_{\mathrm{v}} = [1, \cdots, 1]^{\mathrm{T}}$ denotes an appropriately dimensioned vector composed of ones, and $\varpi \gg 0 \in \mathbb{R}$ is used to replace $+\infty$ numerically. In this equation, $u \in \mathbb{R}^m$ is the dual decision variable vector corresponding to the equality constraint (11.11). The augmented coefficient matrix N and vector $\mathbf{p}$ are defined, respectively, as

$$Q = \begin{bmatrix} I & -J^{\mathrm{T}} \\ J & 0 \end{bmatrix} \in \mathbb{R}^{(n+m)\times(n+m)}, \ \mathbf{p} = \begin{bmatrix} \mathbf{z} \\ -\mathbf{b} \end{bmatrix} \in \mathbb{R}^{n+m}.$$

Then, θ can be obtained by exploiting the Euler's difference method, which is given as Algorithm 1. The detailed explanations on Algorithm 1 are presented as follows.

Algorithm 11.1. Euler difference method for kth update	
1	Define the sampling gap $h = 0.01$ s
2	Initialize the variables, e.g., $\mathbf{y}^0 = 0$
3	Calculate $\theta^{k+1} = \theta^k + h\dot{\theta}^k$ with $\dot{\theta}^k$ being a vector made of the first n elements of $\mathbf{y}^k$

As the motors of the robot joints are driven by pulse commands transmitted from the host computer, the resultant joint variables (i.e., θ and $\dot{\theta}$) need to be converted into pulses per second (PPS), as done and shown in Algorithm 2.

Algorithm 11.2. Conversion from θ and $\dot{\theta}$ to PPS	
1	For the 1st joint, set the parameter $\sigma = 3.2 \times 10^5$
2	Calculate $\mathrm{PPS}_1 = \sigma\dot{\theta}_1/(2\pi)$ per sampling gap
3	For the other joints, set the stepping angles as 0.01π and the subdividing multiples as 32
4	For $i = 1, 2, \cdots, 5$ per sampling gap Calculate $\mathrm{PPS}_{i+1} = (2\pi/(0.01\pi/32))v_{i+1}$ $= 6400a_{i+1}b_{i+1}\dot{\theta}_{i+1}\cos\theta_{i+1}/\left(s_{i+1}\sqrt{a_{i+1}^2 + b_{i+1}^2 + 2a_{i+1}b_{i+1}\sin\theta_{i+1}}\right)$

11.4 Computer Simulations and Physical Experiments

In this section, the presented self-motion schemes [i.e., QP (11.7)-(11.9) without imposing the zero-initial-velocity constraint and QP (11.10)-(11.12) with zero-initial-velocity constraint imposed] and numerical algorithm E47 are simulated and implemented

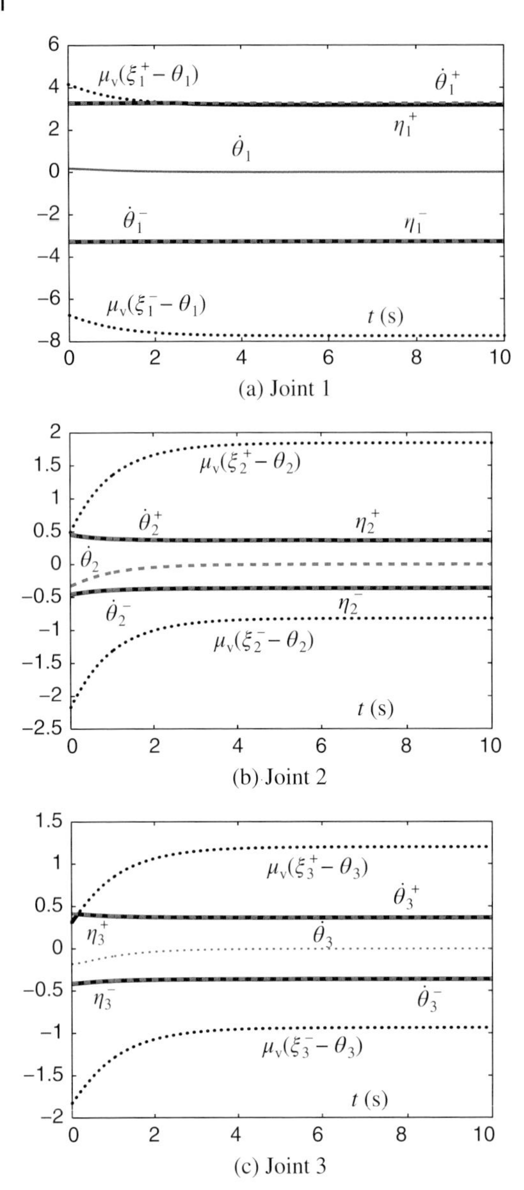

Figure 11.4 Profiles of joint-velocity limits $\dot{\theta}_i^{\pm}$, velocity-level angle-converted bounds $\mu_v(\xi_i^{\pm} - \theta_i)$ and corresponding bounds $\eta_i^{\pm}$ with $i \in \{1, 2, 3\}$ of a six-DOF planar robot manipulator synthesized by QP (11.7)–(11.9) without imposing a zero-initial-velocity constraint.

on a practical six-DOF planar robot manipulator. Without loss of generality, in the simulations and experiment, the task duration $T = 10$ s, and the prescribed error criterion $\|e(\mathbf{y}^k)\|_2 < 10^{-6}$, which is accurate enough for actual application, is used for terminating numerical algorithm E47 throughout this chapter. The self-motion task of the manipulator is to adjust its configuration from the initial state $\theta(0) = [0.27585, 0.6, 0.48469, 0.302, 0.035, 0.0098]^{\mathrm{T}}$ in radians to the desired state $\theta_{\mathrm{d}} = [\pi/6, \pi/12, \pi/12, \pi/6, \pi/12, \pi/8]^{\mathrm{T}}$ in radians.

To lay the ground for the discussion of the realization and implementation of the presented self-motion scheme, simulation results of the joint velocities constrained

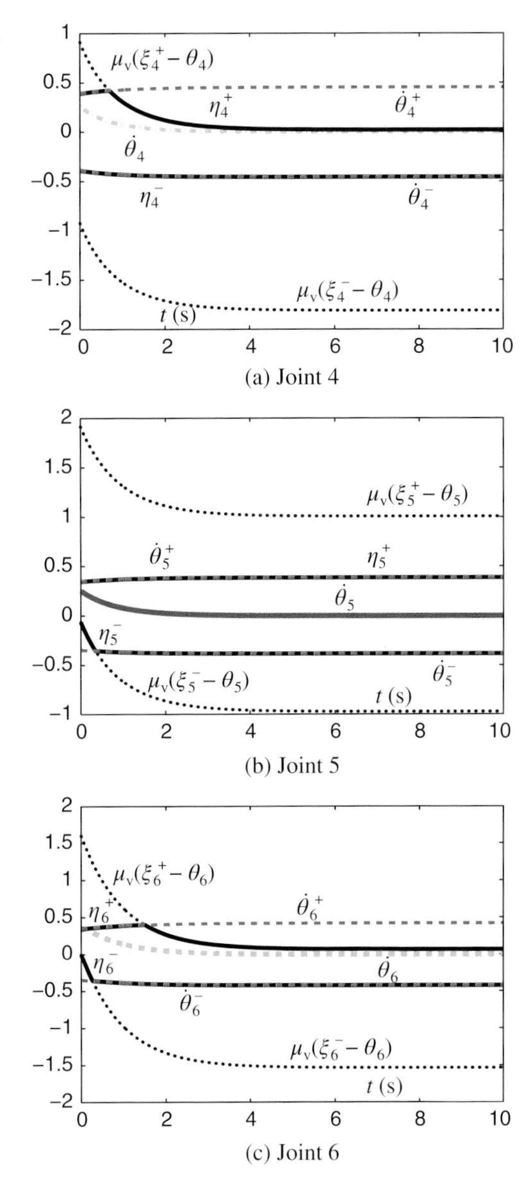

Figure 11.5 Profiles of joint-velocity limits $\dot{\theta}_i^{\pm}$, velocity-level angle-converted bounds $\mu_v(\xi_i^{\pm} - \theta_i)$ and corresponding bounds $\eta_i^{\pm}$ with $i \in \{4, 5, 6\}$ of a six-DOF planar robot manipulator synthesized by QP (11.7)–(11.9) without imposing a zero-initial-velocity constraint.

by their desired bounds $\eta_i^{\pm}$ $(i = 1, 2, \cdots, 6)$ are given in Figure 11.4 and Figure 11.5. The profiles of the joint-velocity limits $\dot{\theta}_i^{\pm}$ and velocity-level angle-converted bounds $\mu_v(\xi_i^{\pm} - \theta_i)$ are also presented in this figure. As seen in Figure 11.4 and Figure 11.5, the joint-velocity variables $\dot{\theta}_i$ are constrained by $\eta_i^{\pm}$ successfully. These results prove that the joint-angle-limit handling technique [i.e., (11.5)] and the joint physical limit conversion technique [i.e., (11.6)] are effective and feasible. Furthermore, for better realization and implementation of the presented self-motion scheme, the zero-initial-velocity constraint [i.e., (11.12)] is considered and imposed. Thus, the resultant $\dot{\theta}^{\pm}\sin(\pi t/2T)$ lead to new smooth zero-initial-velocity constraints to replace the original joint-velocity

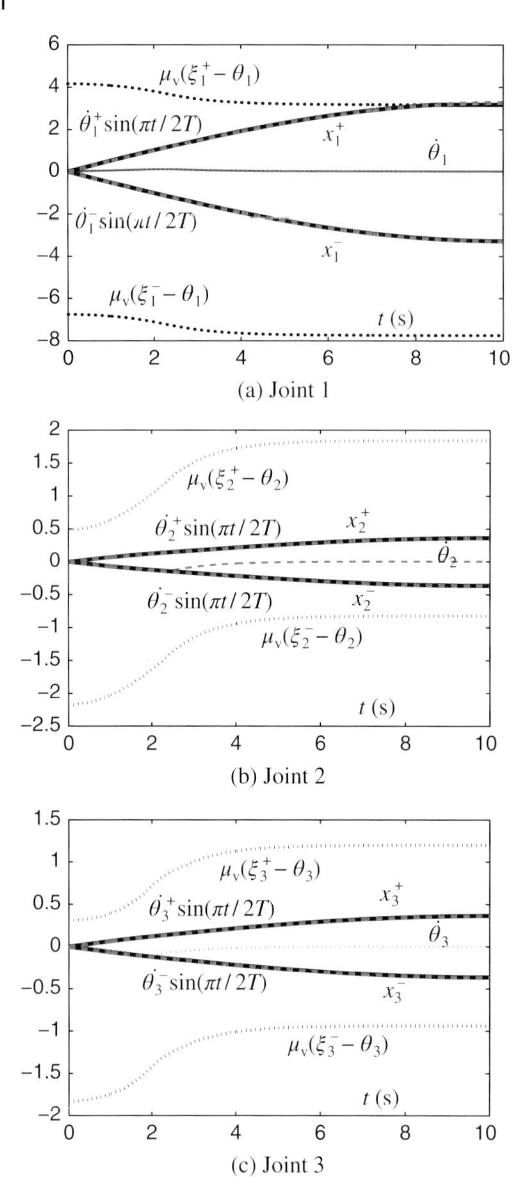

Figure 11.6 Profiles of joint-velocity limits $\dot{\theta}_i^{\pm}$, velocity-level angle-converted bounds $\mu_v(\xi_i^{\pm} - \theta_i)$ and final corresponding bounds $x_i^{\pm}$ with $i \in \{1, 2, 3\}$ of a six-DOF planar robot manipulator synthesized by QP (11.10)–(11.12) with a zero-initial-velocity constraint imposed.

limits $\dot{\theta}^{\pm}$, as shown in Figure 11.6 and Figure 11.7. Correspondingly, the bounds $\eta^{\pm}$ are also replaced by the new desired bounds $\mathbf{x}^{\pm}$, which are smooth and achieve zero initial velocity as well. In Figure 11.6 and Figure 11.7, the initial joint velocity (i.e., $\dot{\theta}(0)$) is constrained to 0 by imposing (11.12). For further investigation and comparison, the θ profiles with and without imposing the zero-initial-velocity constraint (11.12) are given in Figure 11.8. In the figure, θ synthesized by imposing (11.12), especially at the beginning, is smoother than that without imposing (11.12). Moreover, the final state $\theta(T) = [0.523318919705518, 0.262162454370328, 0.262460852147783,$

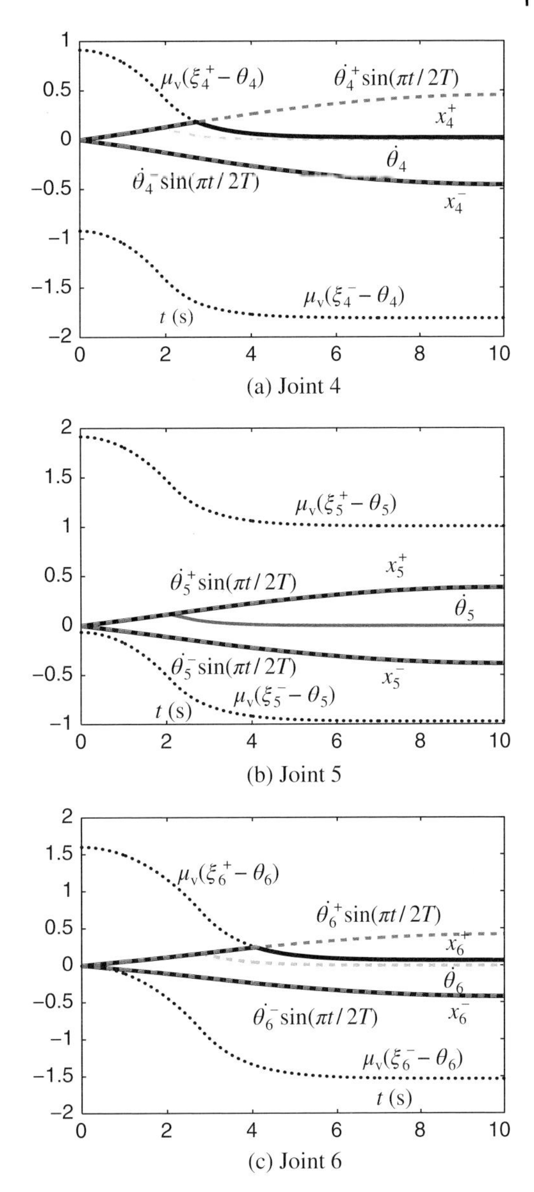

Figure 11.7 Profiles of joint-velocity limits $\dot{\theta}_i^{\pm}$, velocity-level angle-converted bounds $\mu_{\mathrm{v}}(\xi_i^{\pm} - \theta_i)$, and final corresponding bounds $x_i^{\pm}$ with $i \in \{4, 5, 6\}$ of a six-DOF planar robot manipulator synthesized by QP (11.10)–(11.12) with a zero initial velocity constraint imposed.

$0.524302754911326, 0.262301967190858, 0.392804472746665]^{\mathrm{T}}$ rad. Evidently, these simulation results illustrate that the zero-initial-velocity constraint (11.12) eliminates the phenomenon of abrupt and drastic increase in the joint velocity at the beginning of the self-motion task execution, compared with the results of [182]. Thus, it becomes more favorable to the physical realization and implementation.

To control the six-DOF planar robot manipulator, θ and $\dot{\theta}$ are converted to PPS using Algorithm 2, as shown in Figure 11.9 and Figure 11.10. Corresponding to $\dot{\theta}(0)$, the initial PPS is 0 as well, as seen from the figure. That is, the speeds of the motors increase from 0.

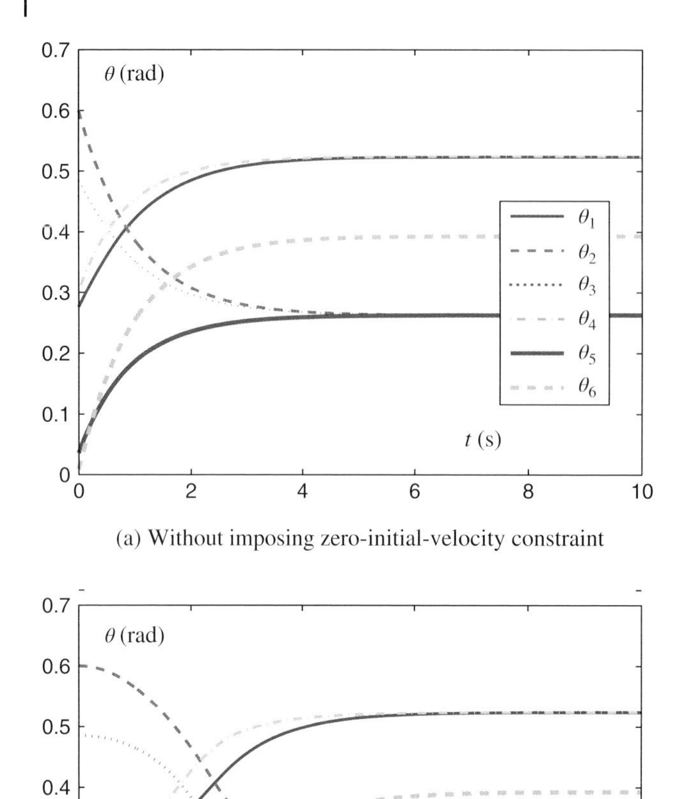

(a) Without imposing zero-initial-velocity constraint

(b) With zero-initial-velocity constraint imposed

Figure 11.8 Joint-angle profiles of a six-DOF planar robot manipulator synthesized by self-motion schemes.

Figure 11.11 illustrates the self-motion task execution of the manipulator synthesized by the presented self-motion scheme (11.10)–(11.12). Specifically, Figure 11.11(a) shows the motion trajectory and Figure 11.11(b) shows the initial and final states of the manipulator. In Figure 11.11(a), the change in joint velocity is from slow to fast and then to slow (i.e., the sparser the trajectories of the manipulator, the faster the manipulator moves because all the sampling gaps between every two adjacent trajectories are the same). In Figure 11.11(b), the solid green lines, dotted blue lines, and solid black lines indicate the initial state, transient states, and final state, respectively. The purpose

Figure 11.9 (a) PPS_1, (b) PPS_2, and (c) PPS_3 for controlling a six-DOF planar robot manipulator.

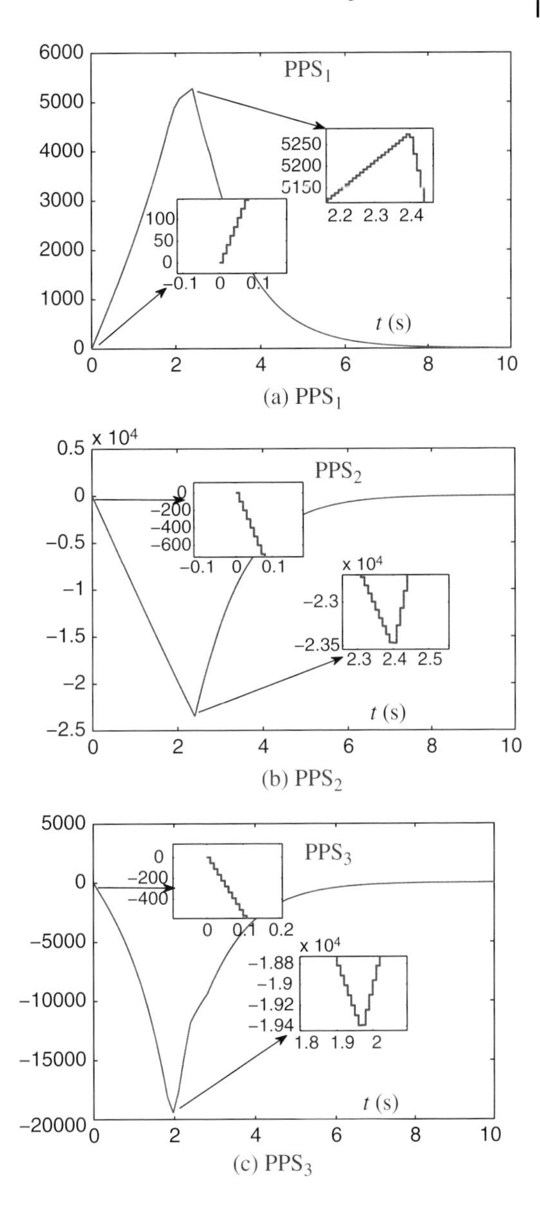

of self-motion is achieved successfully, demonstrating the effectiveness of the self-motion scheme (11.10)–(11.12) and its solver (i.e., numerical algorithm E47). As the corresponding computer simulation shows, the final error $|E(T)|$ between $\theta(T)$ and θ_{d} is $10^{-4} \times [2.79855892780789, 3.63066571178572, 6.61464348633567, 7.03979313027236, 5.02579391708613, 1.05391047940884]^{\mathrm{T}}$ rad, and the maximal joint-angle error is $7.03979313027236 \times 10^{-4}$ rad, which is small and acceptable in practice. The error analysis illustrates the accuracy of the presented zero-initial-velocity self-motion

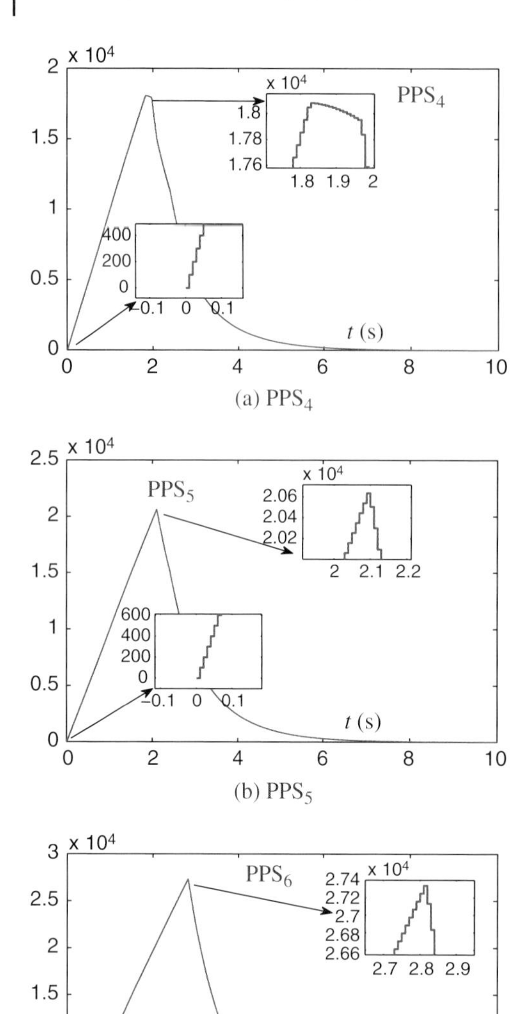

Figure 11.10 (a) PPS_4, (b) PPS_5, and (c) PPS_6 for controlling a six-DOF planar robot manipulator.

scheme (11.10)–(11.12). That is, the presented scheme can make the six-DOF planar robot manipulator successfully realize self-motion.

In summary, these results (i.e., Figures 11.4 through 11.11) substantiate the efficacy and superiority of the presented zero-initial-velocity self-motion scheme (11.10)–(11.12) compared with the self-motion scheme of [182].

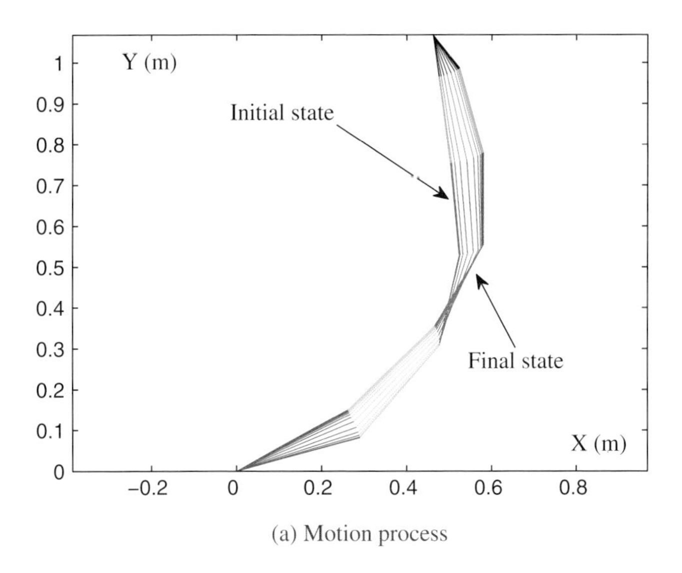

(a) Motion process

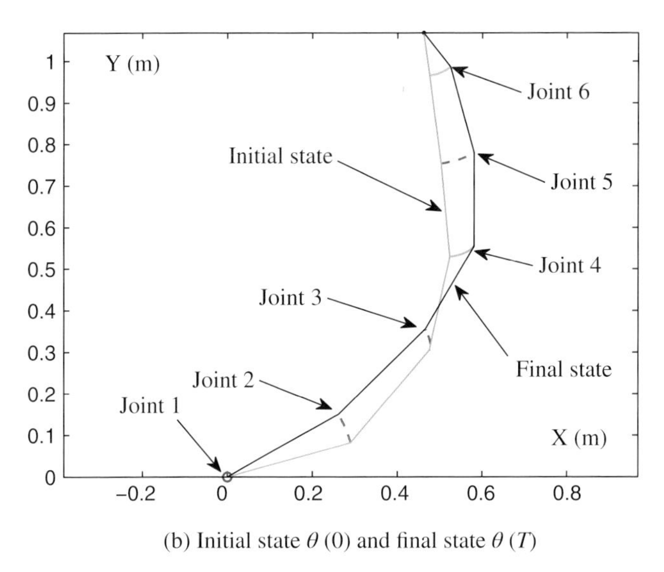

(b) Initial state $\theta(0)$ and final state $\theta(T)$

Figure 11.11 Self-motion task execution of a manipulator synthesized by the self-motion scheme (11.10)–(11.12).

11.5 Chapter Summary

In this chapter, a zero-initial-velocity QP-based self-motion scheme has been presented and investigated to achieve a self-motion task for the six-DOF planar robot manipulator. This scheme has incorporated and unified the joint physical limits at the joint-velocity level and has been formulated as a solvable QP. As a velocity-continuation technique, the zero-initial-velocity constraint (11.12) has been discussed and employed for the presented self-motion scheme. Moreover, a so-called numerical algorithm E47 has been developed and applied to solve the presented self-motion scheme (11.10)–(11.12) at the joint-velocity level. Simulative and experimental results based on the six-DOF planar robot manipulator have illustrated the realizability, effectiveness, and accuracy of the presented zero-initial-velocity QP-based self-motion scheme (11.10)–(11.12).

Part VI

Manipulability Maximization

12

Manipulability-Maximizing SMP Scheme

12.1 Introduction

Studies on and applications of a redundant robot manipulator are the topical research issues in the robotic community [171, 201, 203, 220]. In particular, the inverse-kinematics problem, which finds the joint motion for a given manipulator's end-effector task is one of the important and challenging issues in operating the redundant manipulators. Note that there are an infinite number of joint configurations of a manipulator with respect to a specific task, due to the redundancy [171, 203]; and redundancy resolution is to make a suitable selection from the configurations. This kind of selection is usually used to accomplish some secondary tasks for robot manipulators or generate an optimal solution to achieve some performance criteria, besides finishing the main end-effector path-tracking task. The redundancy-resolution problem can be readily solved by optimization techniques, especially QP [94, 171, 203, 251]. Note that the quadratic program (QP) can be converted to a linear variational inequality (LVI), and the resultant LVI is equivalent to a piecewise-linear equation (PLE) that can be solved by many algorithms and techniques efficiently, such as numerical algorithm E47 and numerical algorithm 94LVI presented in previous chapters as well as recurrent neural networks [94, 168, 171, 203].

An important issue in controlling an actual robot manipulator is the self-motion redundancy-resolution problem [203], that is, the manipulator moves from one joint configuration (or to say, state) to another without affecting the end-effector. That is, keeping the end-effector at certain position and/or orientation (i.e., with $\mathbf{r}_{\mathrm{d}}(t) = 0 \in \mathbb{R}^m$), the manipulator can adjust its configuration in joint space from one state (e.g., an initial joint state, $\theta(0) \in \mathbb{R}^n$) to another state (e.g., a desired joint state, $\theta_{\mathrm{d}} \in \mathbb{R}^n$). By performing suitable self-motion, the mechanism can obtain better maneuverability and achieve different secondary tasks, such as avoiding joint limits and obstacles, and achieving cyclicity. Thus, self-motion planning is evidently very essential in redundant robots' control and it also has significant physical and biological meanings.

To make the robot manipulator execute tasks more readily and efficiently, or for the purposes of avoiding singularities [220], a scheme of self-motion with manipulability maximization (i.e., MMSMP scheme) is presented for redundant robot manipulators,

Robot Manipulator Redundancy Resolution, First Edition. Yunong Zhang and Long Jin.

where no target-configuration should be assigned. Then, the scheme is unified into a QP. The numerical algorithm E47 presented in Chapter 8 is employed for solving such an MMSMP scheme.

12.2 Scheme Formulation

In reference [203], a quadratic performance index of self-motion for approaching middle-value from any initial state $\theta(0)$ for a manipulator is given as

$$(\dot{\theta} + \mathbf{c})^{\mathrm{T}}(\dot{\theta} + \mathbf{c})/2, \quad \text{with} \quad \mathbf{c} = \lambda(\theta - \theta_{\mathrm{m}}), \tag{12.1}$$

where $\dot{\theta} \in \mathbb{R}^n$ denotes the joint-velocity vector, $\theta_{\mathrm{m}} = (\theta^- + \theta^+)/2$ denotes the middle-value of joint vector (with $\theta^{\pm}$ denoting the upper and lower joint-angle limits, respectively), and $\lambda > 0 \in \mathbb{R}$ is a positive design parameter used to scale the magnitude of the manipulator response to the joint drift $\theta - \theta_{\mathrm{m}}$. In addition, superscript ^{T} denotes the transpose operator of a matrix or vector. In this chapter, for better comparison and discussion, the manipulability measure $\sqrt{\det(JJ^{\mathrm{T}})}$ [220] is introduced into the quadratic performance index of self-motion, with $\det(\cdot)$ denoting the determinant of a matrix and J being the Jacobian matrix. Note that, for better mathematical tractability, $\det(JJ^{\mathrm{T}})$ is used in the performance index instead of $\sqrt{\det(JJ^{\mathrm{T}})}$. Considering an actual manipulator is physically constrained by its joint-angle limits and joint-velocity limits, the joint-physical limits are thus incorporated into the scheme formulation. Therefore, an MMSMP (i.e., self-motion with manipulability maximization) scheme constrained by joint-physical limits is presented as

$$\text{minimize} \quad \dot{\theta}^{\mathrm{T}}\dot{\theta}/2 - \gamma\mathbf{q}^{\mathrm{T}}\dot{\theta}, \text{with } \mathbf{q} = \partial\det(JJ^{\mathrm{T}})/\partial\theta \tag{12.2}$$

$$\text{subject to} \quad J\dot{\theta} = \dot{\mathbf{r}}_{\mathrm{d}} = 0, \tag{12.3}$$

$$\theta^- \leqslant \theta \leqslant \theta^+, \tag{12.4}$$

$$\dot{\theta}^- \leqslant \dot{\theta} \leqslant \dot{\theta}^+, \tag{12.5}$$

where $\gamma > 0 \in \mathbb{R}$ is a constant design parameter and $\dot{\mathbf{r}}_{\mathrm{d}} \in \mathbb{R}^m$ is the desired end-effector velocity vector. In addition, $\dot{\theta}^+$ and $\dot{\theta}^-$ denote the upper and lower limits of the joint-velocity vector, respectively.

12.2.1 Derivation of Manipulability Index

To avoid or remedy the degenerate of Jacobian matrix J [note that J is regarded numerically singular when $\det(JJ^{\mathrm{T}})$ is small (or to say, near zero)], the manipulability measure of a robot manipulator [220], that is, $M(\theta) = \sqrt{\det(JJ^{\mathrm{T}})}$, can be considered part of a performance index, which should be maximized. In other words, by maximizing $M(\theta)$, the robot redundancy-resolution scheme selects an optimal configuration for the manipulator during the motion-task execution, and thus the nondegenerate of J is guaranteed to the maximal extent. According to [220], we have the following pseudoinverse-type solution for maximizing the manipulability measure (here we use $\det(JJ^{\mathrm{T}})$ instead of $\sqrt{\det(JJ^{\mathrm{T}})}$ as mentioned):

$$\dot{\theta} = J^{\dagger}\dot{\mathbf{r}}_{\mathrm{d}} + \gamma(I - J^{\dagger}J)\frac{\partial\det(JJ^{\mathrm{T}})}{\partial\theta},$$

where $J^{\dagger}$ is the pseudoinverse of J and I is the identity matrix. According to the GD method presented in Chapter 1, if $\dot{\theta} = \gamma(\partial \det(JJ^{\mathrm{T}})/\partial\theta)$, then the redundant DOF of the manipulator can be utilized to achieve high manipulability during end-effector task execution. In other words, $M(\theta)$ can be maximized when $\dot{\theta} = \gamma(\partial \det(JJ^{\mathrm{T}})/\partial\theta)$. Note that such a dynamic equation is achieved only theoretically.

Based on the QP technique for redundancy resolution presented in previous chapters, we define $\upsilon = \dot{\theta} - \gamma\mathbf{q}$. Therefore, minimizing $\|\upsilon\|_2^2/2$ appears to be more feasible, rather than forcing $\dot{\theta} - \gamma\mathbf{q} = 0$ directly in its exact form. It is worth pointing out here that the minimization of $\|\upsilon\|_2^2/2$ is equivalent to the maximization of $M(\theta)$. Expanding $\|\upsilon\|_2^2/2$, we have $\|\upsilon\|_2^2/2 = \dot{\theta}^{\mathrm{T}}\dot{\theta}/2 - \gamma\mathbf{q}^{\mathrm{T}}\dot{\theta} + \gamma^2\mathbf{q}^{\mathrm{T}}\mathbf{q}/2$. Since the redundancy problem is resolved at the joint-velocity level, the decision variable vector is joint-velocity $\dot{\theta}$. Subsequently, the function $\mathbf{q}$ (i.e., $\mathbf{q} = \partial \det(JJ^{\mathrm{T}})/\partial\theta$) is viewed as a constant in the performance index (with respect to $\dot{\theta}$). In addition, $\gamma^2\mathbf{q}^{\mathrm{T}}\mathbf{q}/2$ is positive and viewed as a constant (with respect to $\dot{\theta}$ as well), which is thus set aside from the performance index. So, we have the new performance index (12.2) for (12.2)–(12.5) to achieve the maximization of the manipulability measure.

12.2.2 Handling Physical Limits

In light of the design requirement for the actual manipulator, protective margins are set for the physical joint-angle limits. That is, a new bound constraint is given for the joint-angle limits as $\xi^- \leqslant \theta \leqslant \xi^+$, where the ith elements of $\xi^{\pm}$ (with $i = 1, 2, \cdots, n$) are defined as $\xi_i^- = \theta_i^- + \mu_{\mathrm{p}}|\theta_i^-|$ and $\xi_i^+ = \theta_i^- - \mu_{\mathrm{p}}|\theta_i^+|$, with symbol $|\cdot|$ denoting the absolute value of a scalar and $\mu_{\mathrm{p}} = 0.1$ (as an example) scaling the width of the protect margins.

In addition, to remedy the non-zero initial velocity weakness of self-motion planning, and especially, to prevent the occurrence of too large initial-joint-velocity portions [203], we exploit the continuation technique by imposing a zero-initial-velocity constraint on the joint-velocity limits $\dot{\theta}^{\pm}$, that is, $\dot{\theta}^{\pm}\sin(\pi t/2T)$ with $T > 0$ being the task duration and time $t \in [0, T]$. Furthermore, the joint-physical limits (12.4)–(12.5) have to be converted into a bound constraint in terms of joint velocity $\dot{\theta}$ [203], since the redundancy problem is resolved at the joint-velocity level. The new bound constraint is thus $\eta^- \leqslant \dot{\theta} \leqslant \eta^+$, where the ith elements of $\eta^{\pm}$ ($i = 1, 2, \cdots, n$) are $\eta_i^+ = \min\{\dot{\theta}_i^+\sin(\pi t/2T), \mu_{\mathrm{v}}(\xi_i^+ - \theta_i)\}$ and $\eta_i^- = \max\{\dot{\theta}_i^-\sin(\pi t/2T), \mu_{\mathrm{v}}(\xi_i^- - \theta_i)\}$, with μ_{v} used to scale the feasible region of $\dot{\theta}$ and being theoretically larger than $2\max_{1\leqslant i\leqslant n}\{\dot{\theta}_i^+/(\theta_i^+ - \theta_i^-), -\dot{\theta}_i^-/(\theta_i^+ - \theta_i^-)\}$. In this chapter, μ_{v} is set at 4 in the computer simulations and physical experiments performed on the six-DOF planar robot manipulator.

12.2.3 QP Formulation

With the decision variable vector $\mathbf{x} = \dot{\theta} \in \mathbb{R}^n$ [203], the presented MMSMP scheme (12.2)–(12.5) is unified as the following QP:

$$\text{minimize} \quad \mathbf{x}^{\mathrm{T}}W\mathbf{x}/2 + \mathbf{z}^{\mathrm{T}}\mathbf{x} \tag{12.6}$$

$$\text{subject to} \quad J\mathbf{x} = \mathbf{b}, \tag{12.7}$$

$$\eta^- \leqslant \mathbf{x} \leqslant \eta^+, \tag{12.8}$$

where coefficients $W = I$, $\mathbf{z} = -\gamma\mathbf{q}$, and $\mathbf{b} = 0$.

The mechanism of the six-DOF planar robot manipulator has been presented in previous chapters and can be also found in reference [171]. For completeness of this chapter and readers' convenience, we provide again the following content. With $i = 1, 2, \cdots, 5$, we have its joint-physical limits as

$$-25\pi/24 \leqslant \dot{\theta}_1 \leqslant 25\pi/24, \quad \text{and} \quad \dot{\theta}^{-}_{i+1} \leqslant \dot{\theta}_{i+1} \leqslant \dot{\theta}^{+}_{i+1},$$

where

$$\dot{\theta}^{\pm}_{i+1} = \frac{v^{\pm}_{i+1} s_{i+1} \sqrt{a^2_{i+1} + b^2_{i+1} + 2a_{i+1} b_{i+1} \sin\theta_{i+1}}}{a_{i+1} b_{i+1} \cos\theta_{i+1}},$$

with v^{-}_{i+1} and v^{+}_{i+1} denoting the negative and positive rotation-rate limits of the $(i+1)$th stepper motor, respectively, and s_{i+1} denoting the elongation rate of the $(i+1)$th push rod (i.e., the elongation length when the motor moves a full turn). In this manipulator system, $-v^{-}_{i+1} = v^{+}_{i+1} = \max(|v_{i+1}|) = 10$ rot/s and $s_{i+1} = 2.5 \times 10^{-3}$ m/rot.

In this chapter, numerical algorithm E47 presented in Chapter 8 is employed as an efficient QP solution for handling the presented QP problem (12.6)–(12.8) as well as the MMSMP scheme (12.2)–(12.5).

12.3 Computer Simulations and Physical Experiments

In this section, based on planar five- and seven-link robot manipulators, computer simulations are performed to illustrate the efficacy of the presented MMSMP scheme (12.2)–(12.5), as compared with the SMMVA scheme [203]. For further investigation, the presented MMSMP scheme is implemented on the practical six-DOF planar robot manipulator [171].

12.3.1 Computer Simulations

In this subsection, the effectiveness of the presented MMSMP scheme (12.2)–(12.5) is verified based on planar five- and seven-link robot manipulators, which are compared with the results of the SMMVA scheme given in reference [203]. With the manipulator parameters and simulation conditions being the same as those of reference [203], the manipulability measures obtained by the MMSMP and SMMVA schemes are shown in Figure 12.1. Specifically, the manipulability measures of the five- and the seven-link robot manipulators are shown in Figure 12.1(a) and Figure 12.1(b), respectively. From Figure 12.1(a), we see that, from near 5, the manipulability measures of both presented schemes increase to some extent. In addition, the maximal manipulability measure synthesized by the SMMVA scheme is 9.20158; by contrast, the maximal manipulability measure synthesized by the MMSMP scheme is 10.09455, clearly larger than the former one. Similar to Figure 12.1(a), from Figure 12.1(b), we see again that the manipulability measures increase; and the maximal manipulability measure synthesized by the SMMVA scheme is 23.20176, while the maximal manipulability measure synthesized by the MMSMP scheme is 24.97755, clearly better than the former one. In summary, these comparative results illustrate the efficacy of the presented MMSMP scheme.

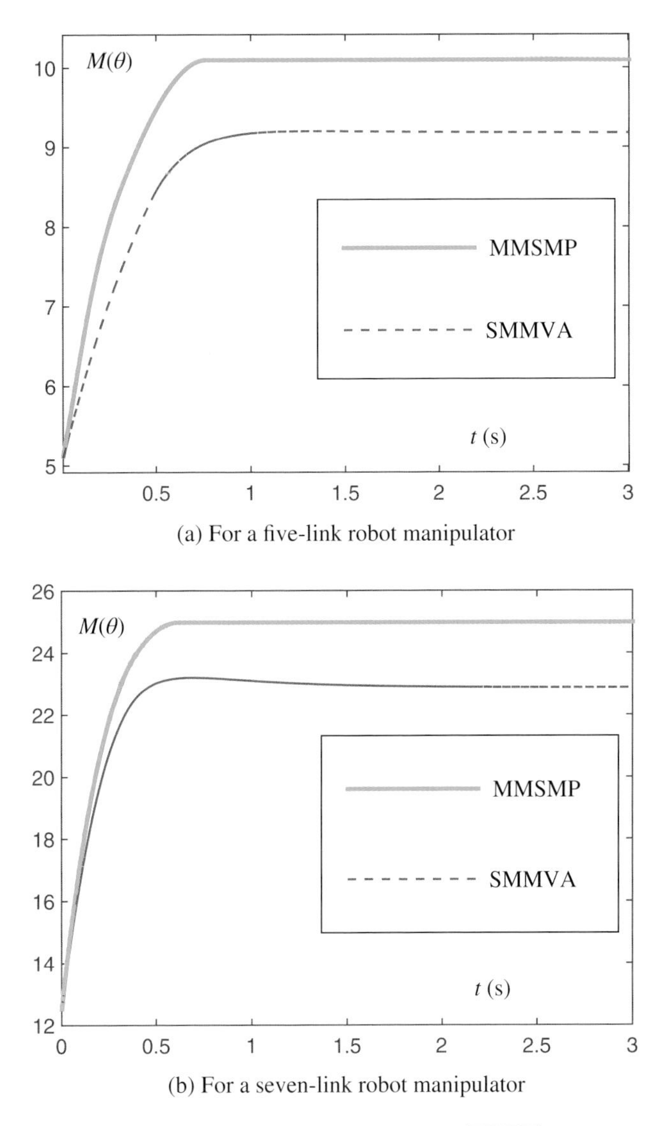

Figure 12.1 Manipulability measures $M(\theta) = \sqrt{\det(JJ^{\mathrm{T}})}$ synthesized by MMSMP scheme (12.2)–(12.5) and SMMVA scheme [203].

12.3.2 Physical Experiments

To further investigate the effectiveness and physical realizability of the presented MMSMP scheme, it is implemented on the practical six-DOF planar robot manipulator [171]. That is, the presented MMSMP scheme (12.2)–(12.5) [corresponding to the QP (12.6)–(12.8)] and numerical algorithm E47 are coded to control the six-DOF planar robot manipulator. Specifically, in the simulation and experiment, the task of the manipulator is to achieve a suitable configuration with best manipulability

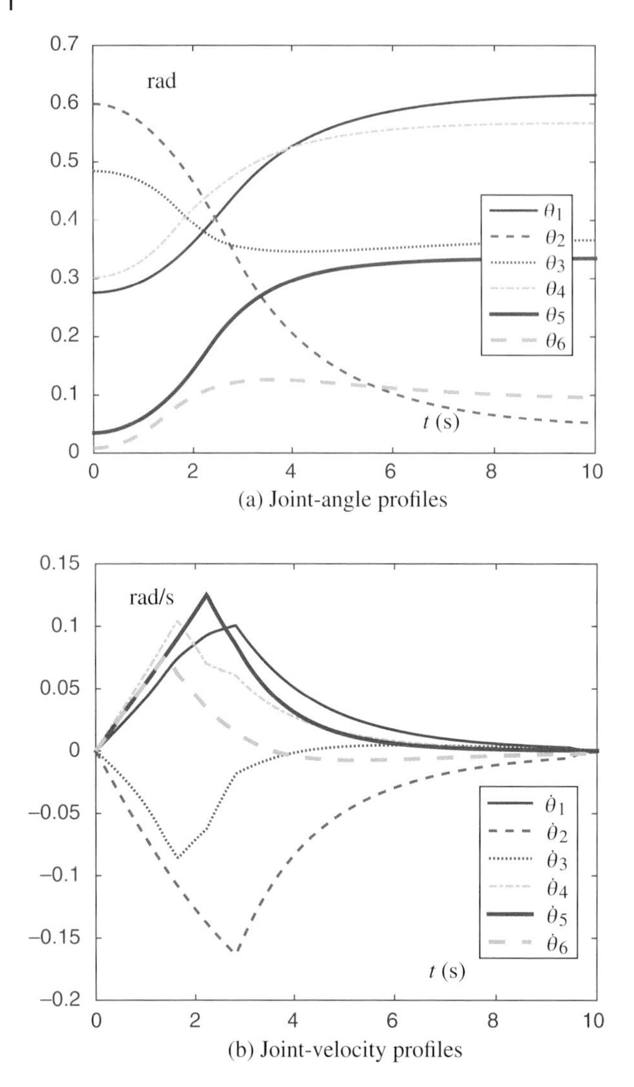

Figure 12.2 (a) Joint-angle and (b) joint-velocity profiles of a six-DOF planar robot manipulator synthesized by the MMSMP scheme (12.2)–(12.5).

(still with no target-configuration assigned). In addition, the initial state is set as $\theta(0) = [0.27585, 0.60000, 0.48469, 0.30200, 0.03500, 0.00980]^{\mathrm{T}}$ rad. Note that the initial state $\theta(0)$ can be set as any value within the joint-angle limits in practice. The corresponding simulative and experimental results are shown in Figures 12.2 through 12.4.

Figure 12.2 illustrates the transient behaviors of joint-angle and joint-velocity variables of the six-DOF planar robot manipulator synthesized by the presented MMSMP scheme (12.2)–(12.5). As seen from Figure 12.2(a), the final state $\theta_{\mathrm{f}} = [0.60757, 0.05767, 0.39573, 0.51842, 0.36185, 0.09829]^{\mathrm{T}}$ rad. In addition, from Figure 12.2(b), which illustrates the profiles of the joint-velocity variable $\dot{\theta}$, we see that, by imposing the zero-initial-velocity constraint, the manipulator motion starts with

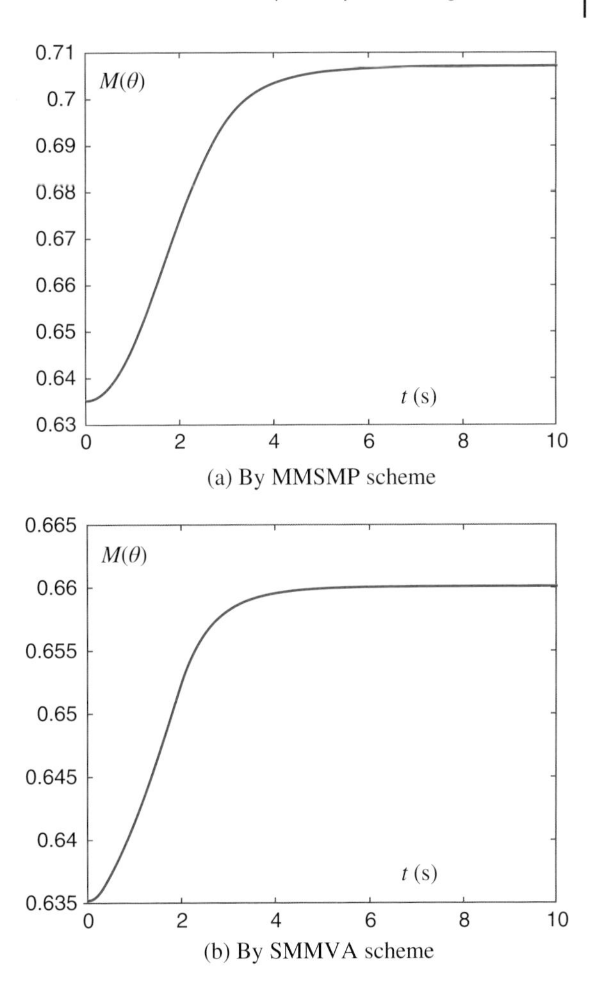

Figure 12.3 Manipulability measures $M(\theta)$ of a six-DOF planar robot manipulator synthesized by the (a) MMSMP and (b) SMMVA schemes.

zero initial velocity, that is, $\dot{\theta}(0) = 0$. It is worth pointing out here that, without the zero-initial-velocity constraint considered and imposed, the initial velocity synthesized by the self-motion scheme can be very large or even arrive at the joint-velocity limits, as shown in Figure 1(d) and Figure 3(d) of reference [203]. Such an undesirable phenomenon will generate a sudden impact on the joints and may cause physical damage (in view of the infinitely large joint acceleration and torque/force). Therefore, the continuation technique (i.e., imposing a zero-initial-velocity constraint on the joint-velocity limits) protects the manipulator effectively.

For comparison, the performance index of the SMMVA [i.e., Equation (12.1)] is also coded and simulated based on the model of the six-DOF planar robot manipulator. The design parameters and the initial state $\theta(0)$ are set the same as those in the MMSMP scheme's simulation and experiment. The desired state $\theta_{\mathrm{d}} = \theta_{\mathrm{m}} = (\theta^{+} + \theta^{-})/2 = [-0.05250, 0.41850, 0.31850, 0.32100, 0.28800, 0.22700]^{\mathrm{T}}$ rad. Note that, due to $J(\theta)\dot{\theta} = \dot{\mathbf{r}}_{\mathrm{d}} = 0$ for the self-motion (i.e., keeping the end-effector at a certain position), the final state $\theta(T)$ approaches the desired state θ_{d} with joint angles adjusted accordingly. Besides, the final state $\theta(T)$ is $[0.31707, 0.58359, 0.34713, 0.28923, 0.24117, 0.20212]^{\mathrm{T}}$

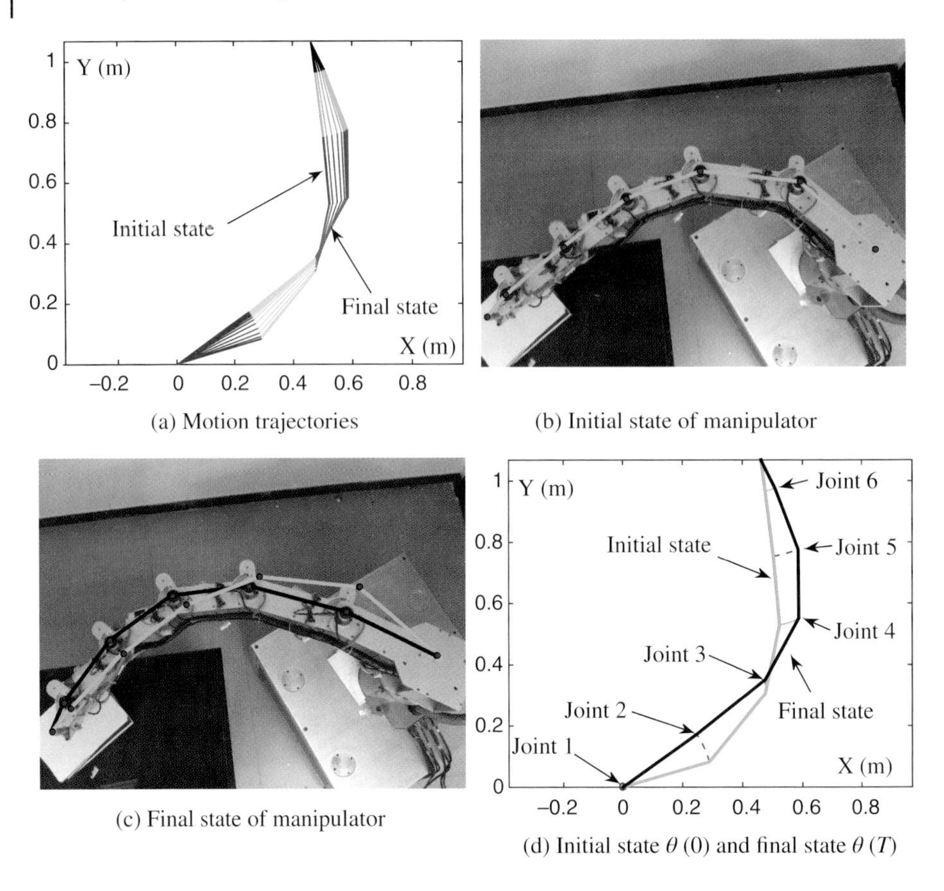

Figure 12.4 Self-motion task execution of a six-DOF planar robot manipulator synthesized by the MMSMP scheme (12.2)–(12.5). *Source*: IEEE 2012. Reproduced with permission of IEEE.

rad, which is the configuration nearest to the desired state θ_{d} (i.e., θ_{m} in the SMMVA simulation). The manipulability indices $M(\theta)$ synthesized by the MMSMP and SMMVA schemes are thus shown in Figure 12.3. Note that both of the two self-motion tasks start with the same initial manipulability index $M(\theta) = 0.63533$. As seen from the figure, the maximal manipulability value $\max\{M(\theta)\} = 0.70702$ synthesized by the MMSMP scheme is clearly larger than that synthesized by the SMMVA scheme (i.e., 0.66017). This substantiates again the efficacy of the presented MMSMP scheme (12.2)–(12.5).

For the actual control of the physical robot manipulator, the resultant joint variables (i.e., θ and $\dot{\theta}$ shown in Figure 12.2) are converted into the pulses per second (PPS), for example, by using Equation (8.15) and Equation (8.16) of Chapter 8. By sending the corresponding PPS to drive the motors, the self-motion of the robot manipulator is achieved. Figure 12.4 illustrates the motion transients of the actual six-DOF planar robot manipulator during the self-motion task execution from the initial state $\theta(0)$ with task duration $T = 10$ s. As shown in Figure 12.4, all joints of the robot manipulator move toward their final states. These results further verify the physical realizability of the presented MMSMP scheme.

12.4 Chapter Summary

In this chapter, an MMSMP scheme has been presented and investigated to achieve the best manipulability for redundant robot manipulators. Such a scheme has incorporated and unified the joint physical limits at the joint-velocity level; and, as a velocity-continuation technique, the zero-initial-velocity constraint has been employed for the presented self-motion scheme to remedy the non-zero initial velocity weakness of previous self-motion schemes. In addition, the MMSMP scheme (12.2)–(12.5) has been unified as a solvable QP (12.6)–(12.8). Moreover, numerical algorithm E47 presented in Chapter 8 has been exploited to solve the QP as well as the presented MMSMP scheme at the joint-velocity level. Simulative and experimental results based on different planar multi-link robot manipulators have further illustrated the effectiveness, superiority, and physical realizability of the presented MMSMP scheme (12.2)–(12.5).

13

Time-Varying Coefficient Aided MM Scheme

13.1 Introduction

The issue of optimal motion control in robotics is becoming a more and more appealing topic as robots are widely applied in scientific research, military and civilian areas, of which the operating environment is much more complicated than that in the past [205–209]. Redundant manipulators can achieve subtasks such as obstacle avoidance [210], repetitive motion planning [128], joint limits, and singularity avoidance [211–214], or optimization of various performance criteria [215], since they have more degrees of freedom (DOF) than required to perform a given end-effector primary task [92]. A fundamental issue in controlling such redundant manipulators is to design suitable redundancy-resolution approaches [92, 215]. The conventional solution to such a redundancy-resolution problem is the pseudoinverse-based formulation [211, 212].

In recent years, to avoid online matrix inversion and achieve higher computation efficiency, much effort has been made to find the solution of the redundancy-resolution problem via optimization techniques [92]. Generally speaking, such techniques can reformulate different schemes as a unified quadratic program (QP) that is subject to equality, inequality, and bound constraints. Since the QP is equivalent to a linear variational inequality (LVI) problem, it can be solved by many methods and techniques efficiently, for example, numerical algorithm 94LVI[136] and/or recurrent neural networks [92, 128].

A singularity problem can refer to the situation that the manipulator's Jacobian matrix is ill-conditioned and loses rank at a kinematic singular point [216], and nearby a singular point of the kinematic mapping, joint velocities, accelerations, and torques may become too large when the end-effector moves in certain directions [217]. To keep the manipulator away from singular points, kinematic manipulability, as a well-established tool in the design or motion control of redundant robot manipulators, which can reflect the manipulator's ability to affect the end-effector's velocities and accelerations, was first introduced with explicit expression by Yoshikawa [218] and were investigated by many researchers [213, 214, 219]. Different from previous methods and techniques, this chapter adopts the QP-technique to solve the redundancy-resolution problem of redundant robot manipulators (note that the 6-DOF planar robot manipulator is specifically considered) and simultaneously maximize the kinematic manipulability. This is realized

Robot Manipulator Redundancy Resolution, First Edition. Yunong Zhang and Long Jin.

by utilizing the redundant DOF of the robot to adjust its configuration in joint space during the end-effector task execution.

Among those existing methods and techniques, the physical constraints such as the joint-velocity limits are usually considered to be constant. This, however, may not be applicable to some kinds of manipulators, such as the robot manipulators with push-rod-type joints (also called push-rod-joint manipulators), of which the joint-velocity limits change with the end-effector and joints movement (i.e., joint-velocity limits are functions of joint angles). In this chapter, a time-varying coefficient aided manipulability-maximizing (TVCMM) scheme subject to varying joint-velocity limits (VJVL) is presented and analyzed. Furthermore, such a scheme is implemented on the six-DOF planar robot manipulator presented in previous chapters.

13.2 Manipulability-Maximization with Time-Varying Coefficient

This section discusses the time-varying coefficient aided manipulability maximizing (TVCMM) scheme. Firstly, this scheme with joint physical limits considered is presented and reformulated into a QP. Secondly, numerical algorithm 94LVI presented in previous chapters is employed to solve effectively such a time-varying QP problem, which is subject to equality and bound constraints. Based on the solutions (i.e., the computed $\dot{\theta}$ and θ) of the presented optimal-control scheme, the six-DOF planar robot manipulator can fulfill the user-specified end-effector task. It is worth mentioning that, during the task execution, the manipulability can keep maximizing, which means that the manipulator maintains its optimality and is away from singular points.

13.2.1 Nonzero Initial/Final Joint-Velocity Problem

The relation between end-effector position-and-orientation vector $\mathbf{r}_\mathrm{e}(t) \in \mathbb{R}^m$ in Cartesian space and the joint angle vector $\theta(t) \in \mathbb{R}^n$ in joint space can be described as $\mathbf{r}_\mathrm{e} = \mathbf{f}(\theta)$ and $\dot{\mathbf{r}}_\mathrm{e} = J(\theta)\dot{\theta}$, where $\mathbf{f}(\cdot)$ is a nonlinear mapping from $\mathbb{R}^n$ to $\mathbb{R}^m$, $\dot{\mathbf{r}}_\mathrm{e} \in \mathbb{R}^m$ is the time-derivative of $\mathbf{r}_\mathrm{e}$ (i.e., the end-effector velocity, or termed Cartesian velocity), while $J(\theta) = \partial \mathbf{f}(\theta)/\partial\theta \in \mathbb{R}^{m\times n}$ denotes the Jacobian matrix. The variable $\mathbf{r}_\mathrm{d} \in \mathbb{R}^m$ denotes the desired or user-defined Cartesian path of the manipulator's end-effector. Based on the user-defined Cartesian path $\mathbf{r}_\mathrm{d}$, we need to generate the corresponding joint trajectory $\theta(t) \in \mathbb{R}^n$ in real time t so that $\mathbf{f}(\theta) = \mathbf{r}_\mathrm{e} \rightarrow \mathbf{r}_\mathrm{d}$. Note that $\mathbf{r}_\mathrm{d}(t)$, $\theta(t)$ and $J(\theta)$ are sometimes written as $\mathbf{r}_\mathrm{d}$, θ and J, respectively, for presentation convenience. In this chapter, $m = 2$ since we consider only the end-effector position, and $n = 6$ in view of the manipulator having six DOF.

During recent decades, kinematic manipulability has been used in optimal control of robot manipulators [212–214, 218–221], and the quantitative measure of manipulability at configuration θ can be given as $w = \det(JJ^\mathrm{T})$. Here, "det(·)" denotes the determinant of a matrix, and superscript$^\mathrm{T}$ denotes the transpose of a matrix or vector. Note that w is quite helpful for the detailed evaluation of the manipulation ability of a robotic mechanism; for example, w can give an overall measure of the directional uniformity of the ellipsoid and the upper bound of the velocity with which the end-effector can be moved in any directions [202]. Therefore, it can be used as an important reference factor on configuration adjustment of redundant manipulators via self-motion. Inspired by the gradient method in optimization [222–224], for maximizing the manipulability w, we can design

$$\dot{\theta} = p\frac{\partial \det(JJ^{\mathrm{T}})}{\partial \theta}, \tag{13.1}$$

where, conventionally speaking, $p \in \mathbb{R}$ is a positive design parameter. In the pseudoin-verse-based formulation,

$$\dot{\theta} = \dot{\theta}_{\mathrm{p}} + \dot{\theta}_{\mathrm{h}} = J^{\dagger}\dot{\mathbf{r}}_{\mathrm{d}} + (I - J^{\dagger}J)p\frac{\partial \det(JJ^{\mathrm{T}})}{\partial \theta},$$

where I denotes an appropriately-dimensioned identity matrix and $J^{\dagger} \in \mathbb{R}^{n\times m}$ denotes the pseudoinverse of J. Projection matrix $(I - J^{\dagger}J) \in \mathbb{R}^{n\times n}$ is utilized to project $p(\partial \det(JJ^{\mathrm{T}})/\partial\theta)$ onto the null-space of J to keep the manipulator away from singularities, without having an effect on the end-effector task execution [1, 225, 226]. Here, $\dot{\theta}_{\mathrm{p}} = J^{\dagger}\dot{\mathbf{r}}_{\mathrm{d}}$ denotes the particular solution (i.e., the minimum Euclidean-norm solution) to $\dot{\mathbf{r}}_{\mathrm{d}} = J(\theta)\dot{\theta}$, and $\dot{\theta}_{\mathrm{h}} = (I - J^{\dagger}J)p(\partial \det(JJ^{\mathrm{T}})/\partial\theta)$ denotes the homogeneous solution, which is referred to as the self-motion of the manipulator and can keep the manipulator away from the singularities by maximizing the manipulability w.

Note that the particular solution $\dot{\theta}_{\mathrm{p}} = J^{\dagger}\dot{\mathbf{r}}_{\mathrm{d}}$ generates zero initial and final joint velocities, when $\dot{\mathbf{r}}_{\mathrm{d}}(t)$ starts from a zero initial value and ends with a zero final value. However, the homogeneous solution $\dot{\theta}_{\mathrm{h}} = (I - J^{\dagger}J)p(\partial \det(JJ^{\mathrm{T}})/\partial\theta)$ would lead to nonzero initial and final joint velocities, since the gradient, $\partial \det(JJ^{\mathrm{T}})/\partial\theta$, may not equal zero at the initial and final time instants, which is usually the case. As a result, $\dot{\theta} = \dot{\theta}_{\mathrm{p}} + \dot{\theta}_{\mathrm{h}}$ can cause a nonzero initial/final joint-velocity phenomenon. The phenomenon may induce a problem that large (or even infinitely large) joint-accelerations arise at the beginning of the task duration. It is not suitable for engineering application and can even cause damage to the hardware system.

For the remedy of the nonzero initial/final joint-velocity problem, this chapter presents using a time-varying coefficient $p(t)$ instead of a constant coefficient p. We simply design $p(t) = p_1 \sin(\pi t/T)$, where constant parameter $p_1 \in \mathbb{R}$ is used to scale the maximum magnitude of coefficient $p(t)$ and T is the task duration. It is worth mentioning that time-varying coefficient $p(t)$ is nonnegative, continuous, and differentiable, in addition to $p(0) = p(T) = 0$.

13.2.2 Scheme Formulation

To improve the manipulability of the robot manipulator and fulfill the desired end-effector task, according to (13.1) and $p(t) = p_1 \sin(\pi t/T)$, we define the following performance index to be minimized: $\rho(t) = \|\dot{\theta}(t) - p(t)\mathbf{q}(\theta(t))\|_2^2/2$, or $\rho = \|\dot{\theta} - p\mathbf{q}\|_2^2/2$ in short, where $\|\cdot\|_2$ denotes the two-norm of a vector, and $\mathbf{q} = \mathbf{q}(\theta) = \partial \det(JJ^{\mathrm{T}})/\partial\theta \in \mathbb{R}^n$. Note that, as the robot's end-effector task requirement $J\dot{\theta} = \dot{\mathbf{r}}_{\mathrm{d}}$ should be considered in the scheme formulation, it is better to minimize ρ rather than force $\dot{\theta} = p\mathbf{q}$ (i.e., (13.1)) directly. By minimizing ρ, the manipulability of the manipulator can be maximized during the task execution. Besides, ρ can be expanded as

$$\rho = \|\dot{\theta} - p\mathbf{q}\|_2^2/2 \tag{13.2}$$

$$= (\dot{\theta} - p\mathbf{q})^{\mathrm{T}}(\dot{\theta} - p\mathbf{q})/2$$
$$= (\dot{\theta}^{\mathrm{T}}\dot{\theta} - \dot{\theta}^{\mathrm{T}}p\mathbf{q} - p\mathbf{q}^{\mathrm{T}}\dot{\theta} + p^2\mathbf{q}^{\mathrm{T}}\mathbf{q})/2$$
$$= \dot{\theta}^{\mathrm{T}}\dot{\theta}/2 - p\mathbf{q}^{\mathrm{T}}\dot{\theta} + p^2\mathbf{q}^{\mathrm{T}}\mathbf{q}/2.$$

Via (13.3), minimizing ρ is equivalent to minimizing $\dot{\theta}^{\mathrm{T}}\dot{\theta}/2 - p\mathbf{q}^{\mathrm{T}}\dot{\theta}$ in the velocity-level redundancy resolution scheme, and the TVCMM scheme subject to varying joint-velocity limits is thus presented as

$$\text{minimize} \quad \dot{\theta}^{\mathrm{T}}\dot{\theta}/2 - p(t)\mathbf{q}^{\mathrm{T}}\dot{\theta} \tag{13.3}$$
$$\text{subject to} \quad J\dot{\theta} = \dot{\mathbf{r}}_{\mathrm{d}} + \kappa_{\mathrm{p}}(\mathbf{r}_{\mathrm{d}} - \mathbf{f}(\theta)), \tag{13.4}$$
$$\theta^- \leqslant \theta \leqslant \theta^+, \tag{13.5}$$
$$\dot{\theta}^-(\theta) \leqslant \dot{\theta} \leqslant \dot{\theta}^+(\theta), \tag{13.6}$$
$$\text{with } \mathbf{q} = \frac{\partial \det(JJ^{\mathrm{T}})}{\partial \theta},$$

where θ^+ and θ^- denote, respectively, the upper and lower limits of joint-angle vector θ, and, resulting from $\dot{\theta}_1^{\pm} = \pm 25\pi/24$ and the varying joint-velocity limits, $\dot{\theta}^+(\theta)$ and $\dot{\theta}^-(\theta)$ denote, respectively, the upper and lower limits of joint-velocity vector $\dot{\theta}$, which vary with θ. To achieve higher precision and robustness of end-effector positioning, it is necessary to introduce a feedback of Cartesian position error into the redundancy-resolution equation, that is, $J\dot{\theta} = \dot{\mathbf{r}}_{\mathrm{d}} + \kappa_{\mathrm{p}}(\mathbf{r}_{\mathrm{d}} - \mathbf{f}(\theta))$, where $\kappa_{\mathrm{p}} = 8$ in the simulations and experiment. In addition, with $i = 1, 2, \cdots, 6$, the ith element of $\mathbf{q}$ can be given mathematically [218] as

$$q_i = \frac{\partial \det(JJ^{\mathrm{T}})}{\partial \theta_i} = \det(JJ^{\mathrm{T}}) \operatorname{trace}\left((JJ^{\mathrm{T}})^{-1}\frac{\partial (JJ^{\mathrm{T}})}{\partial \theta_i}\right)$$
$$= \det(JJ^{\mathrm{T}}) \operatorname{trace}\left((JJ^{\mathrm{T}})^{-1}\left(\frac{\partial J}{\partial \theta_i}J^{\mathrm{T}} + J\left(\frac{\partial J}{\partial \theta_i}\right)^{\mathrm{T}}\right)\right).$$

Here, trace$(\cdot)$ denotes the trace of a matrix argument.

Furthermore, as the redundancy is resolved at the joint-velocity level (i.e., in terms of $\dot{\theta}$), joint-angle limit constraint (13.5) has been converted into the following expression about $\dot{\theta}$ [128]:

$$\mu(\theta^- - \theta) \leqslant \dot{\theta} \leqslant \mu(\theta^+ - \theta), \tag{13.7}$$

where $\mu > 0 \in \mathbb{R}$ is used to scale the feasible region of joint-velocity $\dot{\theta}$. Note that, in the presented six-DOF planar robot manipulator in this chapter, the safety devices are used. If a joint of the manipulator approaches or reaches its physical limit, the joint would be locked by the physical safety device throughout the task duration, which would lead to the failure of the end-effector task execution. It is thus necessary to leave some "safety region" to the joint physical limits $\theta^{\pm}$. So, (13.7) can be further improved into the following expression:

$$\mu((\theta^- + \vartheta) - \theta) \leqslant \dot{\theta} \leqslant \mu((\theta^+ - \vartheta) - \theta),$$

where ϑ is a constant vector used to scale the safety region, with each element ϑ_i $(i = 1, 2, \cdots, 6)$ set to be 0.0349 rad (i.e., 2 degrees) in the simulations and experiment in this chapter.

Therefore, the physically-constrained TVCMM scheme can be formulated as a quadratic program:

$$\text{minimize} \quad \mathbf{x}^{\mathrm{T}} W\mathbf{x}/2 + \varsigma^{\mathrm{T}}\mathbf{x} \tag{13.8}$$

$$\text{subject to} \quad J\mathbf{x} = \mathbf{d}, \tag{13.9}$$

$$\zeta^{-} \leqslant \mathbf{x} \leqslant \zeta^{+}, \tag{13.10}$$

$$\text{with } \mathbf{x} = \dot{\theta},\ W = I, \varsigma = -p(t)\mathbf{q}, \mathbf{d} = \dot{\mathbf{r}}_{\mathrm{d}} + \kappa_{\mathrm{p}}(\mathbf{r}_{\mathrm{d}} - \mathbf{f}(\theta)),$$

where the ith elements of ζ^{-} and ζ^{+} (with $i = 1, 2, \cdots, 6$) are

$$\zeta_i^{-} = \max\{\mu((\theta_i^{-} + \vartheta_i) - \theta_i), \dot{\theta}_i^{-}(\theta_i)\}, \zeta_i^{+} = \min\{\mu((\theta_i^{+} - \vartheta_i) - \theta_i), \dot{\theta}_i^{+}(\theta_i)\}.$$

Note that the performance index, (13.8), is for manipulability-maximizing purposes; the equality constraint, (13.9), expresses the linear relation between joint-velocity vector and desired Cartesian velocity of the end-effector, with position-error feedback $\kappa_{\mathrm{p}}(\mathbf{r}_{\mathrm{d}} - \mathbf{f}(\theta))$ included; and that, to keep all joint variables within their mechanical ranges, it is straightforward and concise to use bound constraint (13.10). It is also worth noting that, as well as the end-effector path-tracking primary task (13.9), in this chapter, the additionally-considered secondary tasks for redundancy resolution are the manipulability maximization (MM) and the avoidance of joint physical limits, especially avoiding the varying joint-velocity limits.

13.2.3 94LVI Assisted QP Solution

In view of the duality theory and the authors' previous work [1] and based on the results presented in Chapter 5, QP problem (13.8)–(13.10) can be converted to a linear variational inequality (LVI). That is, to find a primal-dual vector $\mathbf{y}^{*} \in \Omega = \{\mathbf{y} \in \mathbb{R}^{n+m} | \mathbf{y}^{-} \leqslant \mathbf{y} \leqslant \mathbf{y}^{+}\} \subset \mathbb{R}^{n+m}$ such that

$$(\mathbf{y} - \mathbf{y}^{*})^{\mathrm{T}}(Q\mathbf{y}^{*} + \mathbf{p}) \geq 0, \forall \mathbf{y} \in \Omega, \tag{13.11}$$

where

$$\mathbf{y} = \begin{bmatrix} \mathbf{x} \\ \mathbf{u} \end{bmatrix} \in \mathbb{R}^{n+m}, \mathbf{y}^{+} = \begin{bmatrix} \zeta^{+} \\ \varpi 1_{\mathrm{v}} \end{bmatrix} \in \mathbb{R}^{n+m}, \mathbf{y}^{-} = \begin{bmatrix} \zeta^{-} \\ -\varpi 1_{\mathrm{v}} \end{bmatrix} \in \mathbb{R}^{n+m},$$

$$Q = \begin{bmatrix} W & -J^{\mathrm{T}} \\ J & 0 \end{bmatrix} \in \mathbb{R}^{(n+m)\times(n+m)}, \mathbf{p} = \begin{bmatrix} \varsigma \\ -\mathbf{d} \end{bmatrix} \in \mathbb{R}^{n+m}, 1_{\mathrm{v}} = [1, \cdots, 1]^{\mathrm{T}} \in \mathbb{R}^{m}.$$

Here, $\mathbf{u} \in \mathbb{R}^{m}$ is the dual decision variable vector defined for equality constraint (13.9), and $\varpi \gg 0 \in \mathbb{R}$ is defined sufficiently large (e.g., $\varpi = 10^{6}$) to replace $+\infty$ numerically. It is further known that solving LVI (13.11) is equivalent to solving the following piecewise-linear equation:

$$P_{\Omega}(\mathbf{y} - (Q\mathbf{y} + \mathbf{p})) - \mathbf{y} = 0, \tag{13.12}$$

where $P_{\Omega}(\cdot) : \mathbb{R}^{n+m} \to \Omega$ is a projection operator, with the ith element of $P_{\Omega}(\cdot)$ defined as

$$\begin{cases} y_i^{-} & \text{if } y_i < y_i^{-}, \\ y_i & \text{if } y_i^{-} \leqslant y_i \leqslant y_i^{+}, \forall i \in \{1, 2, \cdots, n+m\}. \\ y_i^{+} & \text{if } y_i > y_i^{+}, \end{cases}$$

To solve (13.12) [as well as LVI (13.11) and QP (13.8) through (13.10)], let us firstly define the solution set, $\Omega^* = \{\mathbf{y}^* | \mathbf{y}^*$ is a solution of (13.12)$\}$, and a continuous error function, $\mathbf{e}(\mathbf{y}) = \mathbf{y} - P_\Omega(\mathbf{y} - (Q\mathbf{y} + \mathbf{p}))$. According to the previous chapters as well as reference [136], LVI-converted piecewise-linear equation (13.12) can be solved by the following numerical algorithm. Given the initial value of primal-dual decision variable vector $\mathbf{y}^0 \in \mathbb{R}^{n+m}$, for iteration index $k = 0, 1, 2, 3, \cdots$, if $\mathbf{y}^k \notin \Omega^*$, the numerical iterative formula for $\mathbf{y}^{k+1}$ is

$$\mathbf{y}^{k+1} = \mathbf{y}^k - \frac{\|\mathbf{e}(\mathbf{y}^k)\|_2^2 \delta(\mathbf{y}^k)}{\|\delta(\mathbf{y}^k)\|_2^2}, \tag{13.13}$$

where $\delta(\mathbf{y}^k) = (Q^\mathrm{T} + I)\mathbf{e}(\mathbf{y}^k)$. In addition, as generalized from Theorems 1 through 3 of [136], the sequence $\{\mathbf{y}^k\}$ generated by numerical algorithm 94LVI (13.13) globally linearly converges to solution $\mathbf{y}^*$, of which the first n elements constitute the optimal solution $\mathbf{x}^*$ (i.e., the computed $\dot{\theta}$) to QP (13.8)–(13.10). The QP problem and the original TVCMM scheme are thus solved.

13.3 Computer Simulations and Physical Experiments

In order to test the TVCMM scheme, computer simulations and experiment are performed by using the six-DOF planar robot manipulator to track an “R” path in the two-dimensional horizontal work-plane (i.e., the motion plane of the whole robot is parallel to the X-Y plane). For the “R” path to be tracked, the character height is 0.15 m, the task duration $T = 40$ s, and the initial joint state $\theta(0) = [\pi/4, \pi/12, \pi/12, \pi/12, \pi/36, \pi/36]^\mathrm{T}$ in radians. Besides, design parameters $p_1 = 2$ and $\mu = 4$. Furthermore, throughout this section, the error tolerance $\|\mathbf{e}(\mathbf{y}^k)\|_2 \leqslant 1.0 \times 10^{-6}$ is checked and guaranteed for assuming $\mathbf{y}^k \in \Omega^*$ in numerical algorithm 94LVI (13.13), which is very accurate for the simulations and experiment.

13.3.1 Computer Simulations

Firstly, for comparison, the simulation results synthesized by the constant coefficient aided manipulability maximizing scheme with $p(t) = 1$ are illustrated, which are shown in Figure 13.1. Specifically, Figure 13.1(a) shows the joint-velocity profiles when the end-effector of the six-DOF planar robot manipulator tracks the “R” path. As seen from the figure, initial joint-velocity $\dot{\theta}(0) \neq 0$. That is to say, an abrupt change of joint-velocity happens at the beginning of the task execution, which is undesirable for engineering application. The reason is similar to the one given in Subsection 13.2.1 that the homogeneous solution is nonzero at the beginning of the task. Also Besides, Figure 13.1(b) shows that the corresponding joint-angle profiles.

Secondly, synthesized by the time-varying coefficient aided manipulability maximizing scheme with $p(t) = 2\sin(\pi t/T)$, the simulation results are generated and shown mainly in Figure 13.2 and Figure 13.3. Specifically, Figure 13.2(a) illustrates the simulated six-DOF planar robot manipulator, of which the end-effector tracks the “R” path. As seen from the figure, the end-effector trajectory is very close to the desired path, which validates that the task is completed well. In addition, Figure 13.2(b) illustrates the desired path and the end-effector trajectory, which coincide well with each other. Furthermore,

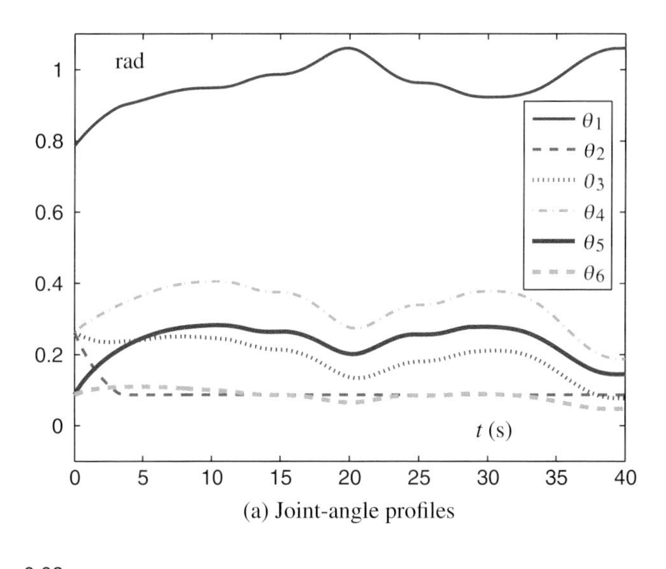

(a) Joint-angle profiles

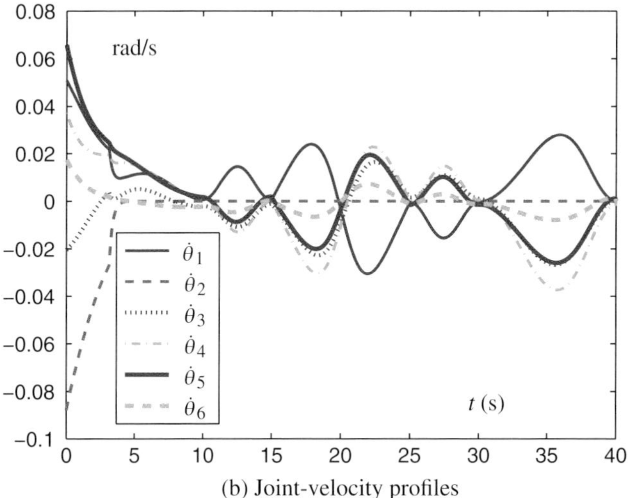

(b) Joint-velocity profiles

Figure 13.1 (a) Joint-angle and (b) joint-velocity profiles of a six-DOF planar robot manipulator synthesized by a manipulability-maximizing scheme with a constant coefficient $p(t) = 1$.

Figure 13.3 shows, respectively, the corresponding joint-velocity and joint-angle profiles of the six-DOF planar robot manipulator when its end-effector tracks the "R" path, as synthesized by the presented TVCMM scheme. Comparing Figure 13.3 with Figure 13.1, we can observe that all joint velocities now start from zero continuously and smoothly and also return to zero. That is to say, by exploiting the time-varying coefficient $p(t) = 2\sin(\pi t/T)$, the nonzero initial/final joint-velocity phenomenon can be remedied well.

Following these two simulations (i.e., shown in Figure 13.1 through Figure 13.3), we give Figure 13.4 showing the comparative results about the profiles of manipulability index $w = \det(JJ^{\mathrm{T}})$, which are synthesized by the schemes with $p(t) = 2\sin(\pi t/T)$, $p(t) = 0$ and $p(t) = 1$. It can be seen from the figure that, by using the time-varying coefficient aided manipulability-maximizing (TVCMM) scheme, the manipulability measure

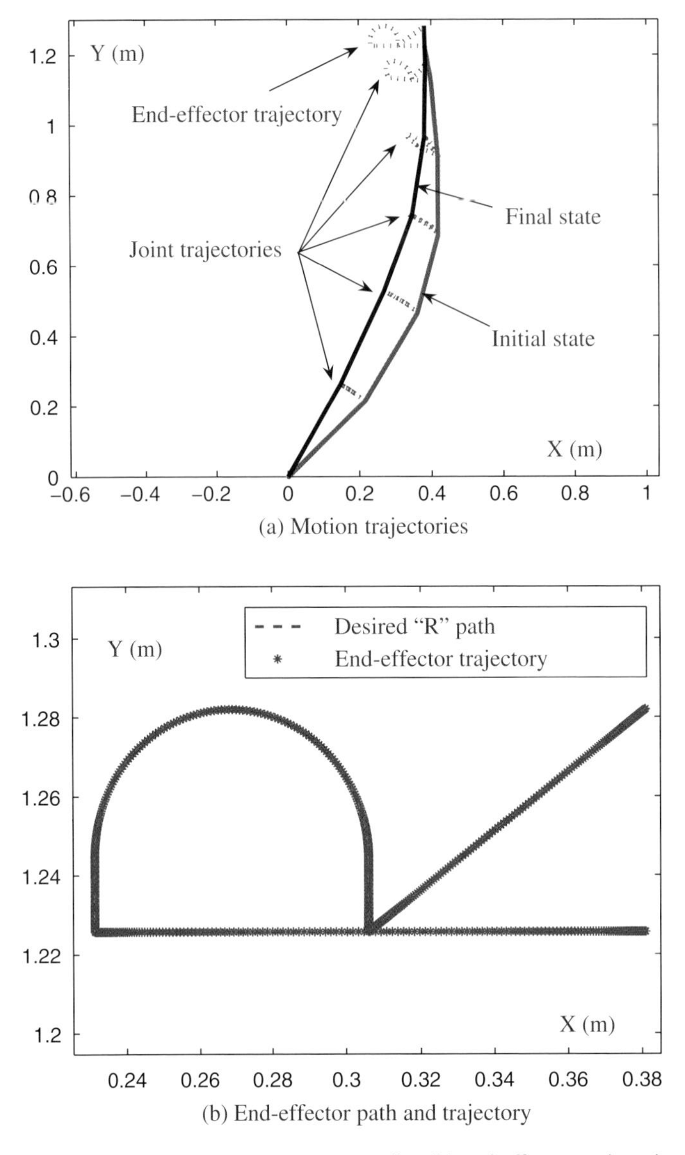

Figure 13.2 (a) Motion trajectories as well as (b) end-effector path and trajectory of the end-effector of a six-DOF planar robot manipulator tracking an "R" path synthesized by time-varying coefficient aided manipulability-maximizing scheme (i.e., $p(t) = 2\sin(\pi t/T)$).

w is always larger than that by using the scheme without a manipulability-maximizing index (i.e., with $p(t) = 0$). That is to say, the manipulability of the manipulator is well improved, and the manipulator is farther away from singularities owing to the TVCMM scheme. It is worth mentioning that, when $p(t) = 0$, the scheme becomes a minimum velocity norm (MVN) scheme [1]. In addition, from the comparison between the profiles of manipulability measure w synthesized by the TVCMM scheme and the scheme with a constant coefficient (i.e., $p(t) = 1$), we can observe that the latter is a little larger

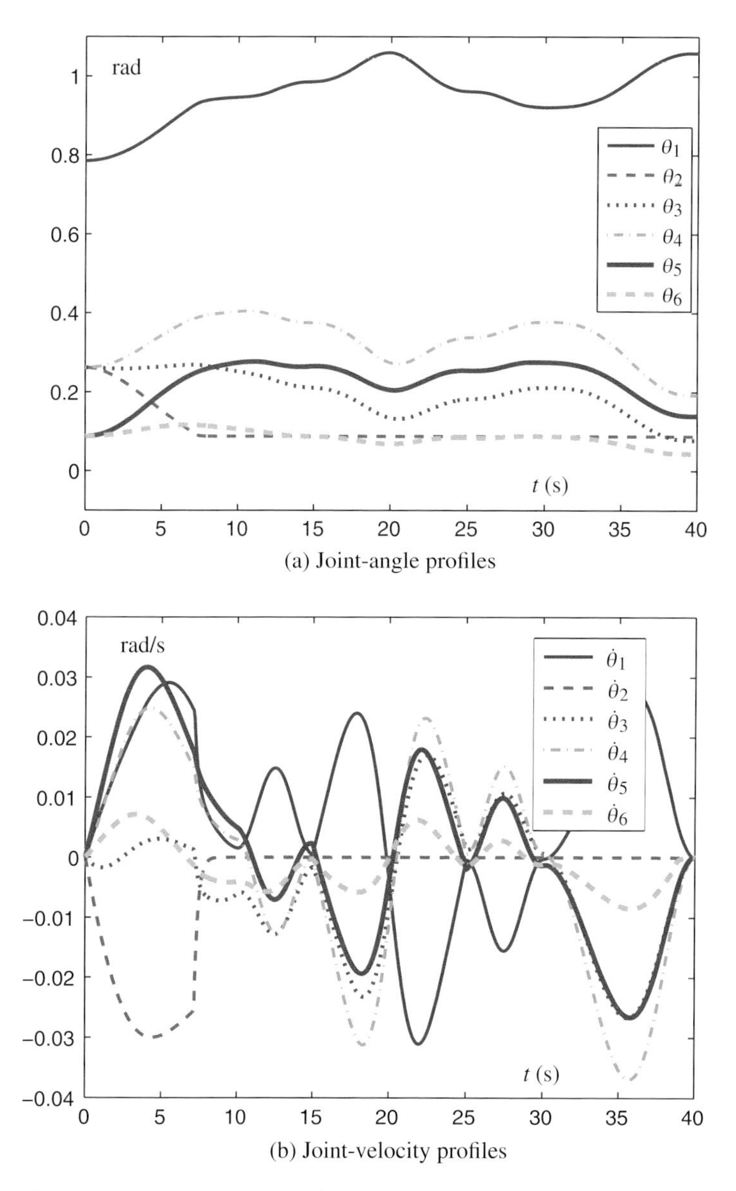

Figure 13.3 (a) Joint-angle and (b) joint-velocity profiles of a six-DOF planar robot manipulator tracking an "R" path synthesized by a time-varying coefficient aided manipulability-maximizing scheme (i.e., $p(t) = 2\sin(\pi t/T)$).

than the former only at the beginning of the task duration, after which (from around $t = 8$ to $t = 40$ s), the w measures are nearly the same. Note that, as shown in Figure 13.1 and discussed in Subsection 13.2.1, the nonzero initial/final joint-velocity problem is caused by employing a constant coefficient (i.e., $p(t) = 1$); and that, by employing a time-varying coefficient (i.e., $p(t) = 2\sin(\pi t/T)$), the nonzero initial/final joint-velocity problem is remedied well. So, it can be concluded that the TVCMM scheme is a better

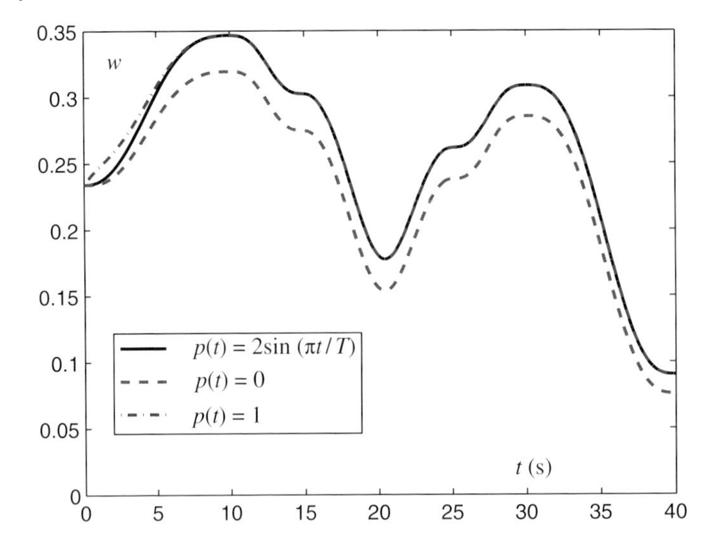

Figure 13.4 Comparison about manipulability index $w = \det(JJ^{\mathrm{T}})$ synthesized by schemes with $p(t) = 2\sin(\pi t/T)$, $p(t) = 0$ (i.e., an MVN scheme) and $p(t) = 1$.

choice for optimal motion planning and control of redundant robots, in view of its better performance in MM and better suitability to engineering application.

Furthermore, corresponding to the aforementioned simulation results synthesized by the TVCMM scheme with $p(t) = 2\sin(\pi t/T)$ (i.e., Figures 13.2 through 13.4), the joint angles and joint-angle limits are shown in more detail in Figures 13.5 through 13.7, which illustrate that all joint-angles are kept within their physical limits. More specifically, none of the joint-angle profiles enters into the "safety region." In addition, the joint velocities, the joint-velocity limits and the combined bounds $\zeta_i^{\pm}$ are shown in Figures 13.8 through 13.10. It can be seen from Figures 13.8 through 13.10 that every joint-velocity $\dot{\theta}_i$ is within its physical limits, or more precisely, within the bounds $\zeta_i^{\pm}$, for $i = 1, 2, \cdots, 6$.

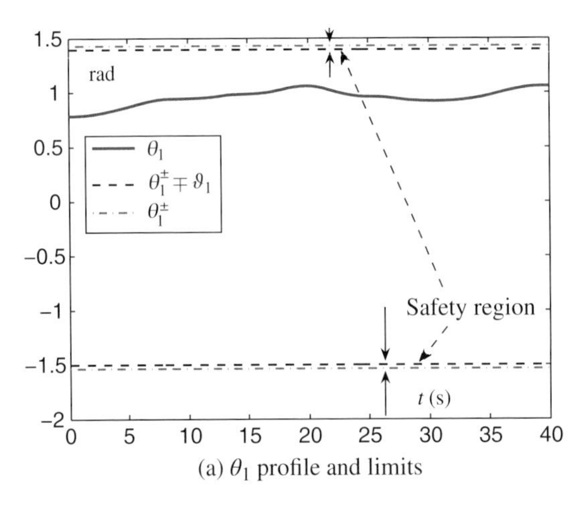

Figure 13.5 Joint-angle profiles (a) θ_1 and (b) θ_2 of corresponding limits of a six-DOF planar robot manipulator tracking an "R" path synthesized by TVCMM scheme with $p(t) = 2\sin(\pi t/T)$.

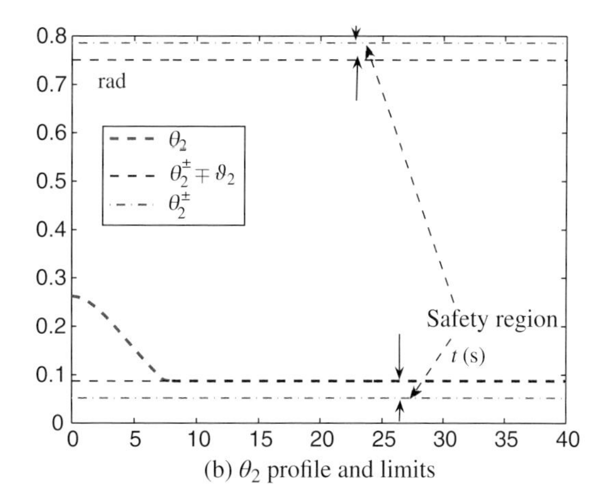

Figure 13.5 (*Continued*)

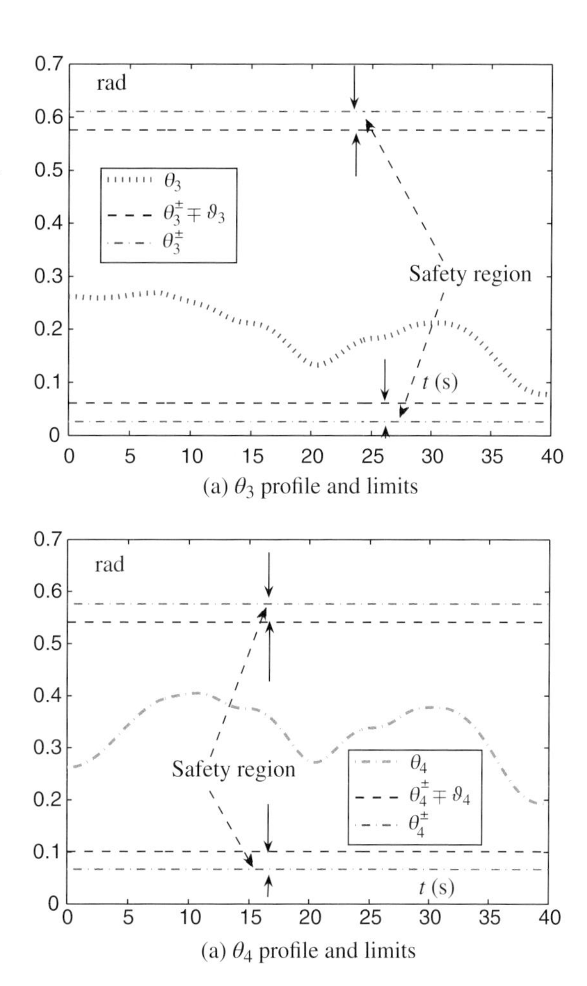

Figure 13.6 Joint-angle profiles (a) θ_3 and (b) θ_4 of corresponding limits of a six-DOF planar robot manipulator tracking an "R" path synthesized by TVCMM scheme with $p(t) = 2\sin(\pi t/T)$.

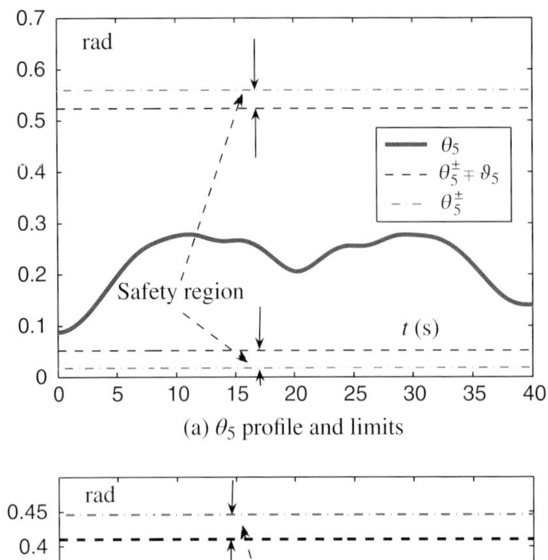

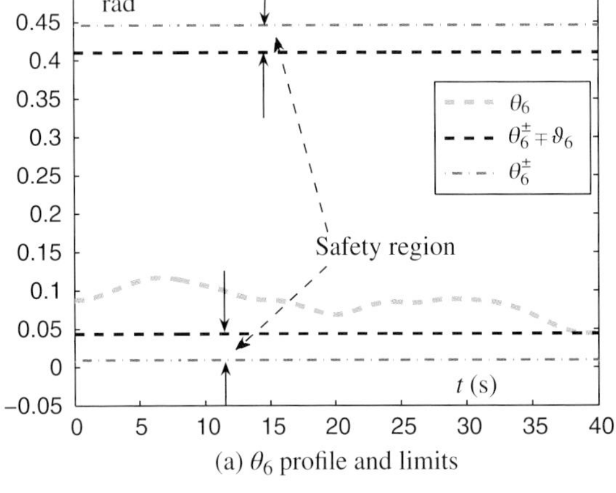

Figure 13.7 Joint-angle profiles (a) θ_5 and (b) θ_6 of corresponding limits of a six-DOF planar robot manipulator tracking an "R" path synthesized by TVCMM scheme with $p(t) = 2\sin(\pi t/T)$.

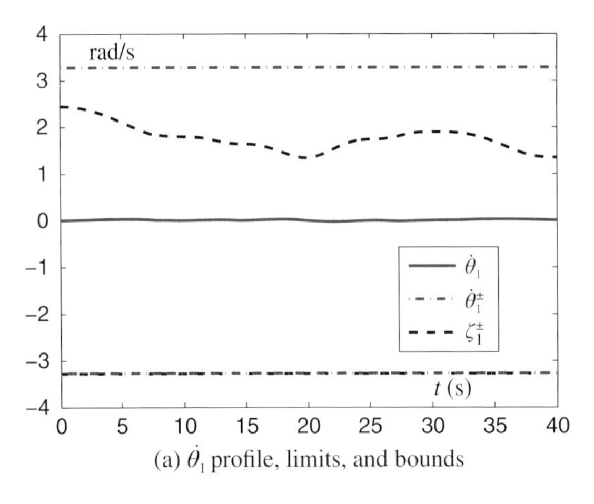

Figure 13.8 Joint-velocity profiles (a) $\dot{\theta}_1$ and (b) $\dot{\theta}_2$ of corresponding limits and bounds of a six-DOF planar robot manipulator tracking an "R" path synthesized by a TVCMM scheme with $p(t) = 2\sin(\pi t/T)$.

Figure 13.8 (*Continued*)

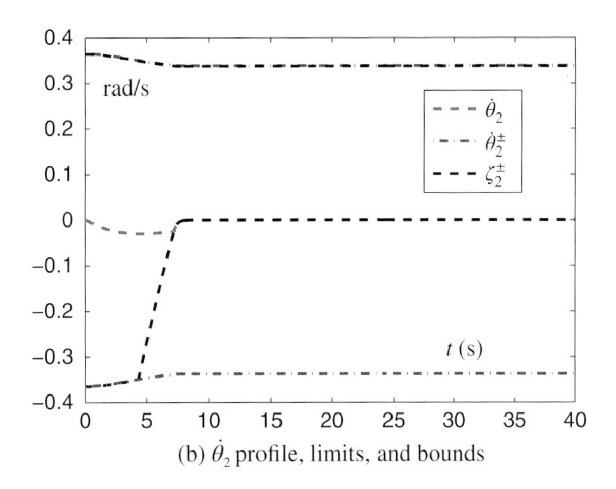

(b) $\dot{\theta}_2$ profile, limits, and bounds

Figure 13.9 Joint-velocity profiles (a) $\dot{\theta}_3$ and (b) $\dot{\theta}_4$ of corresponding limits and bounds of a six-DOF planar robot manipulator tracking an "R" path synthesized by TVCMM scheme with $p(t) = 2\sin(\pi t/T)$.

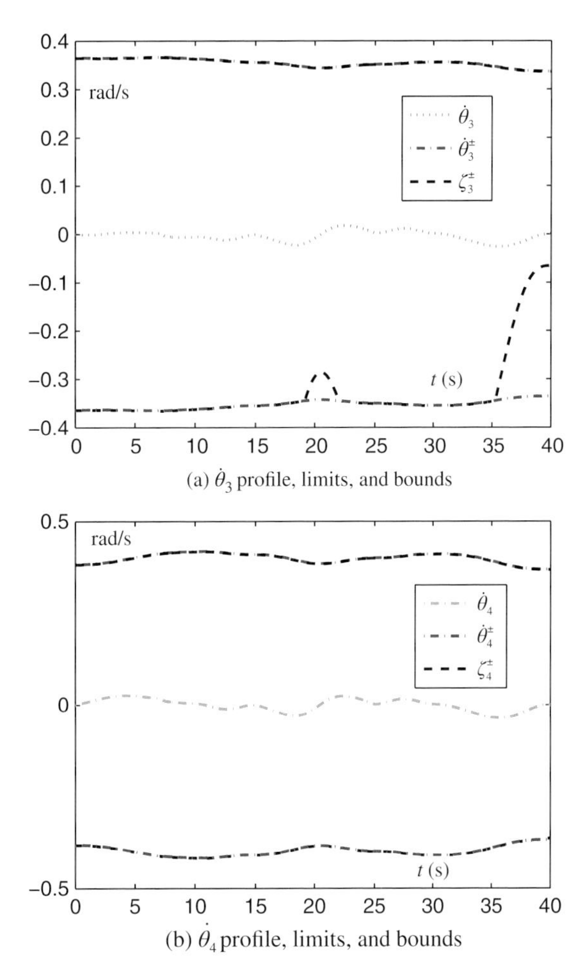

(a) $\dot{\theta}_3$ profile, limits, and bounds

(b) $\dot{\theta}_4$ profile, limits, and bounds

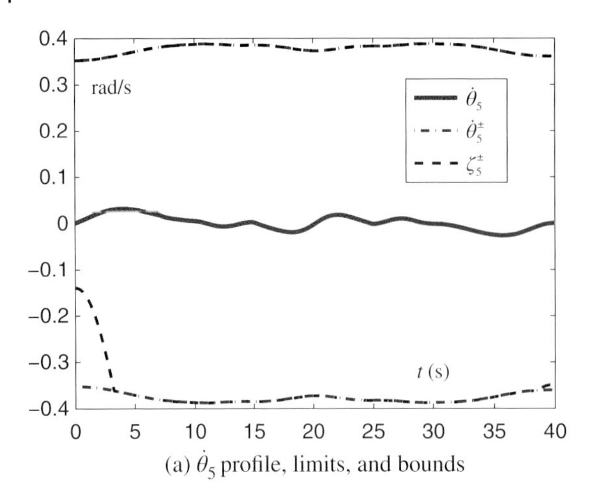

(a) $\dot{\theta}_5$ profile, limits, and bounds

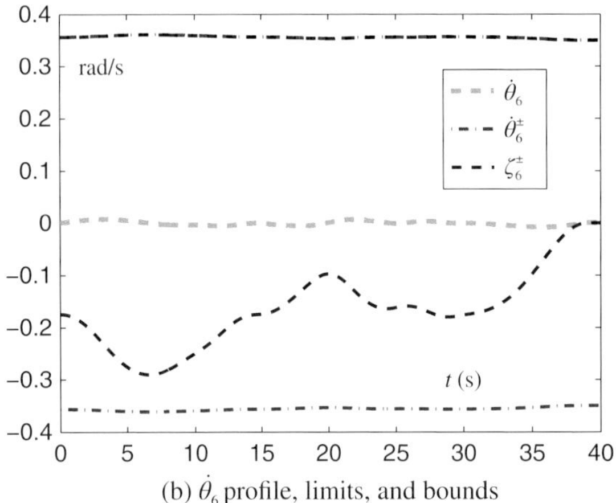

(b) $\dot{\theta}_6$ profile, limits, and bounds

Figure 13.10 Joint-velocity profiles (a) $\dot{\theta}_5$ and (b) $\dot{\theta}_6$ of corresponding limits and bounds of a six-DOF planar robot manipulator tracking an "R" path synthesized by a TVCMM scheme with $p(t) = 2\sin(\pi t/T)$.

In summary, synthesized by the time-varying coefficient aided manipulability-maximizing scheme, the manipulability of the robot manipulator is increased without causing a nonzero initial/final joint-velocity problem, and the manipulator can also fulfill the end-effector task effectively (in addition to being farther away from singularities).

13.3.2 Physical Experiments

In this subsection, by using the TVCMM scheme, an experiment is performed on the six-DOF planar robot manipulator hardware system. The task, the initial joint-angle state and the parameters in the hardware experiment are set the same as those in computer simulations. The experimental results are illustrated in Figure 13.11 and Figure 13.12. Specifically, Figure 13.11 shows the actual task execution of the manipulator synthesized by the TVCMM scheme subject to varying physical limits when the end-effector tracks the "R" path. This illustrates that the task is completed well. In addition, Figure 13.12 shows the top-view result of the "R" trajectory and the positioning errors. As seen from the figure, the height of the letter "R" is nearly 0.15 m.

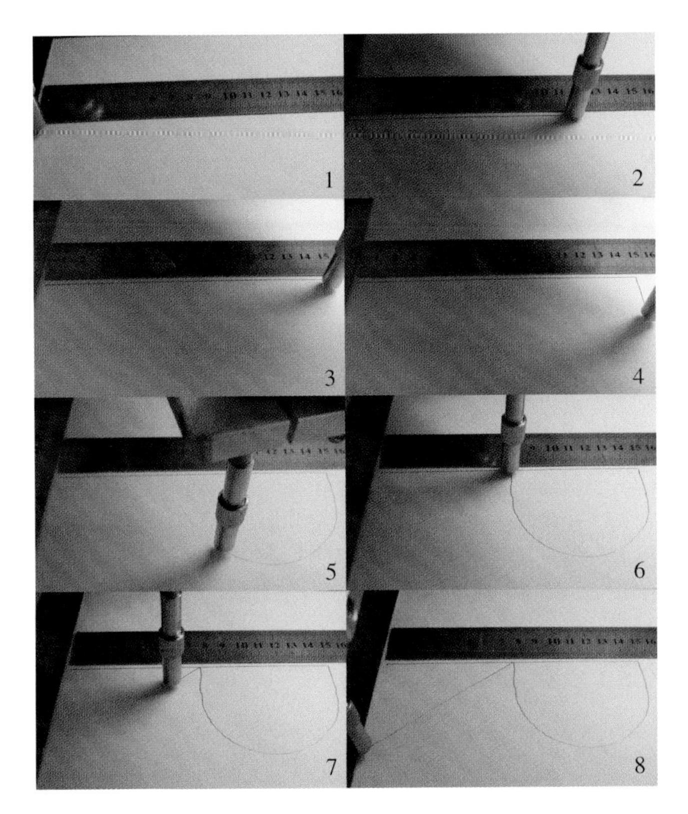

Figure 13.11 Snapshots for actual end-effector task execution of a six-DOF planar robot manipulator tracking an "R" path synthesized by TVCMM scheme with $p(t) = 2\sin(\pi t/T)$. *Source:* Zhang 2013. Reproduced with permission of John Wiley & Sons.

Figure 13.12 (a) Top-view measurement result of "R" trajectory and (b) positioning error of robot end-effector synthesized by TVCMM scheme with $p(t) = 2\sin(\pi t/T)$. *Source:* Zhang 2013. Reproduced with permission of John Wiley & Sons.

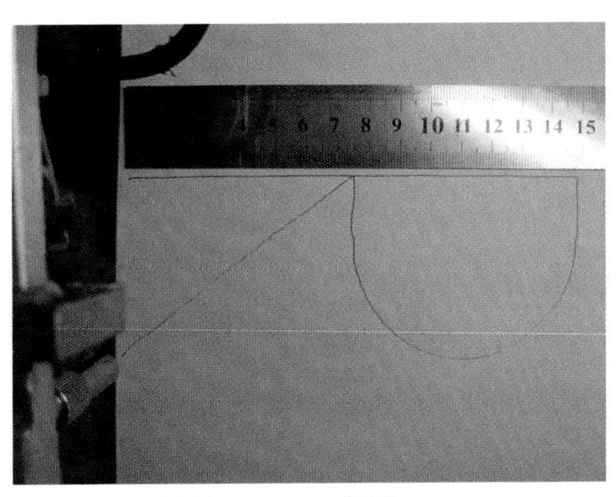

(a) Measurement of "R" trajectory

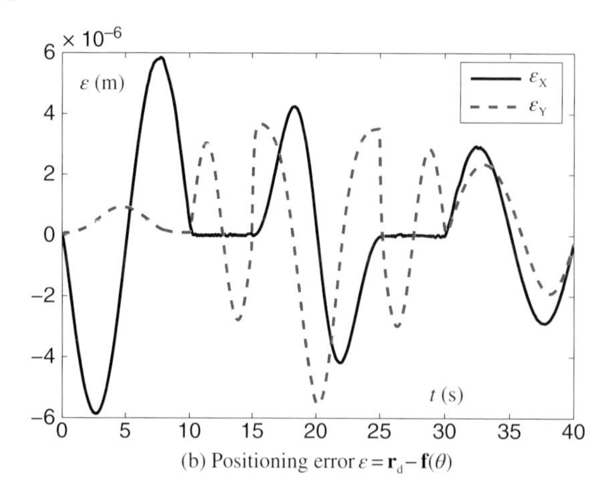

(b) Positioning error $\varepsilon = \mathbf{r}_d - \mathbf{f}(\theta)$

Figure 13.12 (*Continued*)

As for the end-effector positioning errors, we can see from Figure 13.12(b) that the maximum positioning error is less than 6.0×10^{-6} m, which shows the high precision of the end-effector positioning synthesized by the presented TVCMM scheme subject to varying limits.

In summary, this hardware experiment illustrates well the physical realizability and efficacy of the presented TVCMM scheme (which includes the numerical QP-solver (13.13)) subject to VJVL on the optimal motion planning and control of the six-DOF planar robot manipulator.

13.4 Chapter Summary

This chapter designs and presents a novel time-varying coefficient aided manipulability maximizing (TVCMM) scheme for the optimal control of redundant robot manipulators. The scheme is then reformulated into a unified QP problem that can be solved readily by the presented numerical algorithm 94LVI. Furthermore, the scheme is finally implemented on an actual six-DOF planar robot manipulator. As illustrated via comparative simulations and experiment performed on the six-DOF planar robot manipulator, the manipulability of the robot can be maximized as compared with the MVN scheme. Besides, the presented scheme can solve the nonzero initial/fianl joint-velocity problem. Moreover, the experimental results of tracking the "R" path illustrate well the physical realization and efficacy of such a TVCMM scheme subject to varying physical limits. The end-effector positioning-error analysis validates as well the high precision of such a TVCMM scheme.

Part VII

Encoder Feedback and Joystick Control

14

QP Based Encoder Feedback Control

14.1 Introduction

Given they have more DOF than required to perform a given end-effector primary task, redundant robot manipulators can achieve subtasks such as obstacle avoidance [87, 227], fault tolerance [228], repetitive motion planning [128, 229], joint limits, and singularity avoidance [230–232]. Existing optimal control methods for path planning and control of robot manipulators [233] can be roughly categorized into two types: First, optimal control methods that can handle each robot link separately without regard to robot dynamics by computing setpoints for low-level (e.g., position- or velocity-level) single-input single-output controllers [228, 234]; Second, optimal control methods that can handle the robot directly by considering robot dynamics and computing motor torques [235–237] for high-level controllers, specifically, torque-level controllers. The former methods pertain to the description of motion (position, velocity, acceleration, etc.). These methods can perform well if the desired motion is not too fast and does not require large acceleration [238]. The latter methods are related to the explanation of motion in terms of forces and torques and are applicable to numerous practical manipulators. The level at which path planning and control is performed may depend on the type of robot controller [239]. For example, if joints are driven by position- or velocity-controlled servo motors, then a path planning and control scheme at the position or velocity level will generally suffice. However, if controllers can be driven by joint torque or actuator force, then the path planning and control scheme will be designed at the torque level [239]. The computer simulations and experiments in this chapter are based on a six-DOF planar redundant robot manipulator. The manipulator is operated and controlled by pulse signals transmitted from the host computer. Manipulator joints are operated by position- and velocity-controlled servo or stepper motors, and the resultant joint variables (i.e., joint angle and joint velocity) should be converted into pulse per second (PPS) for manipulator control. In consideration of these points, the former optimal control method, which is better suited and much simpler than the latter methods, is exploited to control the six-DOF planar redundant robot manipulator.

Inverse kinematics refers to finding joint-angle vector $\theta(t) \in \mathbb{R}^n$ and/or joint-velocity vector $\dot{\theta}(t)$ in the joint space, with a given end-effector position/orientation vector $\mathbf{r}_\mathrm{d}(t) \in \mathbb{R}^m$ and/or its time derivative $\dot{\mathbf{r}}_\mathrm{d}(t) \in \mathbb{R}^m$ in the Cartesian space [240, 241], which is one of the fundamental and challenging issues in controllings robot manipulators. The existing solutions to the redundancy-resolution problem can be categorized into two methods; global methods [232, 242, 244] and local methods [231, 245]. Global methods (i.e., global optimization methods) usually refer to identifying the extreme value of a given nonconvex function in a certain feasible region. In redundancy resolution, global methods require the formulation of the continuous function (which guarantees the conservative motion over the entire operational space), calculation of variations, and additional time derivatives of variables. The computation required for the global method is complex, thus ruling out its applicability for real-time control of robot manipulators [246, 247]. Note that it may be impossible to guarantee the conservative motion over the entire operational space by formulating the continuous function; for example, in the unknown environment. Such methods can be implemented only as an off-line motion-planning tool in high levels of the manipulator control hierarchy. By contrast, local methods (i.e., local optimization methods) have simpler scheme formulation and the relatively small amount of computation, which makes these methods applicable to the real-time control of robot manipulators. Therefore, it is necessary to investigate the instantaneous motion optimization on the redundancy resolution of robots. For example, [245] presented an online scheme for singularity avoidance in robotic trajectory control, which was realized by limiting the minimum value of the manipulability measure [87] presented an approach for redundant robots to meet the demands of obstacle avoidance in dynamic environments. Most existing solutions, including the aforementioned approaches, are pseudoinverse-based methods [229, 240].

In recent years, to avoid the expensive computation (e.g., online matrix inversion, matrix-matrix multiplication, and high-order nonlinear computation) and to achieve higher computational efficiency for real-time implementation, numerous efforts have been dedicated to solving the redundancy resolution problem via optimization techniques [128, 242, 248]. Generally, such techniques consider the inverse-kinematics problem as a QP subject to equality, inequality, and bound constraints. According to [201, 204], the resultant QP can be converted into a linear variational inequality (LVI) and then into a piecewise-linear equation (PLE) via two "Bridge" theorems. Then, the QP can be efficiently solved by using numerous techniques; for example, numerical algorithm 94LVI [136] and recurrent neural networks [128, 249, 250].

This chapter presents an online motion planning and feedback control (OMPFC) scheme. Such an OMPFC scheme shows an interesting trend of combining motion planning and reactive control methodologically, which is becoming more and more important in the field of automatic control and robotics. This trend results from the real-time requirement and the possibility that unexpected events arise from time to time, for example, in case of time-varying obstacle avoidance. Compared with other methods, the presented method does not require the calculation of the Jacobian matrix inversion, thus avoiding the singularity problem [242]. The original robotic scheme formulation consists of the expression of a QP form that includes an end-effector position-error feedback and joint-physical-limit constraints. Therefore, the OMPFC scheme can ensure the precision of the end-effector positioning owing to the position-error feedback. Meanwhile, such an OMPFC scheme allows to explicitly take into account the joint-angle and joint-velocity limits. The OMPFC scheme is further reformulated as a unified QP. Then, the unified QP is converted into a PLE via

QP conversion techniques. Finally, the resultant QP is solved online by using numerical algorithm 94LVI. This means that, using such an efficient QP solver, the online property of the OMPFC scheme can be guaranteed.

14.2 Preliminaries and Scheme Formulation

This section mainly discusses the presented OMPFC scheme for redundant robot manipulators. In addition, a brief description of the six-DOF planar robot manipulator is given for readers convenience, which has been presented in the previous chapters.

14.2.1 Joint Description

The description of the six-DOF planar robot manipulator has been presented in Chapter 6, so several details about the manipulator system are omitted. Joint 1 is a pivot joint employing a feedback-controlled servo motor to provide a wide operating range. For Joint 2 through Joint 6 of the robot, the rotary motion of each stepper motor without feedback control is converted into the reciprocating motion of the push rod to drive the corresponding joint, as shown in Figure 14.1. Figure 14.1(a) shows the rotary encoders equipped on each joint. These rotary encoders can be used to measure the actual joint angles online. Figure 14.1 and Table 14.1 provide the joint structure and parameters of the employed manipulator system. According to [228] and [94], the upper and lower velocity-limits of Joint 1 through Joint 6 can be obtained by considering the parameters and upper and lower joint-angle limits of the manipulator system (i.e., $\theta^+ = [1.431, 0.785, 0.611, 0.576, 0.559, 0.445]^{\mathrm{T}}$ rad and $\theta^- = [-1.536, 0.052, 0.026, 0.066, 0.017, 0.009]^{\mathrm{T}}$ rad, termed L1 for presentation convenience).

14.2.2 OMPFC Scheme

For manipulators with known structure and parameters, the relation between the desired path $\mathbf{r}_{\mathrm{d}}(t) \in \mathbb{R}^m$ and $\theta(t) \in \mathbb{R}^n$ is described as $\mathbf{f}(\theta) \to \mathbf{r}_{\mathrm{d}}$ and $J(\theta)\dot{\theta} \to \dot{\mathbf{r}}_{\mathrm{d}}$. In this chapter, $\mathbf{f}(\cdot) : \mathbb{R}^n \to \mathbb{R}^m$ is a continuous nonlinear mapping, and $J(\theta) = \partial \mathbf{f}(\theta)/\partial \theta \in \mathbb{R}^{m\times n}$ is the Jacobian matrix; $\mathbf{r}_{\mathrm{d}}(t)$, $\theta(t)$, and $J(\theta)$ are sometimes written as $\mathbf{r}_{\mathrm{d}}$, θ, and J, respectively, to simplify the presentation.

The end-effector of the manipulator hardware system presented in Chapter 6 is a fixed pen-shaped object that is different from the gripper- or hand-shaped end-effector. The orientation of the end-effector has little effect on the hardware system during task execution. Therefore, this study considers the end-effector position only in a 2D horizontal work-plane (i.e., $m = 2$). Here, n is set as 6 in view of the manipulator having six DOF. Following [1] and [128], we provide the following general QP-based scheme formulation to solve the motion planning problem:

$$\text{minimize} \quad \dot{\theta}_{\mathrm{r}}^{\mathrm{T}}\dot{\theta}_{\mathrm{r}}/2 + \mathbf{q}^{\mathrm{T}}\dot{\theta}_{\mathrm{r}} \tag{14.1}$$

$$\text{subject to} \quad J\dot{\theta}_{\mathrm{r}} = \dot{\mathbf{r}}_{\mathrm{d}} + \kappa_{\mathrm{p}}(\mathbf{r}_{\mathrm{d}} - \mathbf{f}(\theta)) \tag{14.2}$$

$$\theta^- \leqslant \theta \leqslant \theta^+ \tag{14.3}$$

$$\dot{\theta}^-(\theta) \leqslant \dot{\theta}_{\mathrm{r}} \leqslant \dot{\theta}^+(\theta) \tag{14.4}$$

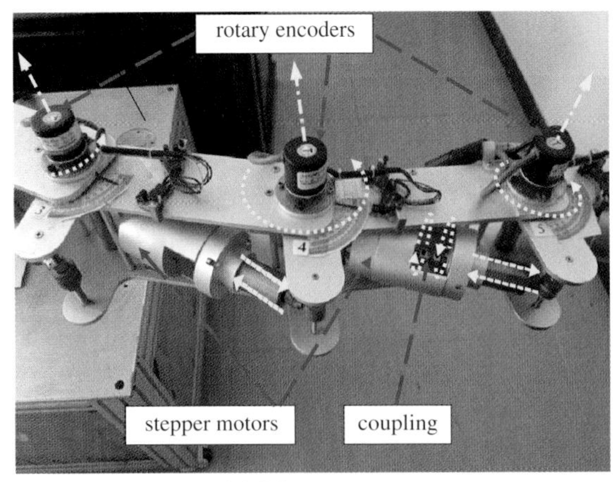

(a) Joint structure

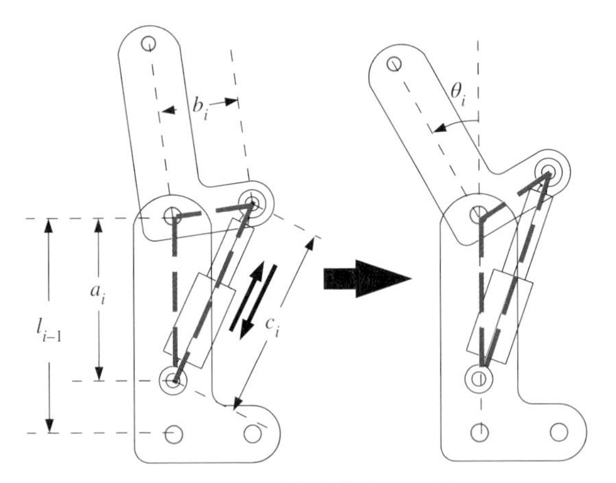

(b) Simplified CAD model

Figure 14.1 Joint structure of Joint 2 through Joint 6. *Source*: Zhang 2013. Reproduced with permission of Elsevier.

Table 14.1 Motor parameters of a six-DOF planar robot manipulator

Motor	Parameter	Meaning	Value
Servo motor (Joint 1)	σ_{se}^{+} (rot/s)	Maximum rotation rate	2500/60
	$\emptyset_{\text{rgr}}$	Reduction gear ratio	80
	δ_{ppr} (pulses/rot)	Pulses per rotation	4000
Stepper motor (Joint 2 through Joint 6)	η_{er} (m/rot)	Elongation rate	2.5×10^{-3}
	σ_{st}^{+} (rot/s)	Maximum rotation rate	10
	ξ_{sm}	Subdivision multiple	32
	θ_{sa} (rad/pulse)	Stepping angle	0.01π

where superscript ^{T} denotes the transpose of a matrix/vector, and coefficient $\mathbf{q} \in \mathbb{R}^n$ is defined accordingly for a specific redundancy-resolution scheme. Symbol $\dot{\theta}_{\mathrm{r}}$ denotes the reference joint-velocity vector in the joint space. Symbols $\mathbf{r}_{\mathrm{d}}$ and $\dot{\mathbf{r}}_{\mathrm{d}}$ denote, respectively, the desired position and velocity vectors of the end-effector for completing the given primary task. Evidently, this scheme formulation considers joint-angle limits and joint-velocity limits as bound constraints. Computational round-off error and force disturbance always exist in real systems because of the modeling error; thus, closed-loop control is necessary in real systems [250–252]. A feedback of the Cartesian position error $\kappa_{\mathrm{p}}(\mathbf{r}_{\mathrm{d}} - \mathbf{f}(\theta))$ can be introduced into the redundancy-resolution equation [253]; that is, $J\dot{\theta}_{\mathrm{r}} = \dot{\mathbf{r}}_{\mathrm{d}} + \kappa_{\mathrm{p}}(\mathbf{r}_{\mathrm{d}} - \mathbf{f}(\theta))$. The feedback-gain coefficient $\kappa_{\mathrm{p}} > 0 \in \mathbb{R}$ is used in this study to scale the magnitude of the manipulator response to the position error $\mathbf{r}_{\mathrm{d}} - \mathbf{f}(\theta)$. Compared with other classical feedback methods, such as velocity, acceleration, and current or voltage feedback method [238], the position-error feedback method is much simpler and more directed in terms of mathematical expressions. The position-error feedback method is incorporated into the OMPFC scheme via Equation (14.2). Theoretical analysis confirms the reasonability of the position-error feedback method and the ensuing simulations and experiments substantiate the efficacy of this method. In joint-limit constraints (14.3) and (14.4), $\theta^{\pm}$ denote, respectively, the upper and lower limits of joint-angle vector θ, whereas $\dot{\theta}^{\pm}(\theta)$ denote, respectively, the upper and lower limits of joint-velocity vector $\dot{\theta}_{\mathrm{r}}$, which varies with θ.

Safety devices are used in the presented six-DOF planar manipulator. If a joint of the manipulator approaches or reaches its physical limit, then the joint would be locked by the physical safety device throughout the task duration, thereby resulting in failure of the end-effector task execution. Therefore, leaving some "safety region" to the joint physical limits $\theta^{\pm}$ is necessary. Given that redundancy is resolved at the joint-velocity level (i.e., in terms of $\dot{\theta}_{\mathrm{r}}$), the joint-angle limit constraint (14.3) has to be converted into a $\dot{\theta}_{\mathrm{r}}$-based expression [128]. On the basis of these two points, Equation (14.3) can be further designed as

$$\mu((\theta^- + \vartheta) - \theta) \leqslant \dot{\theta}_{\mathrm{r}} \leqslant \mu((\theta^+ - \vartheta) - \theta),$$

where $\mu > 0 \in \mathbb{R}$ is used to scale the feasible region of joint-velocity $\dot{\theta}_{\mathrm{r}}$, and ϑ is a constant vector used to scale the safety region, where each element ϑ_i $(i = 1, 2, \cdots, 6)$ is set to be 0.0349 rad (i.e., 2 degrees) in the ensuing simulations and experiments.

By summarizing the aforementioned analysis and by replacing $\dot{\theta}_{\mathrm{r}}$ with the decision variable vector $\mathbf{x}$, the robotic scheme formulation (14.1)–(14.4) can be further reformulated as a unified QP:

$$\text{minimize} \quad \mathbf{x}^{\mathrm{T}} W \mathbf{x}/2 + \mathbf{q}^{\mathrm{T}}\mathbf{x} \tag{14.5}$$

$$\text{subject to} \quad J\mathbf{x} = \mathbf{d} \tag{14.6}$$

$$\zeta^- \leqslant \mathbf{x} \leqslant \zeta^+ \tag{14.7}$$

$$\text{with } \mathbf{x} = \dot{\theta}_{\mathrm{r}},$$

$$W = I \text{ and } \mathbf{d} = \dot{\mathbf{r}}_{\mathrm{d}} + \kappa_{\mathrm{p}}(\mathbf{r}_{\mathrm{d}} - \mathbf{f}(\theta))$$

where I denotes an appropriately-dimensioned identity matrix, and the ith elements of ζ^- and ζ^+ (with $i = 1, 2, \cdots, 6$) are $\zeta_i^- = \max\{\mu((\theta_i^- + \vartheta_i) - \theta_i), \dot{\theta}_i^-(\theta_i)\}$, $\zeta_i^+ = \min\{\mu((\theta_i^+ - \vartheta_i) - \theta_i), \dot{\theta}_i^+(\theta_i)\}$, respectively. Suitable performance indices such as minimum-velocity-norm [1] and repetitive motion index [128] can be incorporated by defining different coefficient $\mathbf{q}$ appropriately. When the end-effector traces a closed path, the joint angles may not return to their initial angles after completing the end-effector task. This condition, which is termed the joint-angle-drift phenomenon or non-repetitive problem, may result in the difficulty of predicting the manipulator behavior. Note that readjusting the configuration of the manipulator via self-motion is less efficient [128]. The parameter $\mathbf{q} = \gamma(\theta - \theta(0))$ with design parameter γ, which is used to scale the magnitude of the manipulator response, can be exploited to solve non-repetitive problems [128]. Thus, setting $\mathbf{q} = \gamma(\theta - \theta(0))$ minimizes the performance index (14.5) and prevents joint displacement in the cyclic motion of the manipulator. This chapter focuses on the online-realization of such a scheme and feedback control.

14.3 Computer Simulations

Computer simulations are performed on the six-DOF planar robot manipulator with two path-tracking examples to test the online property of the presented OMPFC scheme, which is based on numerical algorithm 94LVI presented in Subsection 6.3. In the first example, the end-effector of the manipulator is expected to track a petal-shaped path composed of four semi-circles with diameters of 0.12 m and task duration T of 40 s. The initial joint state $\theta(0)$ of the manipulator is $[\pi/4, \pi/36, \pi/9, \pi/12, \pi/18, \pi/18]^{\mathrm{T}}$ rad. Comparative simulations are further conducted to substantiate the superior efficacy of the OMPFC scheme. The second example is a hexagonal path-tracking task with side length of 0.08 m and task duration T of 24 s. The initial joint state $\theta(0)$ of the manipulator is $[7\pi/36, \pi/9, \pi/9, \pi/12, \pi/12, \pi/12]^{\mathrm{T}}$ rad. Throughout this section, the position-error feedback-gain coefficient κ_{p} is 6, the time-step index $h = 0.01$ s, $\mu = 4$, and $\gamma = 4$. The error tolerance is $\|\mathbf{e}(\mathbf{u}^k)\|_2 < 1 \times 10^{-6}$, and the maximum iteration is 5000. Assuming that $\mathbf{u}^k \in \Omega^*$ in the QP-solving numerical algorithm (13.13), the error tolerance $\|\mathbf{e}(\mathbf{u}^k)\|_2 \leqslant 1.0 \times 10^{-6}$ is considered to pursue the acceptable precision. If the error tolerance is not satisfied, then the maximum iteration count of 5000 is exploited to terminate the procedure. In the simulations, the joint-angle vector θ is obtained by using Euler forward-difference formula, that is, $\theta(t + h) = \theta(t) + h\dot{\theta}_{\mathrm{r}}(t)$. The presented motion planning and control scheme in the experiment is realized in real-time and the joint angles are measured online via rotary encoders of every joint equipped on the manipulator.

14.3.1 Petal-Shaped Path-Tracking Task

A petal-shaped-path-tracking example is illustrated in this subsection to test the online property and efficacy of the presented OMPFC scheme. Simulation results are shown in Figures 14.2 through 14.5.

Figure 14.2(a) presents the simulated manipulator tracking the path, which shows that the end-effector task is completed well. The final joint state coincides with the initial state, demonstrating the effectiveness of the optimization index $\dot{\theta}_{\mathrm{r}}^{\mathrm{T}}\dot{\theta}_{\mathrm{r}}/2 + \mathbf{q}^{\mathrm{T}}\dot{\theta}_{\mathrm{r}}$. Figure 14.2(b) and Figure 14.3(a) illustrate the simulated end-effector trajectory and the corresponding position error $\epsilon = \mathbf{r}_{\mathrm{d}} - \mathbf{f}(\theta)$, with the maximum value of less than 1.5×10^{-5} m. These figures illustrate the high precision of the end-effector positioning

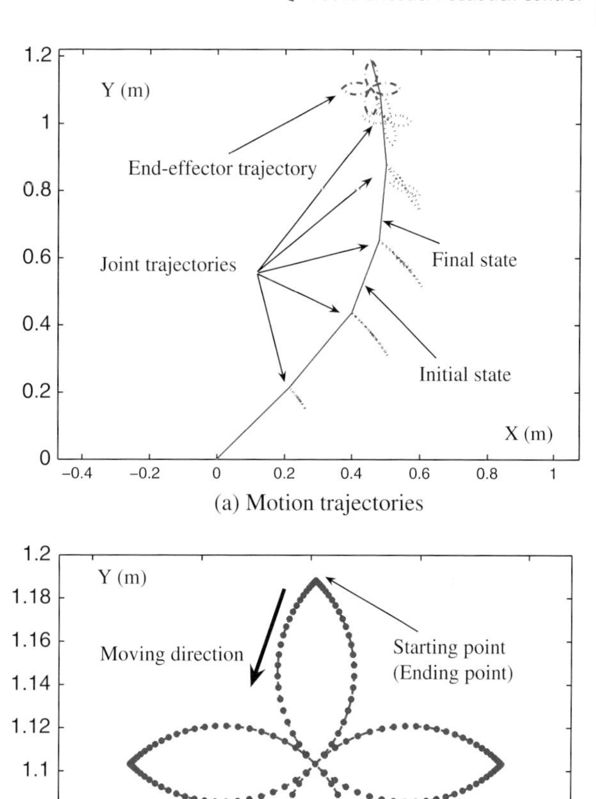

Figure 14.2 (a) Motion trajectories. as well as (b) desired path and actual trajectory of a six-DOF planar robot manipulator tracking a petal-shaped path synthesized by an OMPFC scheme with L1.

synthesized by the OMPFC scheme, where ϵ_X and ϵ_Y denote the x and Y-axis components of ϵ, respectively. Figure 14.3(b) and Figure 14.4(a) show the resolved joint-velocity and joint-angle profiles, respectively. According to Figure 14.4(b), all joint-velocity profiles start from and return to zero continuously and smoothly. Similarly, joint-angle profiles are smooth, and every joint angle returns to its own initial state, which are marked by fine dash lines in Figure 14.4(a). The θ_2 profile and its lower limit/bound are shown in Figure 14.4(b), which illustrates that the θ_2 trajectory is kept within its physical limit/bound (with upper limit/bound omitted in view of that the upper limit/bound is far from the θ_2 trajectory). The joint-angle profile does not enter the "safety region." In the simulation, the resolved θ_i and $\dot{\theta}_i$ profiles are all kept within joint-angle limits $\theta_i^{\pm}$ (specifically, $\theta_i^{\pm} \mp \vartheta_i$) and joint-velocity limits $\dot{\theta}_i^{\pm}$, with $i = 1, 2, \cdots, 6$. These results validate the activation of joint-velocity bounds $\zeta^{\pm}$ as designed in (14.5)–(14.7). However, data are omitted in this chapter because of space limitation.

Corresponding to these simulation results, Figure 14.5 shows the iteration number and running time in each time step in the petal-shaped-path-tracking example of

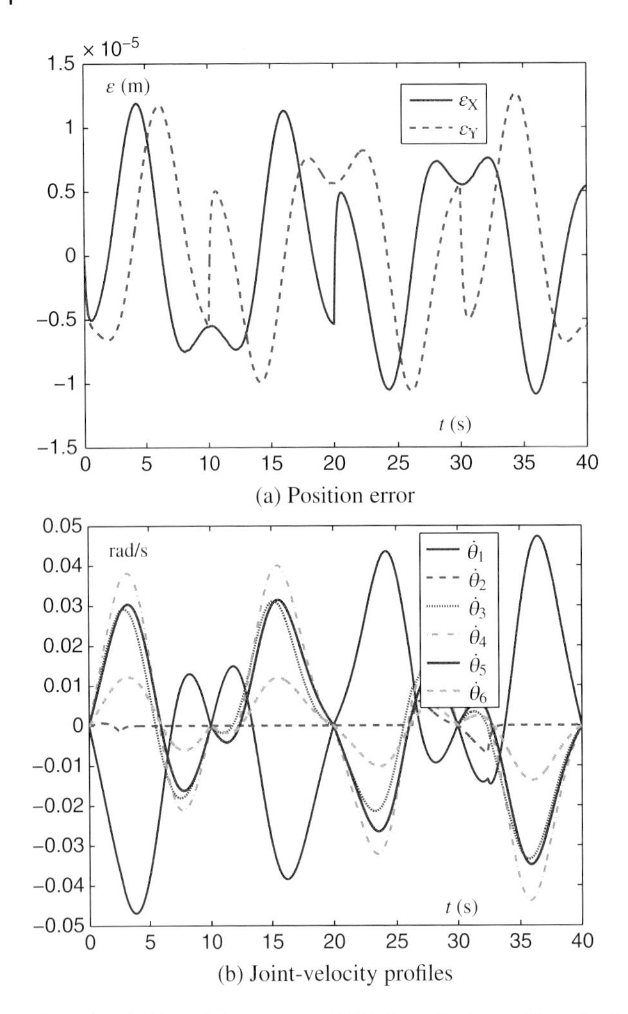

Figure 14.3 (a) Position error and (b) joint-velocity profiles of a six-DOF planar robot manipulator tracking a petal-shaped path synthesized by a OMPFC scheme with L1.

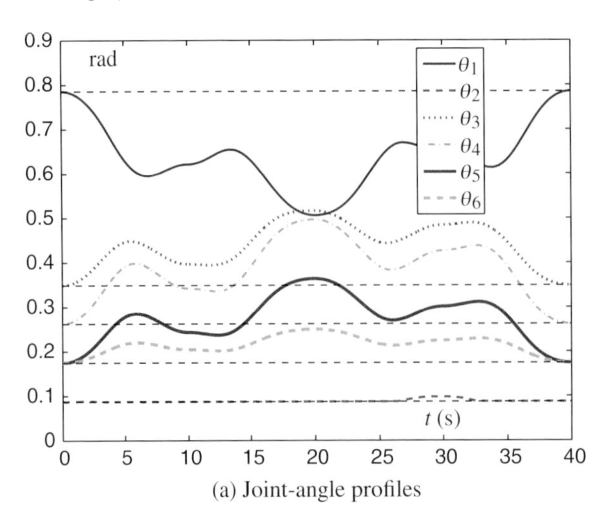

Figure 14.4 (a) Joint-angle profiles and (b) joint-angle-limit constraint (θ_2^-) of a six-DOF planar robot manipulator tracking a petal-shaped path synthesized by a OMPFC scheme with L1.

Figure 14.4 (*Continued*)

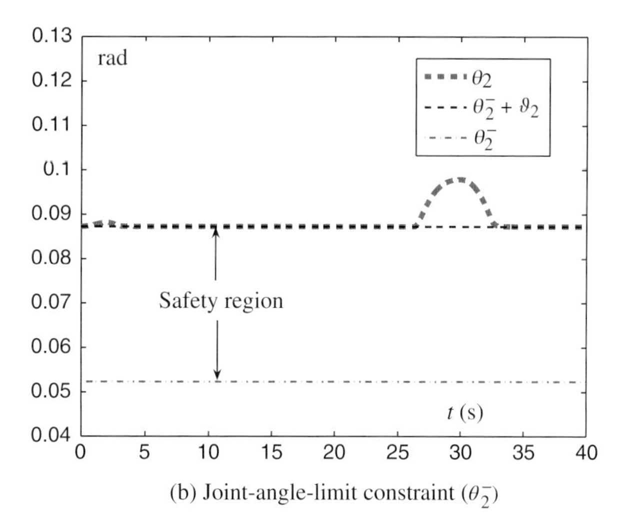

(b) Joint-angle-limit constraint (θ_2^-)

Figure 14.5 (a) Iteration numbers and (b) running times of numerical algorithm 94LVI for a six-DOF planar robot manipulator tracking a petal-shaped path.

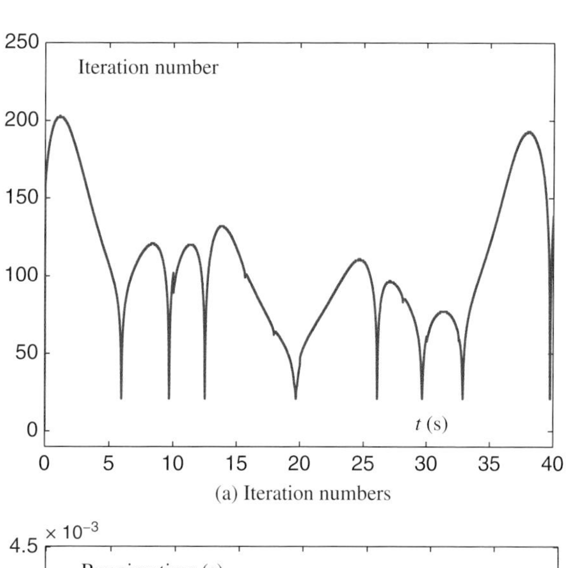

(a) Iteration numbers

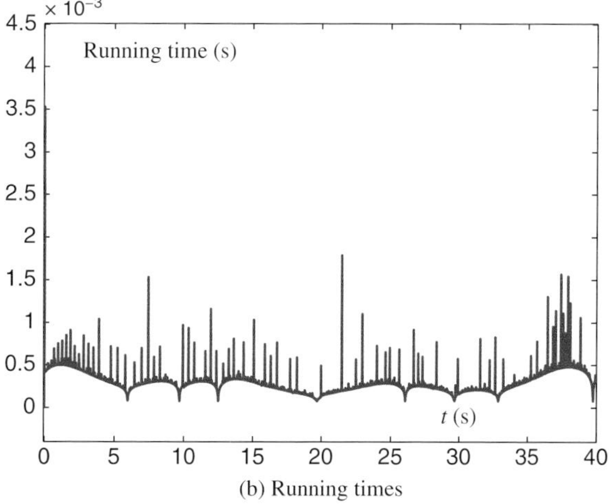

(b) Running times

numerical algorithm 94LVI. On the basis of this numerical algorithm for solving QP (14.5)–(14.7) and the redundancy resolution problem of redundant robots, the maximum iteration number is less than 250. The maximum running time is about 3.5×10^{-3} s, which is significantly less than the time step $h = 0.01$ s. These results indicate that numerical algorithm 94LVI is suitable for online motion planning and control of redundant manipulators.

14.3.2 Comparative Simulations

Two comparative simulations are conducted in this subsection to illustrate further the activation of joint-velocity bounds $\zeta^{\pm}$ and the online property of the OMPFC scheme.

14.3.2.1 Petal-Shaped Path Tracking Using Another Group of Joint-Angle Limits

Another group of joint-angle limits are introduced and considered to perform the same petal-shaped path-tracking task. Following [228], we set the upper and lower joint-angle limits as $\theta^{+} = [4.587, 0.816, 0.621, 0.599, 0.599, 0.445]^{\mathrm{T}}$ rad and $\theta^{-} = [0, 0, 0.035, 0.052, 0.035, 0]^{\mathrm{T}}$ rad (termed L2) in the simulation.

Figures 14.6 and 14.7 present the simulation results. Specifically, Figure 14.6(a) shows the simulated manipulator tracking the path, in which the final joint state coincides with the initial state. Figure 14.6(b) illustrates the corresponding position error with maximum value of less than 1.5×10^{-5} m, thus demonstrating the high precision of the end-effector positioning synthesized by the presented OMPFC scheme. Figure 14.7 shows the θ_2 profile and its lower limit/bound as well as the running time of numerical algorithm 94LVI in each time step during task execution, respectively. As shown in Figure 14.7(a), the θ_2 trajectory is kept within its physical limit/bound (with upper limit/bound omitted in view of that the upper limit/bound is far from the θ_2 trajectory) and does not enter the "safety region." The resolved θ_i and $\dot{\theta}_i$ profiles are all kept within joint-angle limits $\theta_i^{\pm}$ (specifically, $\theta_i^{\pm} \mp \vartheta_i$) and joint-velocity limits $\dot{\theta}_i^{\pm}$, with $i = 1, 2, \cdots, 6$. These results illustrate the activation of joint-velocity bounds $\zeta^{\pm}$. However, data are omitted in this chapter because of space limitation. Figure 14.7(b) shows that the maximum running time is approximately 3.5×10^{-3} s, which is significantly less than the time step $h = 0.01$ s. This result further validates the suitability of numerical algorithm 94LVI for online motion planning and control of redundant manipulators.

14.3.2.2 Petal-Shaped Path Tracking via the Method 4 (M4) Algorithm

The Method 4 (M4) algorithm, which is a numerical algorithm analyzed in [168], is exploited in this part to solve QP (14.5)–(14.7). The M4 algorithm is applied in the same petal-shaped-path-tracking task. Figures 14.8 and 14.9 show the corresponding simulation results, which show that the end-effector task can be completed with tiny position errors, and the maximum iteration number is less than 100. However, the maximum running time is approximately 0.3 s, which is significantly more than the time step $h = 0.01$ s. This condition results in the inapplicability of the M4 algorithm in the real-time control of manipulators. Moreover, the M4 algorithm necessitates matrix inversion operation per iteration, thus taking significant time to resolve redundancy at each step.

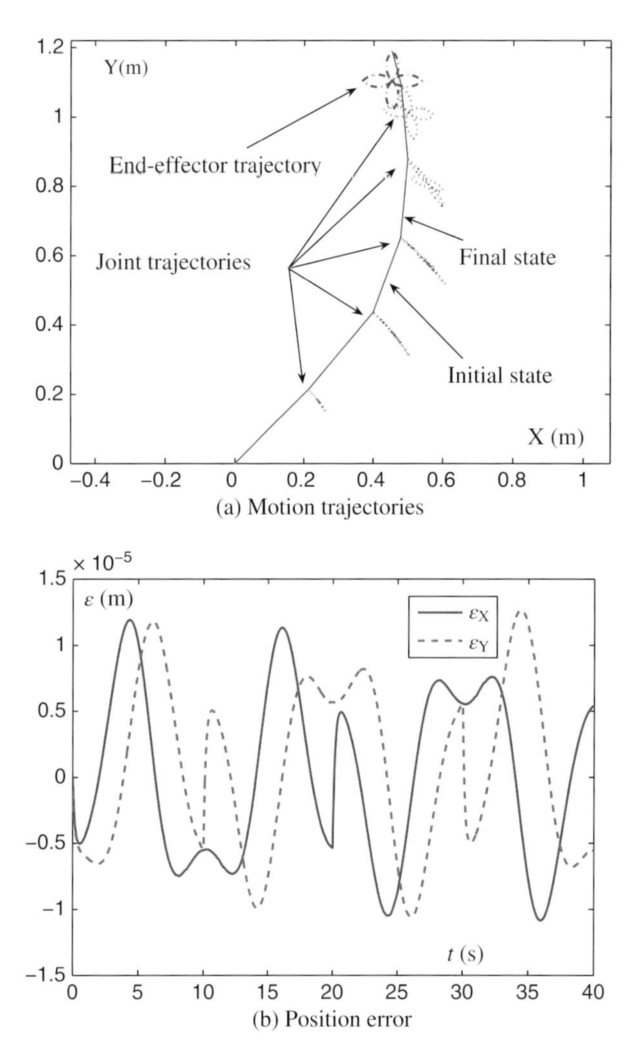

Figure 14.6 (a) Motion trajectories and (b) position error of a six-DOF planar robot manipulator tracking a petal-shaped path synthesized by OMPFC scheme with L2.

14.3.3 Hexagonal-Path-Tracking Task

Figure 14.10 illustrates the simulation results for the hexagonal-path tracking of the second example. Similar to the previous example, Figure 14.10(a) presents the simulated manipulator tracking the end-effector task, which shows that the final joint state coincides well with the initial state. Figure 14.10(b) presents the end-effector position error with a maximum value of under 3×10^{-5} m. These findings further illustrate the efficacy of the drift-free scheme for the motion planning and control of redundant manipulators. Other simulation results are omitted in this chapter because of similarity and space limitation.

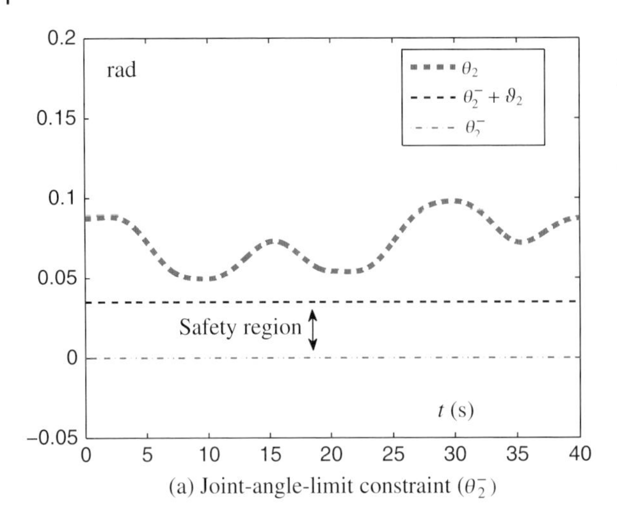

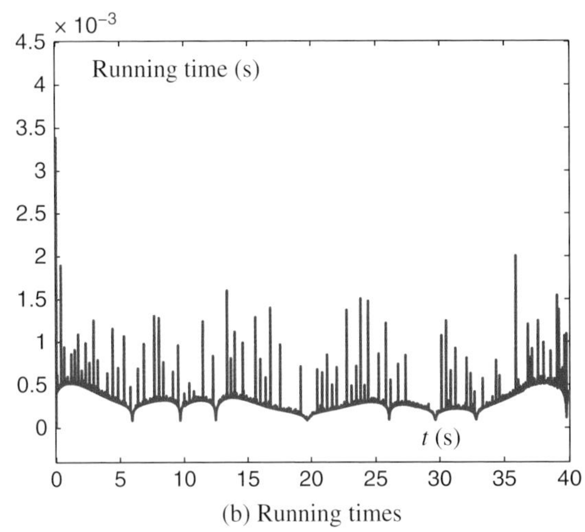

Figure 14.7 (a) Joint-angle-limit constraint (θ_2^-) and (b) running times of a six-DOF planar robot manipulator tracking a petal-shaped path synthesized by a OMPFC scheme with L2.

14.4 Physical Experiments

As synthesized by using the OMPFC scheme, this section presents the comparative experiments performed on the manipulator hardware system presented in Section 14.2.1. The experiments run online have time step h of 0.01 s because of the efficiency of numerical algorithm 94LVI. Each joint-angle θ_i $(i = 1, 2, \cdots, 6)$ is measured online from the corresponding rotary encoder equipped on each joint and then incorporated into the OMPFC scheme by considering the feedback of the Cartesian position error, that is, $J\dot{\theta}_{\mathrm{r}} = \dot{\mathbf{r}}_{\mathrm{d}} + \kappa_{\mathrm{p}}(\mathbf{r}_{\mathrm{d}} - \mathbf{f}(\theta))$.

The tasks, initial joint-angle states, and parameters of the hardware experiments are similar to those in the simulations. According to the recorded joint-angle data, the end-effector position and the corresponding position error can be calculated via forward kinematics $\mathbf{f}(\theta)$ and $\epsilon = \mathbf{r}_{\mathrm{d}} - \mathbf{f}(\theta)$, respectively. Figure 14.11 presents the results of the physical experiment with $\kappa_{\mathrm{p}} = 0$ (no Cartesian position feedback) to show the

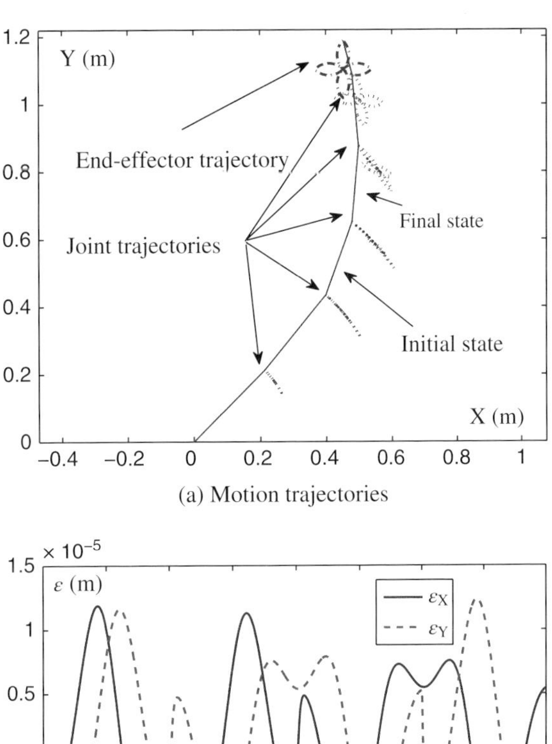

(a) Motion trajectories

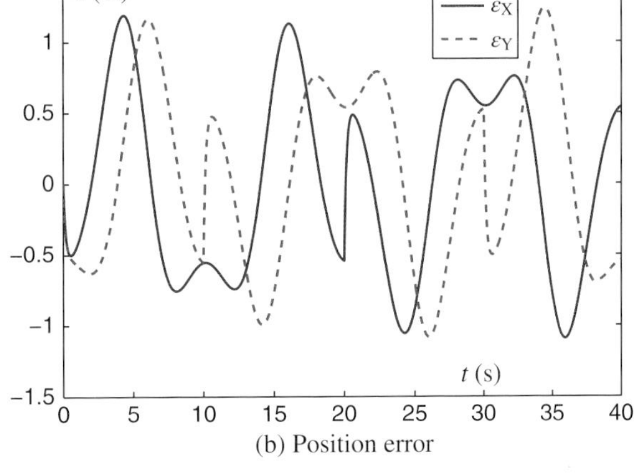

(b) Position error

Figure 14.8 (a) Motion trajectories and (b) position error of a six-DOF planar robot manipulator tracking a petal-shaped path synthesized by a scheme equipped with the M4 algorithm and L1.

importance of Cartesian position feedback within the OMPFC scheme. As illustrated in Figure 14.11(a), the motion trajectory of the manipulator end-effector deviates from the desired petal-shaped path considerably without position error feedback. Figure 14.11(b) shows the corresponding position error [i.e., the distance between the actual position and reference position (or desired position) of the end effector, $\epsilon = \mathbf{r}_\mathrm{d} - \mathbf{f}(\theta)$], in which the maximum error being approximately 1.761×10^{-2} m is considerably large for engineering applications.

This result is attributed to the inability of the pure velocity-level scheme without position-level feedback information to prevent the generation of position errors.

Experiments are performed on the manipulator hardware system on the basis of the feedback-gain coefficient $\kappa_\mathrm{p} = 6$ in Equation (14.2). Figures 14.12 through 14.17 illustrate the experimental results. Figure 14.12 shows the actual task-execution process of tracking the petal-shaped path synthesized via the presented OMPFC scheme, where the arrows denote the directions of the end-effector. According to this figure, the

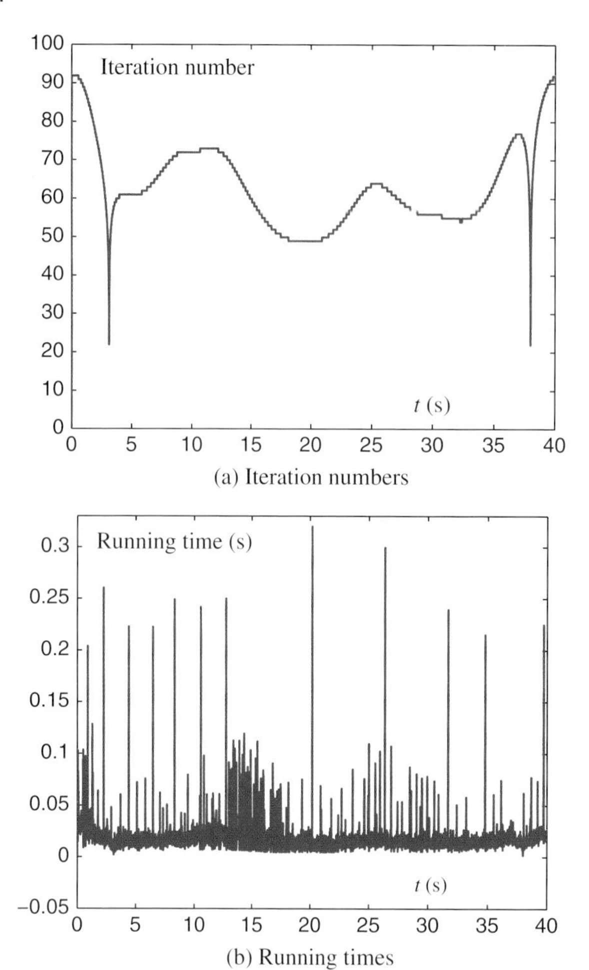

Figure 14.9 (a) Iteration numbers and (b) running times of a six-DOF planar robot manipulator tracking a petal-shaped path synthesized by a scheme equipped with the M4 algorithm and L1.

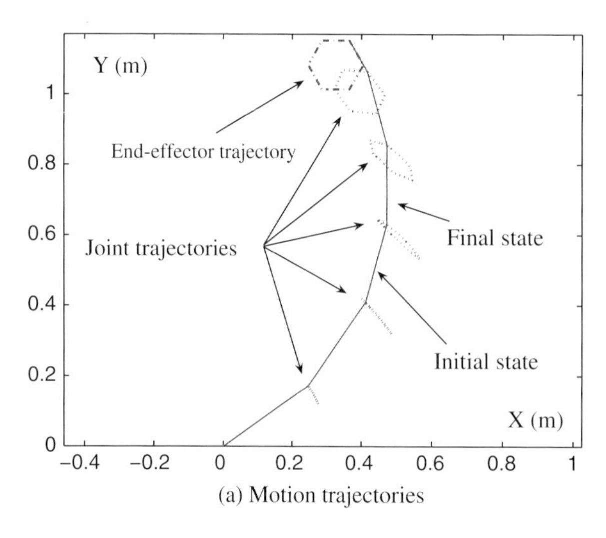

Figure 14.10 Simulation results of a six-DOF planar robot manipulator tracking a hexagonal path synthesized by an OMPFC scheme with L1.

Figure 14.10 (*Continued*)

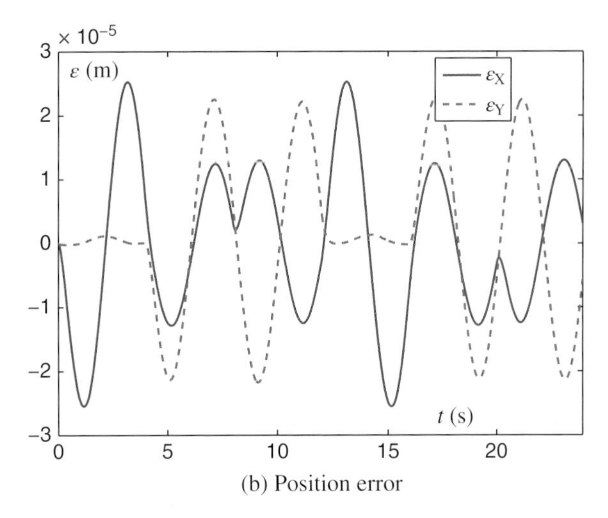

(b) Position error

Figure 14.11 Experimental end-effector position error synthesized by a OMPFC scheme with $\kappa_p = 0$ for tracking a petal-shaped path.

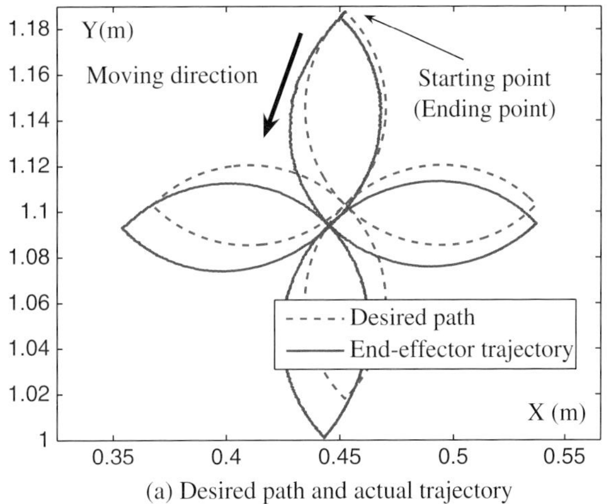

(a) Desired path and actual trajectory

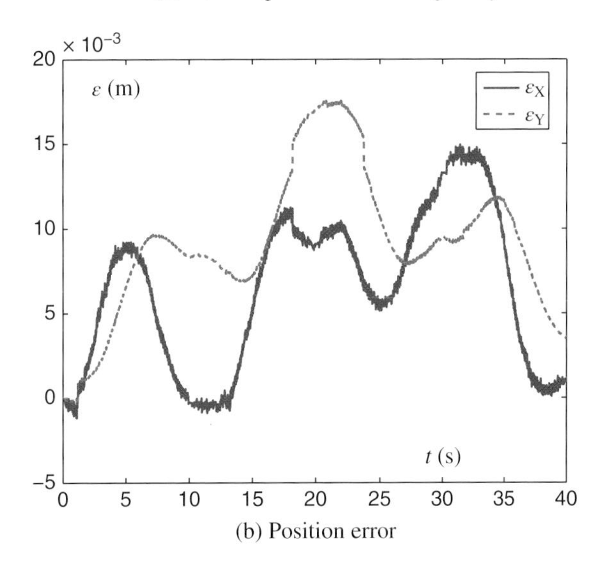

(b) Position error

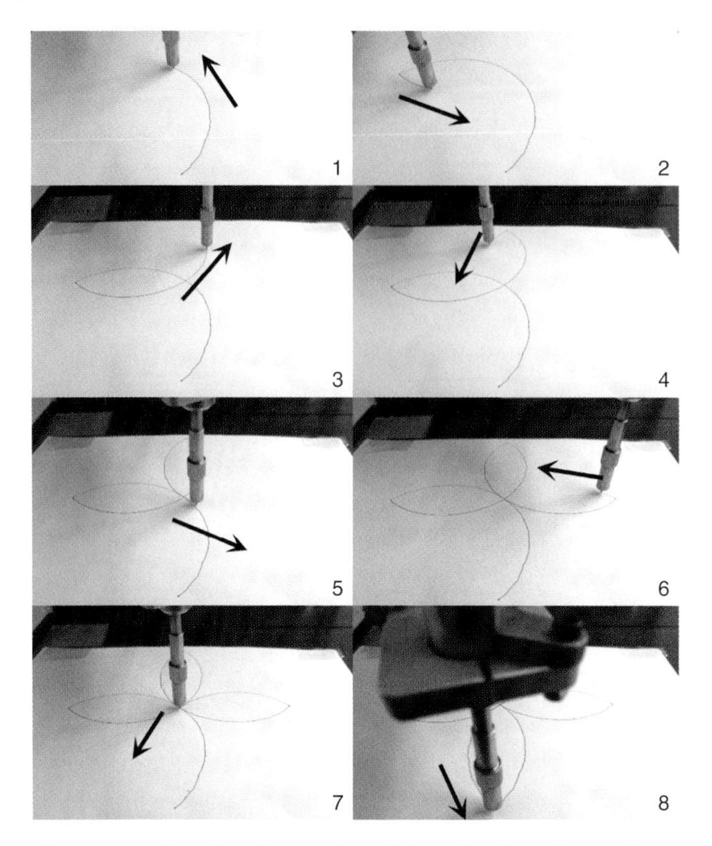

Figure 14.12 Snapshots for actual petal-shaped-path-tracking task execution of manipulator synthesized by OMPFC scheme with $\kappa_p = 6$. *Source*: Zhang 2013. Reproduced with permission of Elsevier.

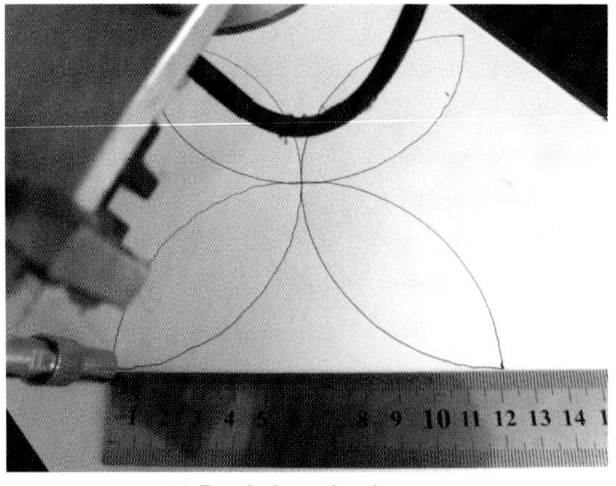

(a) Petal-shaped trajectory

Figure 14.13 Top-view measurement for actual task execution results synthesized by OMPFC scheme with $\kappa_p = 6$. *Source*: Zhang 2013. Reproduced with permission of Elsevier.

Figure 14.13 (*Continued*)

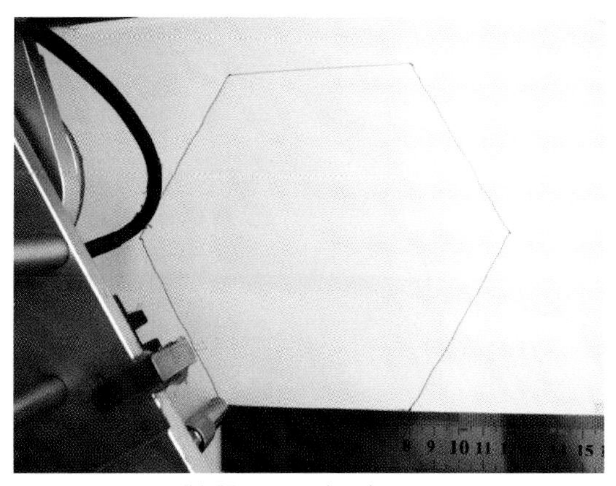

(b) Hexagonal trajectory

Figure 14.14 (a) Experimental end-effector trajectory and (b) position error synthesized by an OMPFC scheme with $\kappa_p = 6$ for tracking a petal-shaped path.

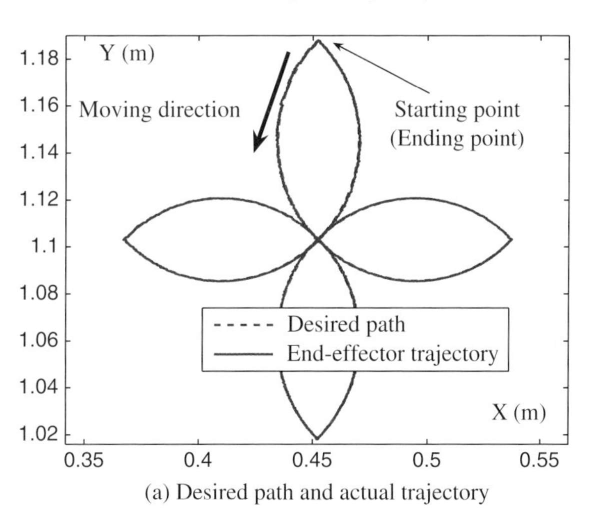

(a) Desired path and actual trajectory

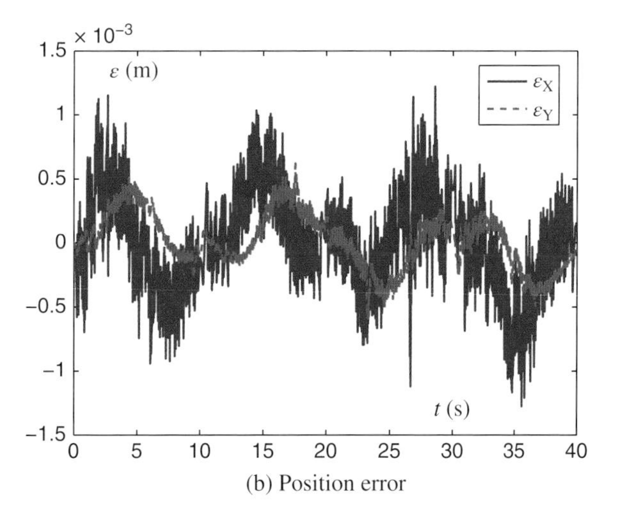

(b) Position error

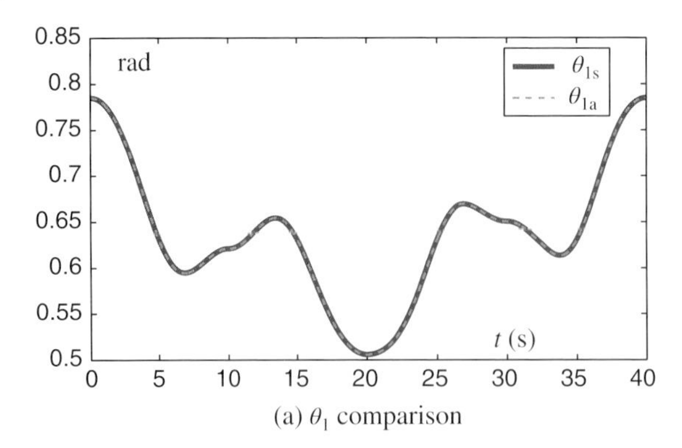

(a) θ_1 comparison

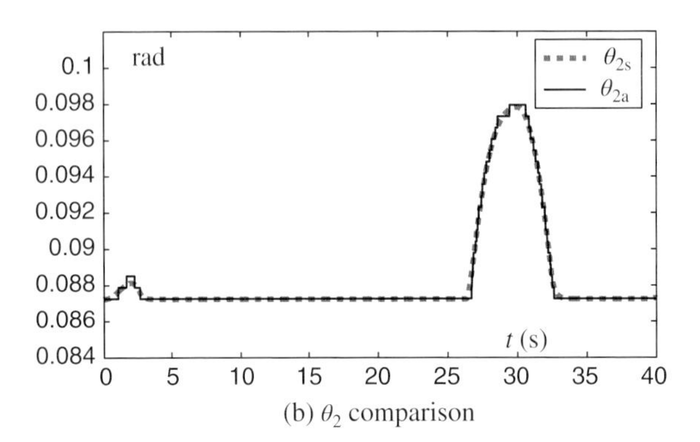

(b) θ_2 comparison

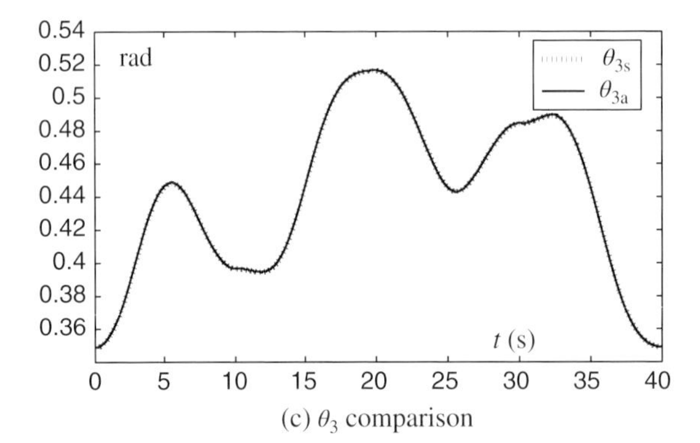

(c) θ_3 comparison

Figure 14.15 Comparison between actual-measured (with subscript $_{ia}$) and simulated (with subscript $_{is}$) θ_i synthesized by an OMPFC scheme with $\kappa_p = 6$ for tracking a petal-shaped path ($i \in \{1, 2, 3\}$).

end-effector task is completed efficiently. Figure 14.13 shows that the diameter of the measured semicircle of the petal-shaped trajectory is 0.12 m, and the side length of the hexagonal trajectory is 0.08 m.

The end-effector trajectory $\mathbf{f}(\theta)$ and corresponding position error $\epsilon = \mathbf{r}_d - \mathbf{f}(\theta)$ are calculated based on the joint angles measured via rotary encoders, as shown in Figure 14.14. By comparing Figure 14.14 with Figure 14.11, we can observe that the end-effector trajectory coincides well with the desired path. As shown in Figure 14.14(b), the maximum position error is 1.278×10^{-3} m, which is less than that

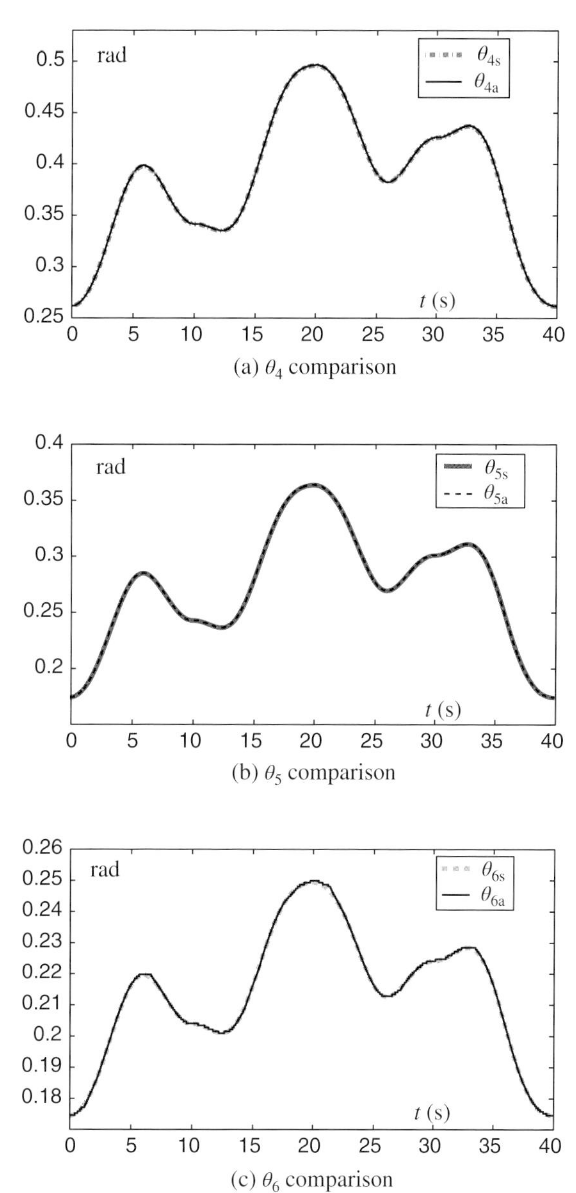

Figure 14.16 Comparison between actual-measured (with subscript $_{ia}$) and simulated (with subscript $_{is}$) θ_i synthesized by an OMPFC scheme with $\kappa_{\mathrm{p}} = 6$ for tracking a petal-shaped path ($i \in \{4, 5, 6\}$).

in Figure 14.11. This difference shows the suitability of the end-effector positioning synthesized by the rotary encoder-based OMPFC scheme. The end-effector positioning precision is significantly improved by exploiting position error feedback. Therefore, the OMPFC scheme with error-feedback of $J\dot{\theta}_{\mathrm{r}} = \dot{\mathbf{r}}_{\mathrm{d}} + \kappa_{\mathrm{p}}(\mathbf{r}_{\mathrm{d}} - \mathbf{f}(\theta))$ in Equation (14.2) is a better choice for optimal motion planning and control of redundant robots compared with the scheme without position error feedback ($\kappa_{\mathrm{p}} = 0$). Figure 14.15 and Figure 14.16 show the corresponding actual measured joint-angle profiles, which are sufficiently close to the desired (simulated) profiles.

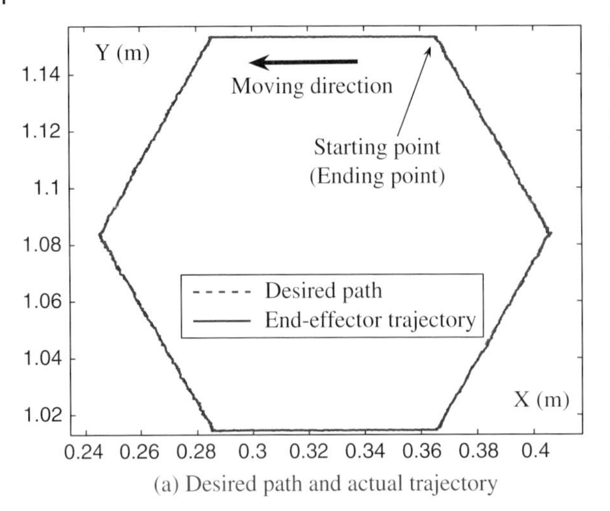

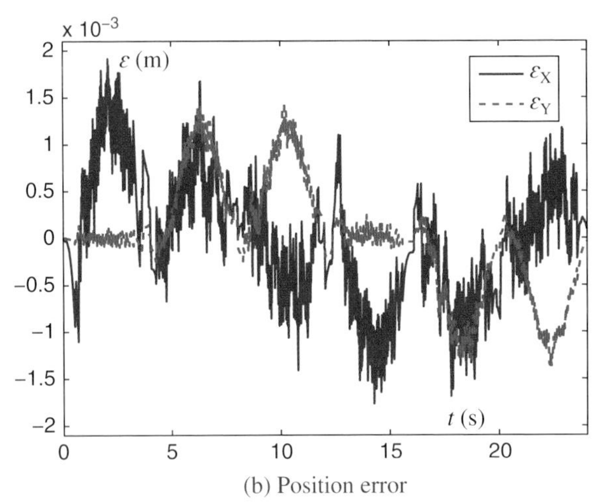

Figure 14.17 (a) Experimental end-effector trajectory and (b) position error synthesized by an OMPFC scheme with $\kappa_{\mathrm{p}} = 6$ for tracking a hexagonal path.

Figure 14.17 shows that the manipulator end-effector trajectory for the hexagonal path-tracking example with $\kappa_{\mathrm{p}} = 6$, coincides well with the desired path, and the maximum position error is 1.917×10^{-3} m. The online measured joint angles are also close to the desired (simulated) angles. However, data are omitted in this chapter because of space limitation. These results further illustrate the efficacy of the presented OMPFC scheme.

In summary, hardware experimental results illustrate the physical realizability and efficacy of the presented rotary encoder-based OMPFC scheme on the drift-free motion planning and control of the six-DOF planar robot manipulator.

14.5 Chapter Summary

In this chapter, the OMPFC scheme for real-time motion planning and control of redundant manipulators has been presented. End-effector path tracking has been considered as the primary task for redundancy resolution, whereas the elimination of joint-drift phenomenon and the avoidance of joint physical limits have been considered as the secondary tasks. Then the QP has been solved by using the presented numerical algorithm 94LVI, thus proving the global linear convergence property of the presented method. The efficacy of the presented method has thus been validated via subsequent simulations. In addition, the presented OMPFC scheme has been implemented on an actual six-DOF planar robot manipulator in real time. Comparative simulations and experiments performed on the six-DOF manipulator have shown that the presented scheme can accurately resolve the redundancy of robot manipulators and remedy the joint-drift phenomenon.

15

QP Based Joystick Control

15.1 Introduction

A robot manipulator is said to be redundant when it has more DOF than necessary to perform a given end-effector task [1, 107]. The fundamental issue to operate such a redundant system is the redundancy-resolution problem [1, 36, 254]. In recent decades, a favorable manner based on online optimization techniques has been applied to solving such redundancy-resolution problem [1, 61] and most of these optimization techniques can be formulated as a (time-varying) quadratic program (QP) subject to equality and inequality constraints [61, 90]. Furthermore, the constrained QP can be transformed into a system of piecewise-linear equations (PLE) [136, 168, 255] and then solved by a number of methods and techniques [1, 36, 61, 65, 97, 104, 128, 136, 168, 255, 256].

In this chapter, to achieve the real-time joystick control of the redundant manipulator, a cosine-aided position-to-velocity mapping subscheme is presented to map the position of the joystick in the motion range to generate the desired velocity of the robot end-effector in real time. On the other hand, a real-time joystick-controlled motion planning (JCMP) subscheme based on minimum-velocity-norm (MVN) is presented to resolve the velocity-specified redundancy-resolution problem of a joystick-controlled manipulator. In addition, such subschemes are implemented on the six-DOF planar robot manipulator hardware system presented in the previous chapters. By using the cosine-aided mapping technique, the joystick can control the manipulator end-effector readily to execute many tasks.

15.2 Preliminaries and Hardware System

The joystick-computer-manipulator control system consists of a joystick, a host computer, and a redundant manipulator hardware system, of which the conceptual block diagram is shown in Figure 15.1. The joystick physically connects to the six-DOF planar robot manipulator hardware system through the host computer so that the operator can control the redundant manipulator via the joystick in real time. Specifically, the joystick can generate and send the end-effector motion command to the host computer according to the operator's action; when receiving the control command, the computer and manipulator hardware system will respond to it by performing redundancy resolution and achieving the corresponding movement and task (e.g., drawing a line segment).

Robot Manipulator Redundancy Resolution, First Edition. Yunong Zhang and Long Jin.

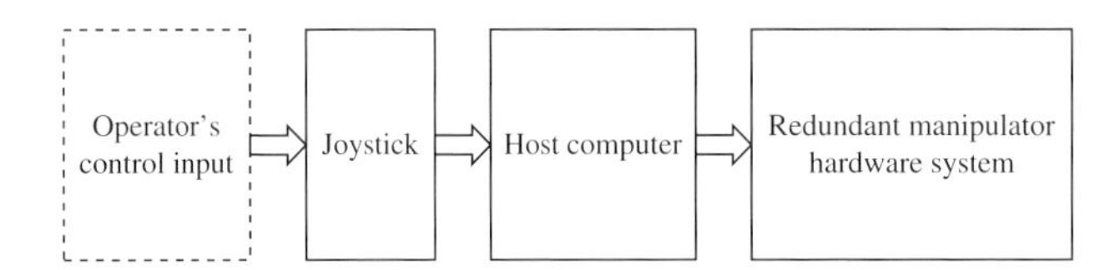

Figure 15.1 Conceptual block diagram of a joystick-controlled robot system.

15.2.1 Velocity-Specified Inverse Kinematics Problem

For the redundant manipulator, the relationship between end-effector position vector $\mathbf{r}_\mathrm{e} \in \mathbb{R}^m$ and joint-angle vector $\theta \in \mathbb{R}^n$ can be represented as $\mathbf{r}_\mathrm{e} = \mathbf{f}(\theta) \rightarrow \mathbf{r}_\mathrm{d}$ [36, 65, 100], where $\mathbf{f}$ is a differentiable function and $\mathbf{r}_\mathrm{d}$ is the desired trajectory. Considering the resolution of the redundancy at velocity level, we get the relationship between end-effector velocity $\dot{\mathbf{r}}_\mathrm{e}$ and joint velocity $\dot{\theta}$ as $J(\theta)\dot{\theta} = \dot{\mathbf{r}}_\mathrm{e}$ where $J(\theta)$ is a Jacobian matrix defined as $J(\theta) = \partial\mathbf{f}/\partial\theta$. It is worth pointing out here that, for the real-time joystick-controlled motion planning, the desired trajectory $\mathbf{r}_\mathrm{d}(t)$ is not necessary (and usually impossible) to be explicitly prescribed in mathematic equations, because the desired end-effector velocity $\dot{\mathbf{r}}_\mathrm{d}$ is generated according to the operator's real-time control input via the joystick directly.

15.2.2 Joystick-Controlled Manipulator Hardware System

The joystick-controlled redundant robot manipulator hardware system, which is designed for the experiment, consists of three parts: a joystick, a six-DOF planar robot manipulator, and a host computer as a linking "bridge".

The joystick shown in Figure 15.2 is employed to control the six-DOF planar robot manipulator directly. It has an eight-way directional pad, 11 buttons (one module button, and 10 control buttons) and two analog sticks. One of the analog sticks is employed for the end-effector motion control, and other buttons are specially designed for manipulator auxiliary control such as joint motion initialization and joint-angle adjustment. In addition, it has a USB cable, which makes the connection convenient between the joystick and the host computer.

It is worth mentioning that each button of the joystick is associated with a hexadecimal number and the host computer can read such a hexadecimal number according to each button's action. For example, when a button is pressed, we can determine which one is pressed by checking the output hexadecimal number of the joystick. In addition, the analog stick moves in a planar region, which means it needs a two-dimensional coordinate to determine its position. Thus, a planar coordinate system is established

Figure 15.2 Joystick. *Source*: IEEE 2011. Reproduced with permission of IEEE.

naturally (termed the original coordinate system of the analog stick). In this coordinate system, the position of the analog stick is associated with a two-dimensional coordinate (p_X, p_Y), where $p_X, p_Y \in [0, 65535]$ are obtained from the joystick. Since the analog stick's initial position is in the center of the motion plane, the default initial coordinate is $(32767, 32767)$ in such an original coordinate system. However, all the coordinate values in such an original coordinate system are positive, which may not be applied to determining the motion direction and the displacement directly.

To remedy the shortage of the expression in the original coordinate system, we can convert (p_X, p_Y) in the original coordinate system of the analog stick to (j_X, j_Y) in a new coordinate system by the following normalization equations:

$$j_X = \frac{p_X - p_{Xmax}/2}{p_{Xmax}/2}, \quad j_Y = \frac{p_Y - p_{Ymax}/2}{p_{Ymax}/2},$$

where $j_X, j_Y \in [-1, 1]$, and $p_{Xmax} = p_{Ymax} = 65535$. The initial position of the analog stick is exactly at the origin of the new coordinate system, that is, $j_X = j_Y = 0$. Thus, once we push the analog stick, it is easy to obtain both the movement direction and the displacement via the new coordinate.

15.3 Scheme Formulation

In this section, a cosine-aided position-to-velocity mapping subscheme is presented for the real-time end-effector velocity generation. With the generated end-effector target velocity, a real-time JCMP subscheme subject to equality and bound constraints is presented, which is then converted into a standard QP. Since such a QP can be equivalent to a system of PLE, numerical algorithm 94LVI presented in previous chapters is employed to solve such a PLE problem as well as the QP problem.

15.3.1 Cosine-Aided Position-to-Velocity Mapping

Since the redundant manipulator hardware system is controlled by a joystick, a cosine-aided position-to-velocity mapping subscheme is presented to map the position of the analog stick of the joystick in its motion range to the desired velocity (or to say, the target velocity, or the specified velocity) of the robot end-effector. The conceptual block diagram of the designed mapping is shown in Figure 15.3. The mapping equations are

$$\dot{r}_X = \begin{cases} j_X(1 - \cos(\pi j_X))/|j_X|, & \text{if } j_X \neq 0 \\ 0, & \text{if } j_X = 0 \end{cases}, \tag{15.1}$$

$$\dot{r}_Y = \begin{cases} j_Y(1 - \cos(\pi j_Y))/|j_Y|, & \text{if } j_Y \neq 0 \\ 0, & \text{if } j_Y = 0 \end{cases}, \tag{15.2}$$

where $\dot{r}_X$ and $\dot{r}_Y$ denote respectively the desired end-effector velocity components in X-axis and Y-axis (or termed, $\dot{r}_{dX}$ and $\dot{r}_{dY}$, respectively, which constitute $\dot{\mathbf{r}}_d$), and $j_X, j_Y \in [-1, 1]$ denote, respectively, the relative position of the analog stick in its motion plane, defined by equation (15.1). It is easy to prove that the $\dot{r}_X(j_X)$ and $\dot{r}_Y(j_Y)$ are differentiable within the range $[-1, 1]$, which illustrates that the cosine-aided position-to-velocity mapping subscheme is feasible.

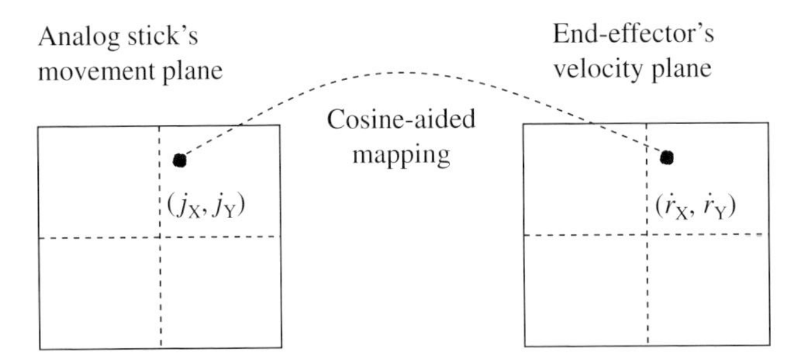

Figure 15.3 Block diagram of cosine-aided position-to-velocity mapping.

15.3.2 Real-Time Joystick-Controlled Motion Planning

According to [1, 65], joint-velocity $\dot{\theta}$ can be obtained by the following subscheme formulation with joint-angle limits and joint-velocity limits considered:

$$\text{minimize} \quad \|\dot{\theta}\|_2^2/2 \tag{15.3}$$
$$\text{subject to} \quad J(\theta)\dot{\theta} = \dot{\mathbf{r}}_\text{d}, \tag{15.4}$$
$$\theta^- \leq \theta \leq \theta^+, \tag{15.5}$$
$$\dot{\theta}^- \leq \dot{\theta} \leq \dot{\theta}^+, \tag{15.6}$$

where $\|\cdot\|_2$ denotes the two norm of a vector, and the superscript $^+$ and $^-$ denote the upper and lower limits of the joint variable vector (e.g., θ or $\dot{\theta}$), respectively.

Based on the previous works [1, 65, 104, 128], the real-time JCMP subscheme presented in (15.3)–(15.6) can be reformulated as a QP-based formulation subject to an equality constraint and a bound constraint:

$$\text{minimize} \quad \mathbf{x}^\text{T}W\mathbf{x}/2 + \mathbf{c}^\text{T}\mathbf{x} \tag{15.7}$$
$$\text{subject to} \quad J\mathbf{x} = \mathbf{b}, \tag{15.8}$$
$$\xi^- \leqslant \dot{\theta} \leqslant \xi^+, \tag{15.9}$$
$$\text{with } W = I, \mathbf{c} = 0, \mathbf{b} = \dot{\mathbf{r}}_\text{d},$$

where $\mathbf{x} \in \mathbb{R}^n$ denotes the joint-velocity vector $\dot{\theta}$ to be solved per sampling period, and the ith elements of ξ^- and ξ^+ are defined as $\xi_i^- = \max\{\dot{\theta}_i^-, \mu(\theta_i^- - \theta_i)\}$ and $\xi_i^+ = \min\{\dot{\theta}_i^+, \mu(\theta_i^+ - \theta_i)$. The QP problem shown in (15.7)–(15.9) is solved by numerical algorithm 94LVI, which is presented in the previous chapters.

15.4 Computer Simulations and Physical Experiments

For the real-time JCMP subscheme, the real-time joint velocity $\dot{\theta}$ is computed by using numerical algorithm 94LVI, which can solve the QP problem (15.7)–(15.9). In order to examine the computational efficiency of the subscheme, we give out the computing time in several tasks in the simulations (i.e. drawing line segments toward six different directions), which is illustrated in Figures 15.4 through 15.6. From these figures, we can see that the computing time is less than 0.01 s. Thus, we can set the sampling period as $\Delta t = 0.01$ s (i.e., 10 ms) in the following hardware experiments.

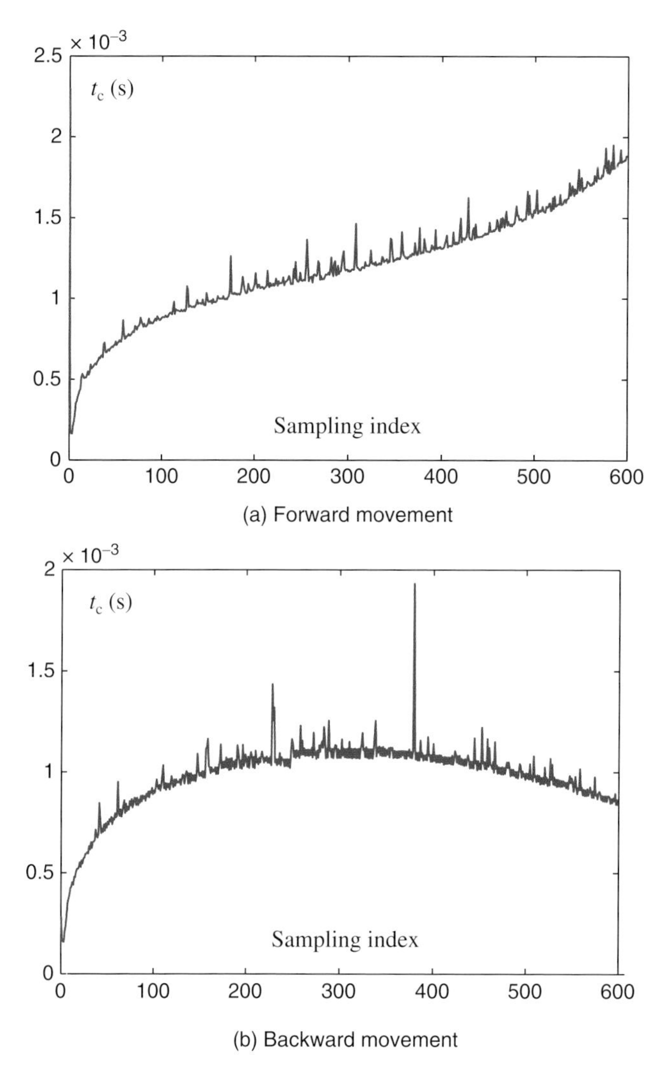

Figure 15.4 Computing times used by numerical algorithm 94LVI for redundancy resolution in simulations of (a) forward and (b) backward movement.

During the experiment, the real-time JCMP subscheme and the cosine-aided position-to-velocity mapping subscheme are implemented on the joystick-controlled redundant robot manipulator hardware system by performing two tasks; that is, drawing line segments toward four different directions and writing "MVN" letters in a two-dimensional plane.

15.4.1 Movement Toward Four Directions

In this experiment, the end-effector of the six-DOF planar robot manipulator is expected to move toward four directions (i.e., move forward, backward, leftward, and rightward) to validate the physical realizability of the robot manipulator controlled by the joystick. Since the position of the analog stick in its motion plane is

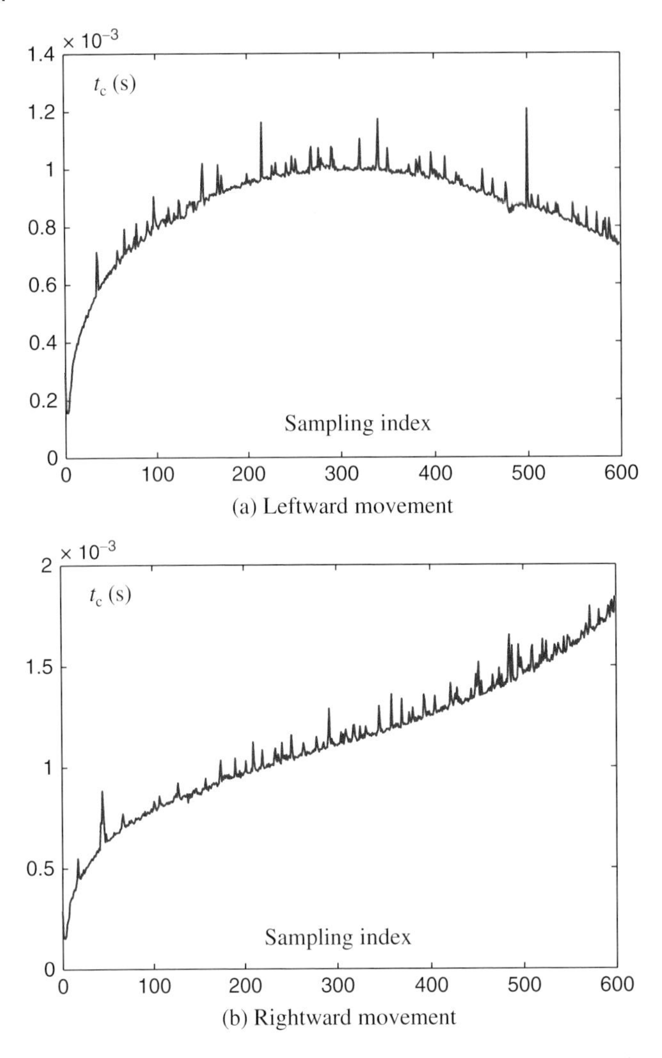

Figure 15.5 Computing times used by numerical algorithm 94LVI for redundancy resolution in simulations of (a) leftward and (b) rightward movement.

associated with the desired velocity of the end-effector via a cosine-aided position-to-velocity mapping, the operator can control the velocity of the end-effector directly by the joystick. The velocity of the end-effector in the x- and y-axis are, respectively, generated by equations (15.1) and (15.2). During the real-time motion control, the JCMP subscheme is employed, which is solved by numerical algorithm 94LVI. Without loss of generality, the tasks start with the initial joint configuration $\theta(0) = [\pi/12, \pi/12, \pi/12, \pi/12, \pi/12, \pi/12]^{\mathrm{T}}$rad.

Firstly, we push the analog stick forward as shown in Figure 15.7(a), and during the task, we take snapshots of the end-effector movement by a digital camera. Then,

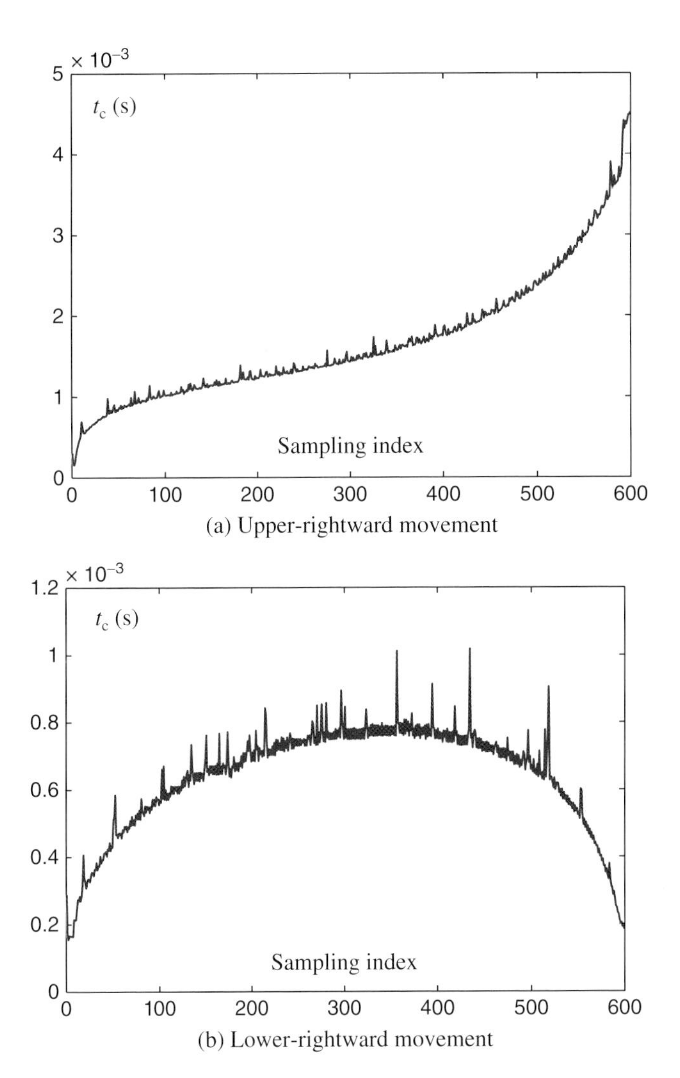

Figure 15.6 Computing times used by numerical algorithm 94LVI for redundancy resolution in simulations of (a) upper-rightward and (b) lower-rightward movement.

we repeat the operation by pushing the analog stick toward other three directions, as shown in Figure 15.7(b) through Figure 15.7(d).

The actual end-effector trajectories for the four directions in the experiment are shown in Figure 15.8, in which the arrows placed in the workspace indicate the expected movement directions of the end-effector. As seen from the figure, the end-effector draws four line segments toward the four expected directions, respectively, illustrating that the tasks are performed successfully. In summary, these experimental results illustrate the realizability and effectiveness of the real-time JCMP subscheme and the cosine-aided position-to-velocity mapping subscheme for the joystick-controlled

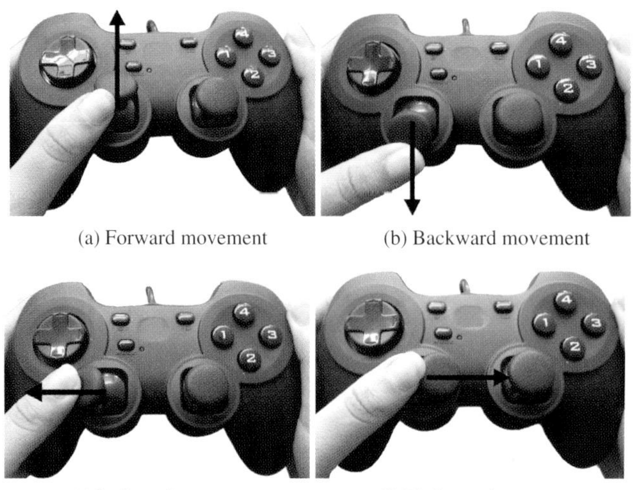

(a) Forward movement (b) Backward movement

(c) Leftward movement (d) Rightward movement

Figure 15.7 Directional manipulation of joystick. *Source*: IEEE 2011. Reproduced with permission of IEEE.

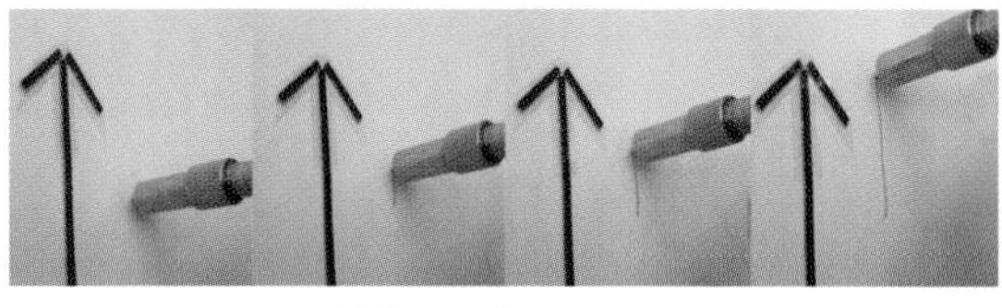

(a) Forward movement

(b) Backward movement

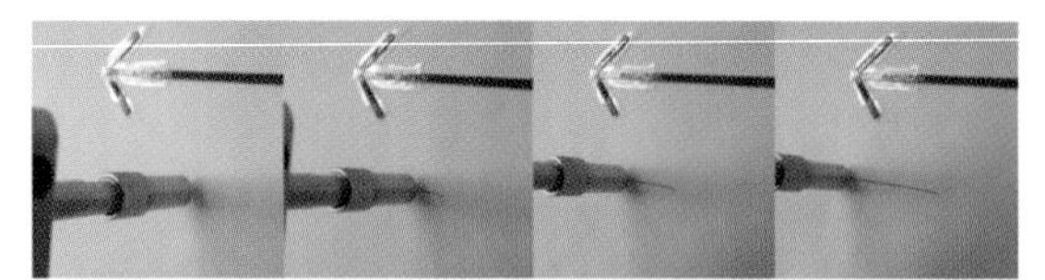

(c) Leftward movement

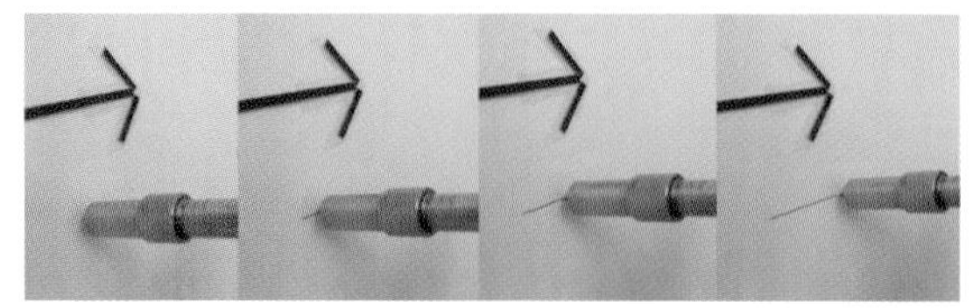

(d) Rightward movement

Figure 15.8 Actual end-effector trajectories corresponding to Figure 15.7. *Source*: Zhang 2011. Reproduced with permission of IEEE.

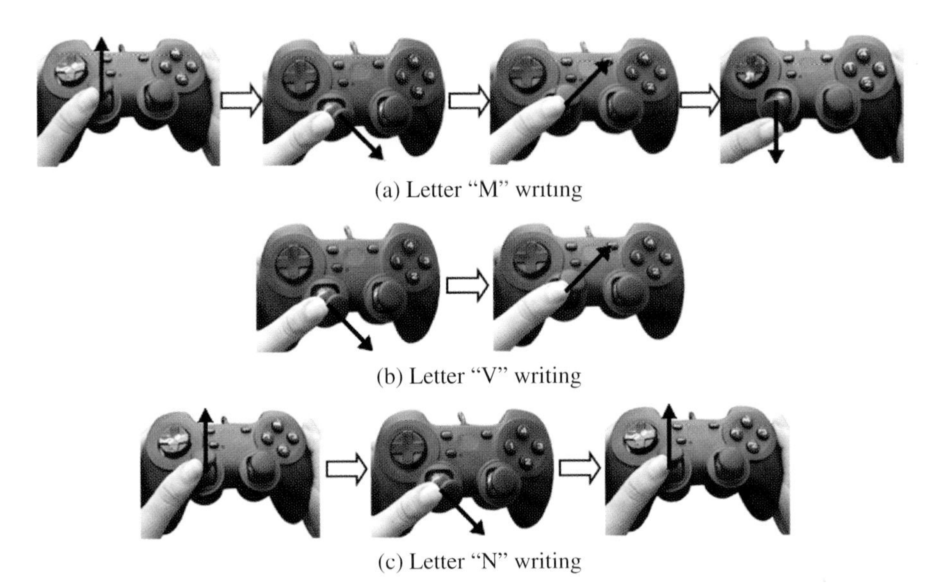
(a) Letter "M" writing

(b) Letter "V" writing

(c) Letter "N" writing

Figure 15.9 Manipulation of joystick for writing "MVN". *Source*: IEEE 2011. Reproduced with permission of IEEE.

six-DOF planar robot manipulator, in addition to the efficacy of numerical algorithm 94LVI presented in the previous chapters on solving QP (15.7)–(15.9).

15.4.2 "MVN" Letter Writing

In this experiment, letters "MVN" are expected to be written by the end-effector of the joystick-controlled robot manipulator. The tasks are also executed with the initial joint state $\theta(0) = [\pi/12, \pi/12, \pi/12, \pi/12, \pi/12, \pi/12]^{\mathrm{T}}$ rad.

During the tasks, we take snapshots to get the actual trajectories of the end-effector's motion in the process of controlling the redundant manipulator via the joystick. Figure 15.9 shows the controlling process of the analog stick to draw the expected letters. In addition, the end-effector actual trajectories of the three letters are shown in Figure 15.10. From the figure, we can see that the end-effector can move according to the operator's control, and fulfill the letter-writing tasks well. The experimental results validate again the realization and effectiveness of the real-time JCMP subscheme and the cosine-aided position-to-velocity mapping subscheme for the joystick-controlled redundant robot manipulator, in addition to the efficacy of numerical algorithm 94LVI on solving QP (15.7)–(15.9).

15.5 Chapter Summary

In this chapter, to operate the joystick-controlled redundant robot manipulator in real time, a cosine-aided position-to-velocity mapping subscheme has been presented to map the position of the joystick in its motion range to the velocity of the robot end-effector for the desired end-effector velocity generation. In addition, to achieve the real-time control of the manipulator, a real-time JCMP subscheme subject to equality

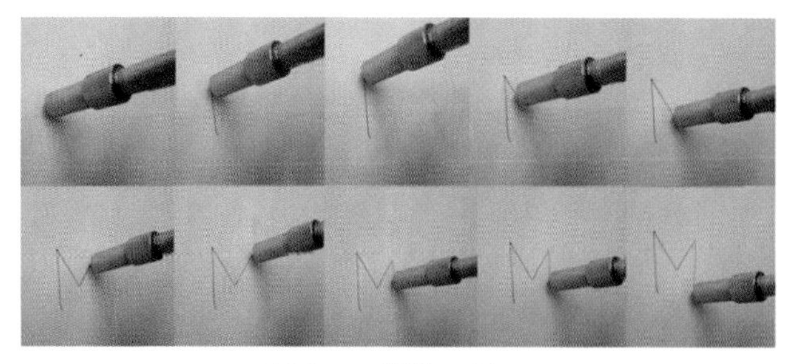

(a) Letter "M" writing

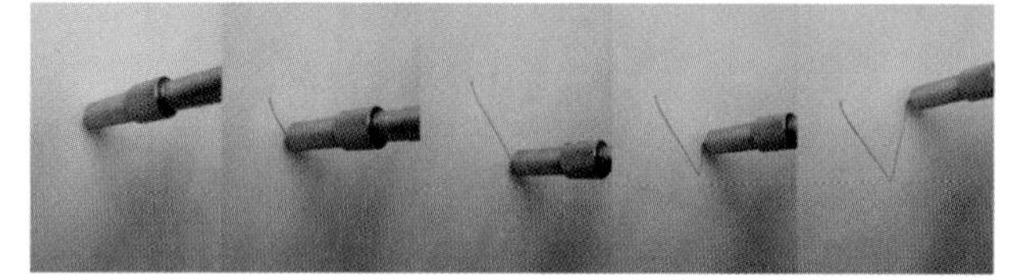

(b) Letter "V" writing

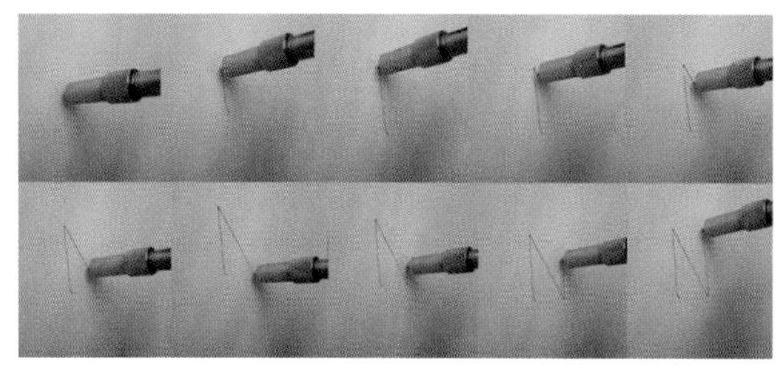

(c) Letter "N" writing

Figure 15.10 Actual end-effector trajectories corresponding to Figure 15.9. *Source*: Zhang 2011. Reproduced with permission of IEEE.

and bound constraints has been presented for operating such a joystick-controlled redundant robot manipulator. The JCMP subscheme can generate the desired joint velocity corresponding to the end-effector velocity given by the joystick user. In addition, the JCMP subscheme has been solved by numerical algorithm 94LVI presented in the previous chapters. Moreover, the experimental results based on the joystick-controlled robot manipulator have verified the realization and effectiveness of the two subschemes by executing two real-time tasks, that is, drawing line segments toward four different directions and writing "MVN" letters in a two-dimensional plane.

References

1 Zhang, Y. (2002) *Analysis and Design of Recurrent Neural Networks and Their Applications to Control and Robotic Systems*, Ph.D. thesis, Chinese University of Hong Kong, Hong Kong, China.

2 Zhang, Y. and Zhang, Z. (2013) *Repetitive Motion Planning and Control of Redundant Robot Manipulators*, Springer-Verlag, New York, USA.

3 Siciliano, B. and Khatib, O. (2008) *Springer Handbook of Robotics*, Springer-Verlag, New York, USA.

4 Jin, L., Zhang, Y., Li, S. and Zhang, Y. (2016) "Modified ZNN for time-varying quadratic programming with inherent tolerance to noises and its application to kinematic redundancy resolution of robot manipulators," *IEEE Trans. Ind. Electron.*, Vol. **63**, No. 11, pp. 6978–6988.

5 Pires, E. J. S., Oliveira, P. B. M. and Machado, J. A. T. (2007) "Manipulator trajectory planning using a MOEA," *Appl. Soft Comput.*, Vol. **7**, No. 3, pp. 659–667.

6 Marcos, M. G., Machado, J. A. T. and Azevedo-Perdicoúlis, T. -P. (2011) "A fractional approach for the motion planning of redundant and hyper-redundant manipulators," *Signal Process.*, Vol. **91**, No. 3, pp. 562–570.

7 Marcos, M. G., Machado, J. A. T. and Azevedo-Perdicoúlis, T. -P. (2012) "A multi-objective approach for the motion planning of redundant manipulators," *Appl. Soft Comput.*, Vol. **12**, No. 2, pp. 589–599.

8 Zhang, Z. and Zhang, Y. (2013) "Variable joint-velocity limits of redundant robot manipulators handled by quadratic programming," *IEEE/ASME Trans. Mechatronics*, Vol. **18**, No. 2, pp. 674–686.

9 Granvagne, I. A. and Walker, I. D. (2000) "On the structure of minimum effort solutions with application to kinematic redundancy resolution," *IEEE Trans. Robot. Automat.*, Vol. **16**, No. 6, pp. 855–863.

10 Kemény, Z. (2003) "Redundancy resolution in robots using parameterization through null space," *IEEE Trans. Ind. Electron.*, Vol. **50**, No. 4, pp. 777–783.

11 Tchoń, K. (2008) "Optimal extended Jacobian inverse kinematics algorithms for robotic manipulators," *IEEE Trans. Robot.*, Vol. **24**, No. 6, pp. 1440–1445.

12 Zhang, Y. and Yi, C. (2011) *Zhang Neural Networks and Neural-Dynamic Method*, Nova Science Publishers, New York, USA.

13 Guo, D. and Zhang, Y. (2014) "Li-function activated ZNN with finite-time convergence applied to redundant-manipulator kinematic control via time-varying Jacobian matrix pseudoinversion," *Appl. Soft Comput.*, Vol. **24**, pp. 158–168.

Robot Manipulator Redundancy Resolution, First Edition. Yunong Zhang and Long Jin.

14 Hopfield, J. J. (1982) "Neural networks and physical systems with emergent collective computational abilities," *PNAS*, Vol. **79**, pp. 2554–2558.

15 Wang, J., Hu, Q. and Jiang, D. (1999) "A Lagrangian network for kinematic control of redundant robot manipulators," *IEEE Trans. Neural Netw.*, Vol. **10**, No. 5, pp. 1123–1132.

16 Hou, Z. G., Cheng, L. and Tan, M. (2010) "Multicriteria optimization for coordination of redundant robots using a dual neural network," *IEEE Trans. Syst., Man, Cybern. B*, Vol. **40**, No. 4, pp. 1075–1087.

17 Guo, D. and Zhang, Y. (2012) "A new inequality-based obstacle-avoidance MVN scheme and its application to redundant robot manipulators," *IEEE Trans. Syst., Man, Cybern. C*, Vol. **42**, No. 6, pp. 1326–1340.

18 Wei, Y. (2000) "Recurrent neural networks for computing weighted Moore–Penrose inverse," *Appl. Math. Comput.*, Vol. **116**, No. 3, pp. 279–287.

19 Ben-Israel, A. and Greville, T. N. E. (2003) *Generalized Inverses: Theory and Applications*, 2nd edn., Springer-Verlag, New York, USA.

20 Huang, F. and Zhang, X. (2006) "An improved Newton iteration for the weighted Moore-Penrose inverse," *Appl. Math. Comput.*, Vol. **174**, No. 2, pp. 1460–1486.

21 Courrieu, P. (2005) "Fast computation of Moore-Penrose inverse matrices," *Neural Inf. Process. Lett. Rev.*, Vol. **8**, No. 2, pp. 25–29.

22 Siciliano, B., Sciavicco, L., Villani, L. and Oriolo, G. (2009) *Robotics: Modelling, Planning and Control*, Springer-Verlag, New York, USA.

23 Zhang, Y., Yang, Y., Tan, N. and Cai, B. (2011) "Zhang neural network solving for time-varying full-rank matrix Moore–Penrose inverse," *Computing*, Vol. **92**, No. 2, pp. 97–121.

24 Li, S., Chen, S. and Liu, B. (2013) "Accelerating a recurrent neural network to finite-time convergence for solving time-varying Sylvester equation by using a sign-bi-power activation function," *Neural Process. Lett.*, Vol. **37**, No. 2, pp. 189–205.

25 Liao, B. and Zhang, Y. (2014) "From different ZFs to different ZNN models accelerated via Li activation functions to finite-time convergence for time-varying matrix pseudoinversion," *Neurocomputing*, Vol. **133**, pp. 512–522.

26 Jin, L. and Zhang, Y. (2014) "Discrete-time Zhang neural network of $O(\tau^3)$ pattern for time-varying matrix pseudoinversion," *Neurocomputing*, Vol. **142**, pp. 165–173.

27 Griffiths, D. F. and Higham, D. J. (2010) *Numerical Methods for Ordinary Differential Equations: Initial Value Problems*, Springer, London, UK.

28 Wei, Y., Cai, J. and Ng, M. K. (2004) "Computing Moore–Penrose inverses of Toeplitz matrices by Newton's iteration," *Math. Comput. Model.*, Vol. **40**, No. 1–2, pp. 181–191.

29 Mathews, J. H. and Fink, K. D. (2004) *Numerical Methods Using MATLAB*, Prentice-Hall, Englewood Cliffs, NJ, USA.

30 Van, M., Kang, H. J. and Ro, Y. S. (2011) "A robust fault detection and isolation scheme for robot manipulators based on neural networks," in: Proceedings of the 7th International Conference on Intelligent Computing, pp. 25–32.

31 Lee, S. C. and Ahn, H. S. (2011) "Multiple manipulator cooperative control using disturbance estimator and consensus algorithm," in: Proceedings of the American Control Conference, pp. 4002–4007.

32 Kim, C. S., Mo, E. J., Jie, M. S., Hwang, S. C. and Lee, K. W. (2007) "Image-based robust control of robot manipulators under jacobian uncertainty," in: Proceedings of the 3rd International Conference on Intelligent Computing, pp. 502–510.

33 Zhou, J., Kang, H. and Ro, Y. (2010) "Comparison of the observability indices for robot calibration considering joint stiffness parameters," in: Proceedings of the 6th International Conference on Intelligent Computing, pp. 372–380.

34 Liu, Y., Yu, Y. and Jiang, C. (2003) "The development of redundant robot manipulators," *Machine Design and Research*, Vol. **19**, No. 1, pp. 24–27.

35 Dubey, R. and Luh, J. (1987) "Redudant robot control for higher flexibility," in: Proceedings of the IEEE International Conference on Robotics and Automation, pp. 1066–1072.

36 Zhang, Y., Wang, J. and Xia, Y. (2003) "A dual neural network for redundancy resolution of kinematically redundant manipulators subject to joint limits and joint velocity limits," *IEEE Trans. Neural Netw.*, Vol. **14**, No. 3, pp. 658–667.

37 Bianco, C. and Ghilardelli, F. (2014) "Real-time planner in the operational space for the automatic handling of kinematic constraints," *IEEE Trans. Autom. Sci. Eng.*, Vol. **11**, No. 3, pp. 730–739.

38 Moreno, J. and Kelly, R. (2003) "Manipulator velocity field control with dynamic friction compensation," in: Proceedings of the 42nd IEEE Conference on Decision and Control, pp. 3834–3839.

39 Cheng, L., Hou, Z. and Tan, M. (2009) "Adaptive neural network tracking control for manipulators with uncertain kinematics, dynamics and actuator model," *Automatica*, Vol. **45**, No. 10, pp. 2312–2318.

40 Cheah, C., Kawamura, S. and Arimoto, S. (2003) "Stability of hybrid position and force control for robotic manipulator with kinematics and dynamics uncertainties," *Automatica*, Vol. **39**, No. 5, pp. 847–855.

41 Cheng, L., Hou, Z., Tan M., Lin, Y. and Zhang, W. (2010) "Neural-network-based adaptive leader-following control for multiagent systems with uncertainties," *IEEE Trans. Neural Networks*, Vol. **21**, No. 8, pp. 1351–1358.

42 Tarokh, M. and Kim, M. (2007) "Inverse kinematics of 7-DOF robots and limbs by decomposition and approximation," *IEEE Trans. Rob.*, Vol. **23**, No. 3, pp. 595–600.

43 Falco, P. and Natale, C. (2011) "On the stability of closed-loop inverse kinematics algorithms for redundant robots," *IEEE Trans. Rob.*, Vol. **27**, No. 3, pp. 780–784.

44 Jin, L. and Zhang, Y. (2015) "G2-type SRMPC scheme for synchronous manipulation of two redundant robot arms," *IEEE Trans. Cybern*, Vol. **45**, No. 2, pp. 153–164.

45 Zhang, Y., Yi, C., Guo, D. and Zheng, J. (2011) "Comparison on Zhang neural dynamics and gradient-based neural dynamics for online solution of nonlinear time-varying equation," *Neural Computing and Applications*, Vol. **20**, No. 1, pp.1–7.

46 Jin, L., Zhang, Y. and Li, S. (2016) "Integration-enhanced Zhang neural network for real-time-varying matrix inversion in the presence of various kinds of noises," IEEE Trans. Neural Netw. Learn. Syst., In Press with DOI 10.1109/TNNLS.2015.2497715.

47 Bowling, M. and Veloso, M. (2001) "Convergence of gradient dynamics with a variable learning rate," in: Proceedings of the 18th International Conference on Machine Learning, pp. 27–34.

48 Ma, X. and Elia, N. (2013) "A distributed continuous-time gradient dynamics approach for the active power loss minimizations," in: Proceedings of the 51st Annual Allerton Conference, pp. 100–106.
49 Taylor, P. and Day, T. (1997) "Evolutionary stability under the replicator and the gradient dynamics," *Evol. Ecol.*, Vol. **11**, No. 5, pp. 579–590.
50 Jin, L. and Zhang, Y. (2016) "Continuous and discrete Zhang dynamics for real-time varying nonlinear optimization," *IEEE Trans. Neural Netw. Learn. Syst.*, Vol. **27**, No. 6, pp. 1525–1531.
51 Jin, L. and Zhang, Y. (2015) "Discrete-time Zhang neural network for online time-varying nonlinear optimization with application to manipulator motion generation," *Numerical Algrithms*, Vol. **73**, No. 1, pp. 115–140.
52 Zhang, Y., Yin, Y., Wu, H. and Guo, D. (2013) Zhang dynamics and gradient dynamics with tracking-control application," in: Proceedings of the 5th International Symposium on Computational Intelligence and Design, pp. 235–238.
53 Cai, B. and Zhang, Y. (2013) *Optimal and Efficient Motion Planning of Redundant Robot Manipulators*, LAMBERT Academic Publishing (LAP), Germany.
54 Zhang, Y. (2002) *Analysis and Design of Recurrent Neural Networks and Their Applications to Control and Robotic Systems*, Ph.D. thesis, Chinese University of Hong Kong, Hong Kong, China.
55 Klein, C. A. and Huang, C. H. (1983) "Review of pseudoinverse control for use with kinematically redundant manipulators," *IEEE Trans. Syst. Man Cybern. B.*, Vol. **13**, No. 3, pp. 245–250.
56 Klein, C. A. and Kee, K. B. (1989) "The nature of drift in pseudoinverse control of kinematically redundant manipulators," *IEEE Trans. Robot. Autom.*, Vol. **5**, No. 2, pp. 231–234.
57 Klein, C. A. and Ahmed, S. (1995) "Repeatable pseudoinverse control for planar kinematically redundant manipulators," *IEEE Trans. Syst. Man Cybern.*, Vol. **25**, No. 12, pp. 1657–1662.
58 Xiao, L. (2013) "Acceleration-level repetitive motion planning and its experimental verification on a six-link planar robot manipulator," *IEEE Trans. Contr. Syst. T.*, Vol. **21**, No. 3, pp. 906–914.
59 Shang, W., Cong, S., Zhang, Y. and Liang, Y. (2009) "Active joint synchronization control for a 2-dof redundantly actuated parallel manipulator," *IEEE Trans. Contr. Syst. T.*, Vol. **17**, No. 2, pp. 416–423.
60 Cai, B. and Zhang, Y. (2012) "Different-level redundancy-resolution and its equivalent relationship analysis for robot manipulators using gradient-descent and Zhang et al.'s neural-dynamic methods," *IEEE Trans. Ind. Electron.*, Vol. **59**, No. 8, pp. 3146–3155.
61 Zhang, Y., Ge, S. S. and Lee, T. H. (2004) "A unified quadratic programming based dynamical system approach to joint torque optimization of physically constrained redundant manipulators," *IEEE Trans. Syst. Man Cybern. Part B Cybern.*, Vol. **34**, No. 5, pp. 2126–2132.
62 Sciavicco, L. and Siciliano, B. (2000) *Modelling and Control of Robot Manipulators*, Springer-Verlag, London, England.
63 Latash, M. L. (1993) *Control of Human Movement*, Human Kinetics Pubisher, Chicago, USA.

64 Ma, S., Ohmameuda, Y. and Inoue, K. (2004) "Dynamic analysis of 3-dimensional snake robots," in: Proceedings of IEEE/RSJ International Conference on Intelligent Robots and Systems, pp. 767–772.

65 Zhang, Y. and Wang, J. (2004) "Obstacle avoidance for kinematically redundant manipulators using a dual neural network," *IEEE Trans. Syst. Man Cybern. Part B Cybern.*, Vol. **34**, No. 1, pp. 752–759.

66 Zhang, Y., Li, Z. and Tan, H. (2006) "Inequality-based manipulator-obstacle avoidance using the LVI-based primal-dual neural network," in: Proceedings of IEEE International Conference on Robotics and Biomimetics, pp. 1459–1464.

67 Hollerbach, J. M. and Suh, K. C. (1987) "Redundancy resolution of manipulators through torque optimization," *IEEE Trans. Robot. Automat.*, Vol. **3**, No. 4, pp. 308–315.

68 Khatib, O. (1987) "A unified approach for motion and force control of robot manipulators: the operational space formulation," *IEEE J. Robot. Automat.*, Vol. **3**, No. 1, pp. 43–53.

69 Ma, S. (1996) "A new formulation technique for local torque optimization of redundant manipulators," *IEEE Trans. Industr. Electro.*, Vol. **43**, No. 4, pp. 462–468.

70 Chen, T., Cheng, F., Sun, Y. and Hung, M. (1994) "Torque optimization schemes for kinematically redundant manipulators," *J. Robot. Syst.*, Vol. **11**, No. 4, pp. 257–269.

71 Cheng, F., Sheu, R. and Chen, T. (1995) "The improved compact QP method for resolving manipulator redundancy," *IEEE Trans. Syst. Man Cybern.*, Vol. **25**, No. 11, pp. 1521–1530.

72 Li, D., Goldenberg, A. A. and Zu, J. W. (1997) "A new method of peak torque reduction with redundant manipulators," *IEEE Trans. Robot. Automat.*, Vol. **13**, No. 6, pp. 845–853.

73 Shim, I. C. and Yoon, Y. S. (1997) "Stabilization constraint method for torque optimization of a redundant manipulator," in: Proceedings of IEEE International Conference on Robotics and Automation, pp. 2403–2408.

74 Park, K. C., Chang, P. H. and Kim, S. H. (1998) "The enhanced compact QP method for redundant manipulators using practical inequality constraints," in: Proceedings of IEEE International Conference on Robotics and Automation, pp. 107–114.

75 Ma, S. and Watanabe, M. (2001) "Time optimal control of kinematically redundant manipulators with limit heat characteristics of actuators," in: Proceedings of IEEE/RSJ International Conference on Intelligent Robots and Systems, pp. 152–157.

76 Ding, H. and Tso, S. K. (1999) "A fully neural-network-based planning scheme for torque minimization of redundant manipulators," *IEEE Trans. Industr. Electro.*, Vol. **46**, No. 1, pp. 199–206.

77 Tang, W. S. and Wang, J. (2001) "A recurrent neural network for minimum infinity-norm kinematic control of redundant manipulators with an improved problem formulation and reduced architectural complexity," *IEEE Trans. Syst. Man Cybern.*, Vol. **31**, No. 1, pp. 98–105.

78 Zhang, Y. and Wang, J. (2002) "A dual neural network for constrained joint torque optimization of kinematically redundant manipulators," *IEEE Trans. Syst. Man Cybern.*, Vol. **32**, No. 5, pp. 654–662.
79 Zhang, S. and Constantinides, A. G. (1992) "Lagrange programming neural networks," *IEEE Trans. Circ. Syst.*, Vol. **39**, No. 7, pp. 441–452.
80 Bouzerdoum, A. and Pattison, T. R. (1993) "Neural network for quadratic optimization with bound constraints," *IEEE Trans. Neural Network*, Vol. **4**, No. 2, pp. 293–304.
81 Zhang, Y. and Wang, J. (2002) "A dual neural network for convex quadratic programming subject to linear equality and inequality constraints," *Phys. Lett. A*, Vol. **298**, No. 4, pp. 271–278.
82 Wang, J. and Zhang, Y. (2004) "Recurrent neural networks for real-time computation of inverse kinematics of redundant manipulators," *Machine Intelligence: Quo Vadis?* Singapore: World Scientific, pp. 299–319.
83 Zhang, Y. and Wang, J. (2003) "Obstacle avoidance of redundant manipulators using a dual neural network," in: Proceedings of IEEE International Conference on Robotics and Automation, pp. 2747–2752.
84 Zhang, Y. and Ma, S. (2007) "Minimum-energy redundancy resolution of robot manipulators unified by quadratic programming and its online solution," in: Proceedings of IEEE International Conference on Mechatronics and Automation, pp. 3232–3237.
85 Guo, J. and Hsia, T. C. (1993) "Joint trajectory generation for redundant robots in an environment with obstacles," *J. Robotic Syst.*, Vol. **10**, No. 2, pp. 199–215.
86 Cheng, F., Lu, Y. and Sun, Y. (1998) "Window-shaped obstacle avoidance for a redundant manipulator," *IEEE Trans. Syst. Man Cybern. Part B Cybern.*, Vol. **28**, No. 6, pp. 806–815.
87 Maciejewski, A. A. and Klein, C. A. (1985) "Obstacle avoidance for kinematically redundant manipulators in dynamically varying environments," *Int. J. Robotic Res.*, Vol. **4**, No. 3, pp. 109–117.
88 Glass, K., Colbaugh, R. Lim, D. and Seraji, H. (1995) "Real-time collision avoidance for redundant manipulators," *IEEE Trans. Rob. Autom.*, Vol. **11**, No. 3, pp. 448–457.
89 Tang, W., Lam, L. and Wang, J. (2001) "Kinematic control and obstacle avoidance for redundant manipulators using a recurrent neural network," in: Proceedings of International Conference on Artificial Neural Networks, pp. 922–929.
90 Cheng, F., Chen, T. and Sun, Y. (1994) "Resolving manipulator redundancy under inequality constraints," *IEEE Trans. Robot. Automat.*, Vol. **10**, No. 1, pp. 65–71.
91 Nedungadi, A. and Kazerounian, K. (1989) "A local solution with global characteristics for joint torque optimization of a redundant manipulator," *J. Robotic Syst.*, Vol. **6**, pp. 631–654.
92 Zhang, Y. (2006) "Inverse-free computation for infinity-norm torque minimization of robot manipulators," *Mechatronics*, vol. **16**, No. 3-4, pp. 177–184.
93 The MathWorks Inc. (2003) *MATLAB Optimization Toolbox.*
94 Zhang, Y. (2005) "On the LVI-based primal-dual neural network for solving online linear and quadratic programming problems," in: Proceedings of American Control Conference, pp. 1351–1356.
95 Ge, S. S., Lee, T. H. and Harris, C. J. (1998) *Adaptive Neural Network Control of Robotic Manipulators*, World Scientific, London, England.

96 Ge, S. S., Lee, T. H., Gu, D. L. and Woon, L.C. (2000) "OpenRob: an open architecture platform for robot control system design," *IEEE Robot. Automat. Mag.*, Vol. **7**, No. 3, pp. 42–54.

97 Xia, Y. and Wang, J. "A recurrent neural network for solving linear projection equations," *Neural Net.*, Vol. **13**, No. 3, pp. 337–350.

98 Zhang, Y., Li, J., Mao, M., Li, W. and Fu, S. (2015) "Complete theory for E47 and 94LVI algorithms solving inequality-and-bound constrained quadratic program efficiently," in: Proceedings of Chinese Automation Congress, pp. 183–189.

99 Zhang, Y., Li, W. Yu, X. Wu, H. and Li, J. (2013) "Encoder based online motion planning and feedback control of redundant manipulators," *Control Eng. Pract.*, Vol. **21**, No. 10, pp. 1277–1289.

100 Zhang, Y., Wang, J. and Xu, Y. (2002) "A dual neural network for bi-criteria kinematic control of redundant manipulators," *IEEE Trans. Robot. Automat.*, Vol. **18**, No. 6, pp. 923–931.

101 O'Neil, K. A. (2002) "Divergence of linear acceleration-based redundancy resolution schemes," *IEEE Trans. Robot. Automat.*, Vol. **18**, No. 4, pp. 625–631.

102 Sangwan, V. and Agrawal, S. K. (2009) "Differentially flat design of bipeds ensuring limit cycles," *IEEE/ASME Trans. Mechatronics*, Vol. **14**, No. 6, pp. 647–657.

103 Kim, B., Li, J. and Tsao, T. (2009) "Two-parameter robust repetitive control with application to a novel dual-stage actuator for noncircular machining," *IEEE/ASME Trans. Mechatronics*, Vol. **9**, No. 4, pp. 644–652.

104 Zhang, Y., Tan, Z., Chen, K., Yang, Z. and Lv, X. (2009) "Repetitive motion of redundant robots planned by three kinds of recurrent neural networks and illustrated with a four-link planar manipulator's straight-line example," *Robot. Auton. Syst.*, Vol. **57**, No. 6-7, pp. 645–651.

105 Zhang, Y., Guo, D., Cai, B. and Chen, K. (2011) "Remedy scheme and theoretical analysis of joint-angled drift phenomenon for redundant robot manipulators," *Robot. Comput. Integr. Manuf.*, Vol. **27**, No. 4, pp. 860–869.

106 Zhang, Y., Zhu, H., Lv, X. and Li, K. (2008) "Joint angle drift problem of PUMA560 robot arm solved by a simplified LVI-based primal-dual neural network," in: Proceedings of IEEE International Conference on Industrial Technology, pp. 1–6.

107 Assal, S. F. M., Watanabe, K. and Izumi, K. (2006) "Neural network-based kinematic inversion of industrial redundant robots using cooperative fuzzy hint for the joint limits avoidance," *IEEE/ASME Trans. Mechatronics*, Vol. **11**, No. 5, pp. 593–603.

108 Koizumi, N., Warisawa, S., Hashizume, H. and Mitsuishi, M. (2008) "Continuous path controller for the remote ultrasound diagnostic system," *IEEE/ASME Trans. Mechatronics*, Vol. **13**, No. 2, pp. 206–218.

109 Cheng, H., Yiu, Y. and Li, Z. (2003) "Dynamics and control of redundantly actuated parallel manipulators," *IEEE/ASME Trans. Mechatronics*, Vol. **8**, No. 4, pp. 483–491.

110 Lee, M. K. and Park, K. W. (2000) "Workspace and singularity analysis of a double parallel manipulator," *IEEE/ASME Trans. Mechatronics*, Vol. **5**, No. 4, pp. 367–375.

111 Angeles, J., Yang, G. and Chen, I. M. (2003) "Singularity analysis of three-legged, six-dof platform manipulators with URS legs," *IEEE/ASME Trans. Mechatronics*, Vol. **8**, No. 4, pp. 469–475.

112 Tan, U-X., Latt, W. T., Shee, C. Y., Riviere, C. N. and Ang, W. T. (2009) "Feedforward controller of ill-conditioned hysteresis using singularity-free Prandtl-Ishlinskii model," *IEEE/ASME Trans. Mechatronics*, Vol. **14**, No. 5, pp. 598–605.

113 Siqueira, A. A. G. and Terra, M. H. (2009) "A fault-tolerant manipulator robot based on H_2, H_∞, and Mixed H_2/H_∞ Markovian Controls," *IEEE/ASME Trans. Mechatronics*, Vol. **14**, No. 2, pp. 257–263.

114 Ma, S., Kobayashi, I., Hirose, S. and Yokoshima, K. (2002) "Control of a multi-joint manipulator 'Moray Arm'," *IEEE/ASME Trans. Mechatronics*, Vol. **7**, No. 3, pp. 304–317.

115 Zergeroglu, E., Dawson, D. D., Walker, I. W. and Setlur, P. (2004) "Nonlinear tracking control of kinematically redundant robot manipulators," *IEEE/ASME Trans. Mechatronics*, Vol. **9**, No. 1, pp. 129–132.

116 Liljebäck, P., Pettersen, K. Y., Stavdahl, Ø. and Gravdahl, J. T. (2012) "Snake robot locomotion in environments with obstacles," *IEEE/ASME Trans. Mechatronics*, Vol. **17**, No. 6, pp. 1158–1169.

117 Andruska, A. M. and Peterson, K. S. (2008) "Control of a snake-like robot in an elastically deformable channel," *IEEE/ASME Trans. Mechatronics*, Vol. **13**, No. 2, pp. 219–227.

118 Peer, A. and Buss, M. B. (2008) "A new admittance-type haptic interface for bimanual manipulations," *IEEE/ASME Trans. Mechatronics*, Vol. **13**, No. 4, pp. 416–428.

119 Zhu, X., Tao, G., Yao, B. and Cao, J. (2008) "Adaptive robust posture control of parallel manipulator driven by pneumatic muscles with redundancy," *IEEE/ASME Trans. Mechatronics*, Vol. **13**, No. 4, pp. 441–449.

120 Caccavale, F., Chiacchio, P., Marino, A. and Villani, L. (2008) "Six -DOF impedance control of dual-arm cooperative manipulators," *IEEE/ASME Trans. Mechatronics*, Vol. **13**, No. 5, pp. 576–586.

121 Takaki, T. and Omata, T. (2011) "High-performance anthropomorphic robot hand with grasping-force-magnification mechanism," *IEEE/ASME Trans. Mechatronics*, Vol. **16**, No. 3, pp. 583–591.

122 Tlalolini, D., Chevallereau, C. and Aoustin Y. (2011) "Human-like walking: optimal motion of a bipedal robot with toe-rotation motion," *IEEE/ASME Trans. Mechatronics*, Vol. **16**, No. 2, pp. 310–320.

123 Ukidve, C. S., McInroy, J. E. and Jafari, F. (2008) "Using redundancy to optimize manipulability of stewart platforms," *IEEE/ASME Trans. Mechatronics*, Vol. **13**, No. 4, pp. 475–479.

124 Huang, J., Fukuda, T. and Matsuno, T. (2008) "Model-based intelligent fault detection and diagnosis for mating electric connectors in robotic wiring harness assembly systems," *IEEE/ASME Trans. Mechatronics*, Vol. **13**, No. 1, pp. 86–94.

125 Patel, R. V., Talebi, H. A., Jayender, J. and Shadpey, F. (2009) "A robust position and force control strategy for 7-DOF redundant manipulators," *IEEE/ASME Trans. Mechatronics*, Vol. **14**, No. 5, pp. 575–589.

126 Wang, L., Wu, J., Wang, J. and You, Z. (2009) "An experimental study of a redundantly actuated parallel manipulator for a 5-DOF hybrid machine tool," *IEEE/ASME Trans. Mechatronics*, Vol. **14**, No. 1, pp. 72–81.

127 Xie, H. and Régnier, S. (2011) "Development of a flexible robotic system for multiscale applications of micro/nanoscale manipulation and assembly," *IEEE/ASME Trans. Mechatronics*, Vol. **16**, No. 2, pp. 266–276.

128 Zhang, Y., Lv, X., Li, Z., Yang, Z. and Chen, K. (2008) "Repetitive motion planning of PA10 robot arm subject to joint physical limits and using LVI-based primal-dual neural network," *Mechatronics*, Vol. **18**, No. 9, pp. 475–485.
129 Bebek, Ö., Hwang, M. J. and RÇavuşoğlu, M. C. (2011) "Design of a parallel robot for needle-based interventions on small animals," *IEEE/ASME Trans. Mechatronics*, Vol. **18**, No. 1, pp. 62–73.
130 Groom, K. N., Maciejewski, A. A. and Balakrishnan, V. (1999) "Real-time failure-tolerant control of kinematically redundant manipulators," *IEEE Trans. Robot. Autom.*, Vol. **15**, No. 6, pp. 1109–1116.
131 Goel, M., Maciejewski, A. A., Balakrishnan, V. and Proctor, R. W. (2003) "Failure tolerant teleoperation of a kinematically redundant manipulator: An experimental study," *IEEE Trans. Syst., Man, Cybern., B, Cybern.*, Vol. **33**, No. 6, pp. 758–765.
132 English, J. D. and Maciejewski, A. A. (2000) "On the implementation of velocity control for kinematically redundant manipulators," *IEEE Trans. Syst., Man, Cybern., A, Syst., Humans*, Vol. **30**, No. 3, pp. 233–237.
133 Ulrey, R. R., Maciejewski, A. A. and Siegel, H. J. (2005) "Parallel algorithms for singular value decomposition," in: Proceedings of IEEE International Parallel and Distributed Processing Symposium, pp. 524–533.
134 Chou, C. C., Lian, F. L. and Wang, C. C. "Characterizing indoor environment for robot navigation using velocity space approach with region analysis and look-ahead verification," *IEEE Trans. Instrum. Meas.*, Vol. **60**, No. 2, pp. 442–451.
135 Seder, M. and Petrovic, I. (2007) "Dynamic window based approach to mobile robot motion control in the presence of moving obstacles," in: Proceedings of IEEE International Conference on Robotics and Automation, pp. 1986–1991.
136 He, B. (1994) "A new method for a class of linear variational inequalities," *Math. Programming*, Vol. **66**, No. 1–3, pp. 137–144.
137 Cai, B. and Zhang, Y. 2010 "Bi-criteria optimal control of redundant robot manipulators using LVI-based primal-dual neural network," *Optim. Control Appl. Meth.*, Vol. **31**, No. 3, pp. 213–229.
138 Zhang, Y., Lv, X., Yang, Z. and Li, Z. (2007) "Repetitive motion planning of kinematically redundant manipulators using LVI-based primal-dual neural network," in Proceedings of IEEE International Conference on Mechatronics and Automation, pp. 3138–3143.
139 Zhang, Y., Yin, J. and Cai, B. (2009) "Infinity-norm acceleration minimization of robotic redundant manipulators using the LVI-based primal-dual neural network," *Robot. Comput. Integr. Manuf.*, Vol. **25**, No. 2, pp. 358–365.
140 Whitney, D. E. (1969) "Resolved motion rate control of manipulators and human prostheses," *IEEE Trans. Man Machine Sys.*, Vol. **10**, No. 2, pp. 47–53.
141 Uchiyama, M., Shimizu, K. and Hakomori, K. (1985) *Performance evaluation of manipulators using the Jacobian and its application to trajectory planning,* MIT Press, Cambridge.
142 Nakamura, Y. and Hanafusa, H. (1987) "Optimal redundancy control of robot manipulators," *Int. J. Robot. Res.*, Vol. **6**, No. 1, pp. 32–42.
143 Kazerounian, K. and Wang, Z. (1988) "Global versus local optimization in redundancy resolution of robotic manipulators," *Int. J. Robot. Res.*, Vol. 7, No. 5, pp. 3–12.

144 Chan, T. F. and Dubey, R. V. (1995) "A weighted least-norm solution based scheme for avoiding joint limits for redundant joint manipulators," *IEEE Trans. Robot. Autom.*, Vol. **11**, No. 2, pp. 286–292.

145 Hasan, G. M. and Nawab, K., and Morteza, D. (2011) "Near time-optimal control of redundant manipulators along a specified path with jerks constraint," *Adv. Robotics*, Vol. **25**, No. 18, pp. 2319–2339.

146 Taghirad, H. D. and Nahon, M. (2008) "Kinematic analysis of a macro-micro redundantly actuated parallel manipulator," *Adv. Robotics*, Vol. **22**, No. 6–7, pp. 657–687.

147 Martins, A. M., Dias, A. M. and Alsina, P. J. (2006) "Comments on manipulability measure in redundant planar manipulators," in: Proceedings of IEEE 3rd Latin American Robotics Symposium, pp. 520–525, pp. 169–173.

148 Xia, Y., and Wang, J. (2001) "A dual neural network for kinematic control of redundant robot manipulators," *IEEE Trans. Syst., Man Cybernet. B*, Vol. **31**, No. 1, pp. 147–154.

149 Choi, H. B., Lee, S. and Lee, J. (2011) "Minimum infinity-norm joint velocity solutions for singularity-robust inverse kinematics," *Int. J. Precis. Eng. Man.*, Vol. **12**, No. 3, pp. 469–474.

150 Ding, H. and Wang, J. (1999) "Recurrent neural networks for minimum infinity-norm kinematic control of redundant manipulators," *IEEE Trans. Syst. Man, Cybern. A*, Vol. **29**, No. 3, pp. 269–276.

151 Kang, H. J. and Freeman, R. A. (1992) "Joint torque optimization of redundant manipulators via the null space damping method," in: Proceedings of 1992 IEEE International Conference on Robotics and Automation, pp. 520–525.

152 Dermatas, E., Nearchou, A. and Aspragathos, N. (1996) "Error-back-propagation solution to the inverse kinematic problem of redundant manipulators," *Robot. Comput. Integr. Manuf.*, Vol. **12**, No. 4, pp. 303–310.

153 Tang, W. S. and Wang, J. (2001) "A recurrent neural network for minimum infinity-norm kinematic control of redundant manipulators with an improved problem formulation and reduced architecture complexity," *IEEE Trans. Syst., Man Cybernet. B*, Vol. **31**, No. 1, pp. 98–105.

154 Xia, Y., Feng, G. and Wang, J. (2005) "A primal-dual neural network for online resolving constrained kinematic redundancy in robot motion control," *IEEE Trans. Syst., Man Cybernet. B*, Vol. **35**, No. 1, pp. 54–64.

155 Zhang, Y, Ma, W. and Cai, B. (2009) "From Zhang neural network to Newton iteration for matrix inversion," *IEEE Trans. on Circ. Syst. I.*, Vol. **56**, No. 7, pp. 1405–1415.

156 De Luca, A. (1992) "Control of redundant robots on cyclic trajectories," in: *Proc. IEEE Int. Conf. on Robotics and Autom., Nice, France*, Vol. **1**, pp. 500–506.

157 Meghdari, A., Naderi, D. and Eslami, S. (2006) "Optimal stability of a redundant mobile manipulator via genetic algorithm," *Robotica*, Vol. **24**, No. 6, pp. 739–743.

158 Puga, J. P. and Chiang, L. E. (2008) "Optimal trajectory planning for a redundant mobile manipulator with non-holonomic constraints performing push-pull tasks," *Robotica*, Vol. **26**, No. 3, pp. 385–394.

159 Ozbay, U., Sahin, H. T. and Zergeroglu, E. (2008) "Robust tracking control of kinematically redundant robot manipulators subject to multiple self-motion criteria," *Robotica*, Vol. **26**, No. 6, pp. 711–728.

160 Allotta, B., Colla, V. and Bioli, G. (1999) "Kinematic control of robots with joint constraints," *ASME J. Dyn. Syst. Meas. Control*, Vol. **121**, No. 3, pp. 433–442.
161 Mao, Z. and Hsia, T. C. (1997) "Obstacle avoidance inverse kinematics solution of redundant robots by neural networks," *Robotica*, Vol. **15**, No. 1, pp. 3–10.
162 Zhang, Y. (2006) "A set of nonlinear equations and inequalities arising in robotics and its online solution via a primal neural network," *Neurocomputing*, Vol. **70**, No. 1, pp. 513–524.
163 Zhang, Y., Tan, Z., Yang, Z. and Lv, X. (2008) "A dual neural network applied to drift-free resolution of five-link planar robot arm," in: Proceedings of the 2008 IEEE International Conference on Information and Automation, pp. 1274–1279.
164 Donelan, P. S. (2007) "Singularity-theoretic methods in robot kinematics," *Robotica*, Vol. **25**, No. 6, pp. 641–659.
165 Padois, V., Fourquet, J.-Y. and Chiron, P. (2001) "Kinematic and dynamic model-based control of wheeled mobile manipulators: a unified framework for reactive approaches," *Robotica*, Vol. **25**, No. 2, pp. 157–173.
166 Lee, J. (2001) "A structured algorithm for minimum l_∞-norm solutions and its application to a robot velocity workspace ananlysis," *Robotica*, Vol. **19**, No. 3, pp. 343–352.
167 Qiu, C., Cao, Q. and Miao, S. (2009) "An on-line task modification method for singularity avoidance of robot manipulators," *Robotica*, Vol. **27**, No. 4, pp. 539–546.
168 He, B. (1994) "Solving a class of linear projection equation," *Numer. Math.*, Vol. **68**, No. 1, pp. 71–80.
169 Wang, J. (1997) "Recurrent neural networks for computing pseudoinverses of rank-deficient matrices," *SIAM J. Sci. Comput.*, Vol. **18**, No. 5, pp. 1479–1493.
170 Zhang, Y., Chen, K. and Ma, W. (2007) "MATLAB simulation and comparison of Zhang neural network and gradient neural network for online solution of linear time-varying equations," in: Proceedings of International Conference on Life System Modeling and Simulation, pp. 450–454.
171 Li, K. and Zhang, Y. (2012) "State adjustment of redundant robot manipulator based on quadratic programming," *Robotica*, Vol. **30**, No. 3, pp. 477–489.
172 Zhang, Y., Ruan, G., Li, K. and Yang, Y. (2010) "Robustness analysis of the Zhang neural network for online time-varying quadratic optimization," *J. Phys. A-Math. Theor.*, Vol. **43**, No. 24, pp. 1–19.
173 Khatibm, O. and Bowling, A. (1996) "Optimization of the inertial and acceleration characteristics of manipulators," in: Proceeding of IEEE International Conference on Robotics and Automation, pp. 2883–2889.
174 Klein, C. A. and Huang, C. H. (1983) "Review of pseudoinverse control for use with kinematically redundant manipulators," *IEEE Trans. Syst., Man, Cybern. B, Cybern.*, Vol. **13**, No. 3, pp. 245–250.
175 Liegeois, A. (1997) "Automatic supervisory control of the configuration and behavior of multibody mechanisms," *IEEE Trans. Syst., Man, Cybern. B, Cybern.*, Vol. 7, No. 12, pp. 868–871.
176 Burdick, J. W. (1989) "On the inverse kinematics of redundant manipulators: characterization of the self-motion manifolds," in: Proceeding of IEEE International Conference on Robotics Automation, pp. 264–270.
177 Oriolo, G. (1994) "Stabilization of self-motion in redundant robots," in: Proceedings of IEEE International Conference on Robotics and Automation, pp. 704–709.

178 Wang, L. and Ma, B. (2003) "Self-motion control of redundant manipulators," *Control Decis.*, Vol. **18**, No. 2, pp. 199–202, (in Chinese).

179 Ma, B. (2005). "Bi-directional self-motion path planning of redundant manipulators", *Control Theory and Applications*, Vol. **22** No. 4, pp. 547–550.

180 Ferris, M. C. and Pang, J. S. (1997) *Complementarity and Variational Problems: State of the Art*, PA: SIAM, Philadephia.

181 Zhang, Y., Zhu H., Tan, Z., Cai, B. and Yang Z. (2008) "Self-motion planning of redundant robot manipulators based on quadratic program and shown via PA10 example," in: Proceedings of 2nd International Symposium on Systems and Control in Aerospace and Astronautics, pp. 1–6.

182 Zhang, Y., Huang, Y., Guo, D. (2009) "Self-motion planning of functionally redundant PUMA560 manipulator via quadratic-program formulation and solution," in: Proceedings of International Conference on Mechatronics and Automation, pp. 2518–2523.

183 Lee, J. K. and Cho, H. S. (1997) "Mobile manipulator motion planning for multiple tasks using global optimization approach," *J. Intell. Rob. Syst. Theor. Appl.*, Vol. **18**, No. 2, pp. 169–190.

184 Shahri, N. R. and Troch, I. (1996) "Collision-avoidance for redundant robots through control of the self-motion of the manipulator," *J. Intell. Rob. Syst. Theor. Appl*, Vol. **16**, No. 2, pp. 123–149.

185 Zhang, Y. and Xiao L. (2012) "QP-based SMP scheme for robots with pseudoinverse method compared and singularities discussed," in: Proceedings of the 10th World Congress on Intelligent Control and Automation, pp. 3583–3588.

186 Luck, C. L. (1997) "Self-motion representation and global path planning optimization for redundant manipulators through topology-based discretization," *J. Intell. Rob. Syst. Theor. Appl.*, Vol. **19**, No. 1, pp. 23–38.

187 Khoogar, A. R., Tehrani, A. K. and Tajdari, M. (2011) "A dual neural network for kinematic control of redundant manipulators using input pattern switching," *J. Intell. Rob. Syst. Theor. Appl.*, Vol. **63**, No. 1, pp. 101–113.

188 Pott, P. P., Wagner, A., Badreddin, E., Weiser, H.-P. and Schwarz, M. L. R. (2011) "Inverse dynamic model and a control application of a novel 6-DOF hybrid kinematics manipulator," *J. Intell. Rob. Syst. Theor. Appl.*, Vol. **63**, No. 1, pp. 3–23.

189 Kanoun, O., Lamiraux, F., Wieber, P., Kanehiro, F., Yoshida, E. and Laumond, J. (2009) "Prioritizing linear equality and inequality systems: application to local motion planning for redundant robots," in: Proceedings of IEEE International Conference on Robotics and Automation, pp. 2939–2944.

190 Lenarcic, J. (1999) "The range of self-motion of redundant robots," in: Proceedings of IEEE/ASME International Conference on Advanced Intelligent Mechatronics, pp. 386–391.

191 Cullen, K. E., Brooks, J. X., Jamali, M., Carriot, J. and Massot, C. (2011) "Internal models of self-motion: computations that suppress vestibular reafference in early vestibular processing," *Exp. Brain. Res.*, Vol. **210**, No. 3–4, pp. 377–388.

192 Scholz, J. P., Dwight-Higgin, T., Lynch, J. E., Tseng, Y. W., Martin, V. and Schoner, G. (2011) "Motor equivalence and self-motion induced by different movement speeds," *Exp. Brain. Res.*, Vol. **209**, No. 3, pp. 319–332.

193 Zhang, Y. and Li, K. (2010) "Bi-criteria velocity minimization of robot manipulators using LVI-based primal-dual neural network and illustrated via PUMA560 robot arm," *Robotica*, Vol. **28**, No. 4, pp. 525–537.

194 Corke, P. I., and Armstrong-Helouvry, B. (1994) "A search for consensus among model parameters reported for the PUMA 560 robot," in: Proceedings of IEEE International Conference on Robotics and Automation, pp. 1608–1613.

195 Bogue, R. (2011). "Robots in the nuclear industry: a review of technologies and applications", *Industrial Robot: An International Journal*, Vol. **38** No. 2, pp. 113–118.

196 Bosscher, P. and Hedman, D. (2011). "Real-time collision avoidance algorithm for robotic manipulators", *Industrial Robot: An International Journal*, Vol. **38** No. 2, pp. 186–197.

197 Li, S., Liu, Y. and Xie, M. (2011). "Implementation of a single motor driven manipulator with multiple joints", *Industrial Robot: An International Journal*, Vol. **38** No. 1, pp. 48–57.

198 Wu, X.J., Li, Q. and Heng, K.H. (2005). "Development of a general manipulator path planner using fuzzy reasoning", *Industrial Robot: An International Journal*, Vol. **32** No. 3, pp. 248–258.

199 Zhang, T., Nakamura, M., Goto, S. and Kyura, N. (2005). "Highly accurate contour control of an articulated robot manipulator using a Gaussian neural network", *Industrial Robot: An International Journal*, Vol. **32** No. 5, pp. 408–418.

200 Wang, L. and Ma, B. (2003). "Self-motion control of redundant manipulators", *Control and Decision*, Vol. **18** No. 2, pp. 199–202.

201 Zhang, Y., Guo, D., Li, K. and Li, J. (2012) "Manipulability-maximizing self-motion planning and control of redundant manipulators with experimental validation," in: Proceedings of the IEEE International Conference on Mechatronics and Automation, pp. 1829–1834.

202 Yoshikawa, T. (1985) "Manipulability of robotic mechanisms," *Int. J. Rob. Res.*, Vol. **4**, No. 2, pp. 3–9.

203 Zhang, Y., Huang, Y., Guo, D. and Ke, Z. (2010) "Quadratic-programming based self-motion planning with no target-configuration assigned for planar robot arms," in: Proceedings of the IEEE International Conference on Control and Automation, pp. 534–539.

204 Zhang, Y., Fu, S., Zhang, Z., Xiao, L. and Li, X. (2011) "On the LVI-based numerical method (E47 algorithm) for solving quadratic programming problems," in: Proceedings of the IEEE International Conference on Automation and Logistics, pp. 125–130.

205 Ji, C. Y., Chen, T. C. and Lee, Y. L. (2008) "Joint control for flexible-joint robot with input-estimation approach and LQG method," *Optim. Control Appl. Meth.*, Vol. **29**, No. 2, pp. 101–125.

206 Vossen, G. and Maurer, H. (2006) "On L1-minimization in optimal control and applications to robotics," *Optim. Control Appl. Meth.*, Vol. **27**, No. 6, pp. 301–321.

207 Khadem, S. E. and Dubey, R. V. (1991) "A global cartesian space obstacle avoidance scheme for redundant manipulators," *Optim. Control Appl. Meth.* Vol. **12**, No. 4, pp. 279–286.

208 Fabien, B. C. (2008) "Direct optimization of dynamic systems described by differential-algebraic equations," *Optim. Control Appl. Meth.* Vol. **29**, No. 6, pp. 445–466.

209 Zhang, Y., Li, J. and Zhang, Z. (2013) "A time-varying coefficient-based manipulability-maximizing scheme for motion control of redundant robots subject to varying joint-velocity limits," *Optim. Control Appl. Meth.* Vol. **34**, No. 2, pp. 202–215.

210 Park, D. H., Hoffmann, H., Pastor, P. and Schaal, S. (2008) "Movement reproduction and obstacle avoidance with dynamic movement primitives and potential fields," in: Proceedings of IEEE-RAS International Conference on Humanoid Robots, pp. 91–98.

211 Komoguchi, Y., Yano, K., Peer, A. and Buss, M. (2008) "Redundancy resolution of a 7 DOF haptic interface considering collision and singularity avoidance," in: Proceedings of IEEE/RSJ International Conference on Intelligent Robots and Systems, pp. 3513–3518.

212 Soylu, S., Buckham, B. J. and Podhorodeski R. P. (2010) "Redundancy resolution for underwater mobile manipulators," *Ocean Eng.*, Vol. **37**, pp. 325–343.

213 Mitsi, S., Bouzakis, K. D., Sagris, D. and Mansour, G. (2008) "Determination of optimum robot base location considering discrete end-effector positions by means of hybrid genetic algorithm," *Rob. Comput. Integr. Manuf.*, Vol. **24**, pp. 50–59.

214 Jun, B. H., Lee, P. M. and Kim, S. (2008) "Manipulability analysis of underwater robotic arms on ROV and application to task-oriented joint configuration," *J. Mech. Sci. Technol.*, Vol. **22**, pp. 887–894.

215 Lee, S. H., Lee, J. H., Yi, B. J., Kim, S. H. and Kwak, Y. K. (2005) "Optimization and experimental verification for the antagonistic stiffness in redundantly actuated mechanisms: a five-bar example," *Mechatronics*, Vol. **15**, pp. 213–238.

216 Schreiber, G., Otter, M. and Hirzinger, G. (1999) "Solving the singularity problem of non-redundant manipulators by constraint optimization," in: Proceedings of International Conference on Intelligent Robots and Systems, Vol. 3, pp. 1482–1488.

217 Donelan, P. S. (2007) *Singularity Theory*, World Scientific: Hackensack, New Jersey, USA.

218 Yoshikawa, T. (1984) *Robotics Research*, MIT Press: Cambridge, Master of Art, pp. 735–747.

219 Huang, J., Yamada, D., Nakamura, Y., Hara, M., and Yabuta, T. (2009) "Cooperative impedance control of a finger-arm robot by regulating finger's manipulability," *J. Syst. Des. Dyn.*, Vol. **3**, No. 5, pp. 756–767.

220 Kim, J. O. and Khosla, P. K. (1991) "Dexterity measures for design and control of manipulators," in Proceedings of IEEE/RSJ International Workshop on Intelligent Robots and Systems, Vol. 2, pp. 758–763.

221 Tang, C. P. and Krovi, V. N. (2007) "Manipulability-based configuration evaluation of cooperative payload transport by mobile manipulator collectives," *Robotica*, Vol. **25**, pp. 29–42.

222 Zhang, Y., Chen, K. and Tan, H. (2009) "Performance analysis of gradient neural network exploited for online time-varying matrix inversion," *IEEE Trans. Autom. Control*, Vol. **54**, No. 8, pp. 1940–1945.

223 Zhang, Y. and Ge, S. S. (2005) "Design and analysis of a general recurrent neural network model for time-varying matrix inversion," *IEEE Trans. Neural Networks*, Vol. **16**, No. 6, pp. 1477–1490.

224 Bazaraa, M. S., Sherali, H. D. and Shetty, C. M. (1993) *Nonlinear Programming-Theory and Algorithms*, John Wiley and Sons: Hoboken, New Jersey.

225 Siciliano, B. (1990) "Kinematic control of redundant robot manipulators: a tutorial," *J. Intell. Rob. Syst.*, Vol. **3**, No. 3, pp. 201–212.

226 Nenchev, D. N. (1995) "Tracking manipulator trajectories with ordinary singularities: a null space-based approach," *Int. J. Rob. Res.*, Vol. **14**, No. 4, pp. 399–404.

227 Morales, R., Chocoteco, J., Feliu, V. and Sira-Ramírez, H. (2013) "Obstacle surpassing and posture control of a stair-climbing robotic mechanism," *Control Eng. Pract.*, Vol. **21**, No. 5, pp. 604–621.

228 Li, K. and Zhang, Y. (2012) "Fault-tolerant motion planning and control of redundant manipulator," *Control Eng. Pract.*, Vol. **20**, No. 3, pp. 282–292.

229 Roberts, R. G. and Maciejewski, A. A. (1994) "Singularities, stable surfaces, and the repeatable behavior of kinematically redundant manipulators," *Int. J. Rob. Res.*, Vol. **13**, No. 1, pp. 70–81.

230 Borangiu, T., Dumitrache, A. and Dogar, A. (2012) "Heuristic solution for constrained 7-DOF motion planning in 3D scanning application," *Control Eng. Pract.*, Vol. **20**, No. 2, pp. 93–101.

231 Hayakawa, M., Hara, K., Sato, D., Konno, A. and Uchiyama, M. (2008) "Singularity avoidance by inputting angular velocity to a redundant axis during cooperative control of a teleoperated dual-arm robot," in: Proceedings of the IEEE International Conference on Robotics and Automation, pp. 2013–2018.

232 Shimizu, M., Kakuya, H., Yoon, W. K., Kitagaki, K. and Kosuge, K. (2008) "Analytical inverse kinematic computation for 7-dof redundant manipulators with joint limits and its application to redundancy resolution," *IEEE Trans. Rob.*, Vol. **24**, No. 5, pp. 1131–1142.

233 Zhang, Y., Li, W., Yu, X., Wu, H. and Li, J., (2013) "Encoder based online motion planning and feedback control of redundant manipulators," *Control Eng. Pract.*, Vol. **21**, No. 10, pp. 1277–1289.

234 Kanoun, O. (2011) "Real-time prioritized kinematic control under inequality constraints for redundant manipulators. In Durrant-Whyte, H.., Roy, N., Abbeel, P. (eds)," *Progress in Robotics: Science and Systems VII*. Cambridge, MA: MIT Press.

235 Hashemi S. M., Abbas H. S. and Werner H. (2012) "Low-complexity linear parameter-varying modeling and control of a robotic manipulator," *Control Eng. Pract.*, Vol. **20**, No. 3, pp. 248–257.

236 Van Willigenburg, L. G. (1993) "Computation and implementation of digital time-optimal feedback controllers for an industrial X-Y robot subjected to path, torque, and velocity constraints," *Int. J. Rob. Res.*, Vol. **12**, No. 5, pp. 420–433.

237 Van Willigenburg, L. G. and Loop, R. P. H. (1991) "Computation of time-optimal controls applied to rigid manipulators with friction," *Int. J. Control*, Vol. **54**, No. 5, pp. 1097–1117.

238 Siciliano, B., Sciavicco, L., Villani, L. and Oriolo, G. (2010) *Robotics: Modelling, Planning and Control, Second Edition*. London: Springer.

239 Kumar, V., Žefran, M. and Ostrowski, J. (2007) Motion Planning and Control of Robots. In Shimon Y N (eds) *Handbook of Industrial Robotics, Second Edition* (pp. 295–315).New York: John Wiley & Sons, Inc.

240 Chiaverini, S., Oriolo, G. and Walker, I.D. (2008) "Kinematically redundant manipulators. In Siciliano B and Khatib O (eds)," *Springer Handbook of Robotics*, pp. 245–268. Heidelberg: Springer-Verlag.

241 Kwon, S. and Hwang, J. (2011) "Kinematics, pattern recognition, and motion control of mask-panel alignment system," *Control Eng. Pract.*, Vol. **19**, No. 8, pp. 883–892.
242 Lin, C. J. (2004) "Motion planning of redundant robots by perturbation method," *Mechatronics*, Vol. **14**, No. 3, pp. 281–297.
243 Stilman, M. (2010) "Global manipulation planning in robot joint space with task constraints," *IEEE Trans. Rob.*, Vol. **26**, No. 3, pp. 576–584.
244 Seereeram, S. and Wen, J. (1995) "A global approach to path planning for redundant manipulators," *IEEE Trans. Rob. Autom.*, Vol. **11**, No. 1, pp. 152–160.
245 Marani, G., Kim, J., Yuh, J. and Chung, W. K. (2002) "A real-time approach for singularity avoidance in resolved motion rate control of robotic manipulators," in: Proceedings of the IEEE International Conference on Robotics and Automation, pp. 1973–1978.
246 Nakamura. Y., Zhang, T. and Minami, M. (2008) "On-line configuration-optimizing control of redundant manipulator based on AMSIP," in: Proceedings of the SICE Annual Conference, pp. 1994–2000.
247 Suh, K. and Hollerbach, J. (1987) "Local versus global torque optimization of redundant manipulators," in: Proceedings of the IEEE International Conference on Robotics and Automation, pp. 619–624.
248 Turetta, A., Casalino, G. and Sorbara, A. (2008) "Distributed control architecture for self-reconfigurable manipulators," *Int. J. Rob. Res.*, Vol. **27**, No. 3–4, pp. 481–504.
249 Hu, X. and Wang, J. (2008) "An improved dual neural network for solving a class of quadratic programming problems and its *k*-winners-take-all application," *IEEE Trans. Neural Netw.*, Vol. **19**, No. 12, pp. 2022–2031.
250 Rossomando, F. G., Soria, C. and Carelli, R. (2011) "Autonomous mobile robots navigation using RBF neural compensator," *Control Eng. Pract.*, Vol. **19**, No. 3, pp. 215–222.
251 Kircanski, M. and Kircanski, N. (1998) "Resolved rate and acceleration control in the presence of actuator constraints," *IEEE Control Syst.*, Vol. **18**, No. 1, pp. 42–47.
252 Luh, J., Walker, M. and Paul, R. (1980) "Resolved-acceleration control of mechanical manipulators," *IEEE Trans. Autom. Control*, Vol. **25**, No. 3, pp. 468–474.
253 Zhang, Y. and Li, Z. (2009) "Zhang neural network for online solution of time-varying convex quadratic program subject to time-varying linear-equality constraints," *Phys. Lett. A*, Vol. **373**, No. 18–19, pp. 1639–1643.
254 Hollerbach, J. and Ki, S. (1987) "Redundancy resolution of manipulators through torque optimization," *IEEE Journal of Robotics and Automation*, Vol. **3**, No. 4, pp. 308–316.
255 He, B. (1992) "A projection and contraction method for a class of linear complementarity problems and its application in convex quadratic programming," *Applied Mathematics and Optimization*, Vol. **25**, No. 3, pp. 247–262.
256 Zhang, Y., Xie, L., Zhang, Z., Li, K. and Xiao, L. (2011) "Real-time joystick control and experiments of redundant manipulators using cosine-based velocity mapping," in: Proceedings of the IEEE International Conference on Automation and Logistics, pp. 345–350.

Index